Probability and Mathematical Statistics
  PUKELSHEIM · Optimal Design
  PURI and SEN · Nonparametric M
  PURI, VILAPLANA, and WERTZ · New Perspectives in Theoretical and Applied
    Statistics
  RAO · Asymptotic Theory of Statistical Inference
  RAO · Linear Statistical Inference and Its Applications, *Second Edition*
  ROBERTSON, WRIGHT, and DYKSTRA · Order Restricted Statistical Inference
  ROGERS and WILLIAMS · Diffusions, Markov Processes, and Martingales,
    Volume II: Ito Calculus
  ROHATGI · A Introduction to Probability Theory and Mathematical Statistics
  ROSS · Stochastic Processes
  RUBINSTEIN · Simulation and the Monte Carlo Method
  RUBINSTEIN and SHAPIRO · Discrete Event Systems: Sensitivity Analysis and
    Stochastic Optimization by the Score Function Method
  RUZSA and SZEKELY · Algebraic Probability Theory
  SCHEFFE · The Analysis of Variance
  SEBER · Linear Regression Analysis
  SEBER · Multivariate Observations
  SEBER and WILD · Nonlinear Regression
  SERFLING · Approximation Theorems of Mathematical Statistics
  SHORACK and WELLNER · Empirical Processes with Applications to Statistics
  STAUDTE and SHEATHER · Robust Estimation and Testing
  STOYANOV · Counterexamples in Probability
  STYAN · The Collected Papers of T.W. Anderson: 1943−1985
  WHITTAKER · Graphical Models in Applied Multivariate Statistics
  YANG · The Construction Theory of Denumerable Markov Processes

Applied Probability and Statistics
  ABRAHAM and LEDOLTER · Statistical Methods for Forecasting
  AGRESTI · Analysis of Ordinal Categorical Data
  AGRESTI · Categorical Data Analysis
  ANDERSON and LOYNES · The Teaching of Practical Statistics
  ANDERSON, AUQUIER, HAUCK, OAKES, VANDAELE, and WEISBERG ·
    Statistical Methods for Comparative Studies
  ASMUSSEN · Applied Probability and Queues
 *BAILEY · The Elements of Stochastic Processes with Applications to the Natural
    Sciences
  BARNETT · Interpreting Multivariate Data
  BARNETT and LEWIS · Outliers in Statistical Data, *Third Edition*
  BARTHOLOMEW, FORBES, and McLEAN · Statistical Techniques for
    Manpower Planning, *Second Edition*
  BATES and WATTS · Nonlinear Regression Analysis and Its Applications
  BELSLEY · Conditioning Diagnostics: Collinearity and Weak Data in Regression
  BELSLEY, KUH, and WELSCH · Regression Diagnostics: Identifying Influential
    Data and Sources of Collinearity
  BHAT · Elements of Applied Stochastic Processes, *Second Edition*
  BHATTACHARYYA and WAYMIRE · Stochastic Processes with Applications
  BIEMER, GROVES, LYBERG, MATHIOWETZ, AND SUDMAN · Measurement
    Errors in Surveys
  BIRKES and DODGE · Alternative Methods of Regression
  BLOOMFIELD · Fourier Analysis of Time Series: An Introduction
  BOLLEN · Structural Equations with Latent Variables
  BOX · R.A. Fisher, the Life of a Scientist
  BOX and DRAPER · Empirical Model-Building and Response Surfaces
  BOX and DRAPER · Evolutionary Operation: A Statistical Method for Process
    Improvement
  BOX, HUNTER, and HUNTER · Statistics for Experimenters: An Introduction to
    Design, Data Analysis and Model Building

Continued on back end papers

* Now available in a lower priced paperback edition in the Wiley Classics Library.

Aspects of Uncertainty

Aspects of Uncertainty
A Tribute to D. V. Lindley

Edited by

P. R. FREEMAN
Consultant Statistician

and

A. F. M. SMITH
Imperial College, London, UK

JOHN WILEY & SONS
Chichester . New York . Brisbane . Toronto . Singapore

Copyright © 1994 by John Wiley & Sons Ltd,
Baffins Lane, Chichester,
West Sussex PO19 1UD, England
National Chichester (0243) 779777
International (+44) 243 779777

All rights reserved.

No part of this book may be reproduced by any means,
or transmitted, or translated into a machine language
without the written permission of the publisher.

Other Wiley Editorial Offices

John Wiley & Sons, Inc., 605 Third Avenue,
New York, NY 10158-0012, USA

Jacaranda Wiley Ltd, 33 Park Road, Milton,
Queensland 4064, Australia

John Wiley & Sons (Canada) Ltd, 22 Worcester Road,
Rexdale, Ontario M9W 1L1, Canada

John Wiley & Sons (SEA) Pte Ltd, 37 Jalan Pemimpin #05-04,
Block B, Union Industrial Building, Singapore 2057

British Library Cataloguing in Publication Data

A catalogue record for this book is available from the British Library

ISBN 0 471 94347 9

Typeset in 10/12 pt Times by Pure Tech Corporation, Pondicherry.
Printed and bound in Great Britain by Biddles Ltd, Guildford and Kings Lynn

Contents

Preface	xiii
List of Contributors	xv

CHAPTER 1   DENNIS LINDLEY: THE FIRST 70 YEARS 1
P. Armitage

Acknowledgements	12
References	12

CHAPTER 2   THE OPERATIONAL BAYESIAN APPROACH 19
R. E. Barlow and M. B. Mendel

2.1 Introduction	19
2.2 The Indifference Principle	20
2.3 Indifference Relative to Transformed Random Quantities	23
2.4 Conclusions	26
References	27

CHAPTER 3   PIVOTAL INFERENCE ILLUSTRATED ON THE DARWIN MAIZE DATA 29
G. A. Barnard

3.1 The Darwin Data	29
3.2 Estimating the Difference δ	32
3.3 Limits for the Ratio θ	34
3.4 Combining Sets of Data	37
3.5 General Logical Considerations	38
References	39

CHAPTER 4   BAYES' THEOREM IN LATENT VARIABLE MODELLING 41
D. J. Bartholomew

4.1 Setting the Scene	41
4.2 A General Formulation	42
4.3 The Normal Linear Model	45

4.4 Binary Data	46
4.5 Linear Structural Relations Model	48
References	49

CHAPTER 5 BAYESIAN SAMPLING SCHEMES FOR AUDITORS 51
J. A. Bather and P. J. Browne

5.1 Introduction	51
5.2 Dynamic Programming	54
5.3 Numerical Illustration	57
5.4 Continuity and Uniqueness	59
References	65

CHAPTER 6 OPTIMIZING PREDICTION WITH HIERARCHICAL MODELS: BAYESIAN CLUSTERING 67
J. M. Bernardo

6.1 Introduction	67
6.2 The Prediction Problem	67
6.3 The Decision Problem	69
6.4 The Clustering Algorithm	70
6.5 An Application to Election Forecasting	71
6.6 A Case Study: State Elections in Mexico	73
6.7 Discussion	75
References	75

CHAPTER 7 INFERENCE FOR A COVARIANCE MATRIX 77
P. J. Brown, N. D. Le and J. V. Zidek

7.1 Introduction	77
7.2 Inverted Wishart Prior Distribution	79
7.3 Some Alternative Prior Distributions	81
7.4 Generalized Inverted Wishart (GIW)	82
7.5 Implementing the GIW Prior	87
7.6 Concluding Remarks	90
Acknowledgements	90
References	90

CHAPTER 8 THE ROLE OF STATISTICAL THEORY IN DECISION AIDING: MEASURING DECISION EFFECTIVENESS IN THE LIGHT OF OUTCOMES 93
R. V. Brown

8.1 Some Past Developments in Prescriptive Statistics	94
8.2 Statistical Development Needed on Evaluating Decision Effectiveness	99

8.3 Conclusions	114
Acknowledgements	114
References	115

CHAPTER 9 THE SIGNATURE AS A COVARIATE IN RELIABILITY AND BIOMETRY 119
S. Campodónico and N. D. Singpurwalla

9.1 Introduction and Overview	119
9.2 The Spectrum and Its Least-squares Estimation	120
9.3 The Signatures of Vibrations and Cardiograms	123
9.4 Bayesian Estimation of the Spectrum	125
9.5 The Spectrum as a Covariate	133
9.6 Application: Service Life of Traction Motors	137
Appendix A	142
Appendix B	143
Appendix C	145
References	146

CHAPTER 10 RESIDUAL ANALYSIS AND OUTLIERS IN BAYESIAN HIERARCHICAL MODELS 149
K. Chaloner

10.1 The Realized Errors	149
10.2 The One-way Model	150
10.3 Unknown Variances	152
10.4 Discussion	156
Acknowledgement	156
References	156

CHAPTER 11 THE ISLAND PROBLEM: COHERENT USE OF IDENTIFICATION EVIDENCE 159
A. P. Dawid

11.1 Introduction	159
11.2 The Island Problem	160
11.3 Which Answer?	162
11.4 Searching for Suspects	165
11.5 The Presumption of Innocence	168
11.6 Conclusion	170
References	170

CHAPTER 12 UTILITY: PROBABILITY'S YOUNGER TWIN? 171
S. French

12.1 Introduction	171

12.2	Two Families of Axiomatizations	172
12.3	Modelling Decision Situations	174
12.4	The Reference Experiment	175
12.5	The Importance of Being both Belief and Preference Analysts	178
12.6	Concluding Remarks	178
	References	179

CHAPTER 13 FULLY BAYESIAN HIERARCHICAL ANANYSIS FOR EXPONENTIAL FAMILIES VIA MONTE CARLO COMPUTATION 181

E. I. George, U. E. Makov and A. F. M. Smith

13.1	Motivation	181
13.2	Monte Carlo Evaluation of the Posterior	186
13.3	Application of Fully Bayesian Hierarchical Analysis	193
	References	196

CHAPTER 14 REVISING EXCHANGEABLE BELIEFS: SUBJECTIVIST FOUNDATIONS FOR THE INDUCTIVE ARGUMENT 201

M. Goldstein

14.1	Introduction	201
14.2	Tossing Coins	203
14.3	Exchangeable Structures	206
14.4	Posterior Expectations	209
14.5	Posterior Expectations for Population Quantities	211
14.6	Learning from Exchangeable Data	213
14.7	Separating Posterior Beliefs for the Exchangeable Model	215
14.8	Exchangeable Posterior Beliefs	217
14.9	Exchangeable Revisions of Beliefs	220
14.10	Concluding Comments	221
	References	222

CHAPTER 15 ON STEINIAN SHRINKAGE ESTIMATORS: THE FINITE/INFINITE PROBLEM AND FORMALISM IN PROBABILITY AND STATISTICS 223

B. M. Hill

15.1	Finite and Infinite	223
15.2	Admissibility and Boundedness	225
15.3	Using the True Sphere	231

15.4 The Risk Functions	238
15.5 Extended Admissibility	242
15.6 Random Effects Models	244
15.7 The Steinian Paradoxes and Assorted Red Herrings	249
15.8 Conclusions	254
References	258

CHAPTER 16 BAYESIAN DECISION THEORY AND THE LEGAL STRUCTURE 261
J. B. Kadane

16.1 Introduction	261
16.2 Abolition of Judgements of Guilt	262
16.3 Admission of all Cost-free Evidence	264
16.4 Abolition of the Adversarial Approach	265
16.5 Conclusion	265
References	266

CHAPTER 17 EXPERIMENTAL DESIGN FROM A SUBJECTIVE UTILITARIAN VIEWPOINT 267
F. Lad and J. Deely

17.1 Introduction	267
17.2 The Problem	269
17.3 Mixing Distributions	271
17.4 Elicitation	272
17.5 Experimental Design as a Decision Problem	275
17.6 A Benchmark Utility Valuation	275
17.7 Related Conceptions of Utility as Information	277
17.8 The Expected Utility Valuation of Each Design	279
17.9 Conclusions and Remarks	280
References	281

CHAPTER 18 THE BAYESIAN ANALYSIS OF CATEGORICAL DATA—A SELECTIVE REVIEW 283
T. Leonard and J. S. J. Hsu

18.1 The Foundations of the 1960s	283
18.2 Numerical Examples	287
18.3 Bayes–Stein Methods of the 1970s	289
18.4 Smoothing Grade Distributions for 40 London High Schools	293
18.5 Computational Techniques of the 1980s	295
18.6 Three-way Tables and Simpson's Paradox	296

18.7	Further Problems with Non-randomized Data	304
	Acknowledgements	306
	References	306

CHAPTER 19 CONFLICTING INFORMATION AND A CLASS OF BIVARIATE HEAVY-TAILED DISTRIBUTIONS 311
A. O'Hagan and H. Le

19.1	Heavy-tailed Bayesian Modelling	311
19.2	Multivariate Heavy-tailed Distributions	313
19.3	A Class of Bivariate Distributions	315
19.4	A Simple Example	317
19.5	A More Complex Example	321
19.6	Final Remarks	324
	Acknowledgements	325
	References	326

CHAPTER 20 APPLICATIONS OF LINDLEY INFORMATION MEASURE TO THE DESIGN OF CLINICAL EXPERIMENTS 329
G. Parmigiani and D. A. Berry

20.1	Introduction	329
20.2	First-order Conditions	332
20.3	Information and Sample Size	334
20.4	Duration and Follow-up Time	336
20.5	Multicentre Clinical Trials	341
20.6	Conclusions	345
	References	346

CHAPTER 21 ON TWO CLASSIC THEOREMS INVOLVING THE CHARACTERISTIC FUNCTION 349
W. L. Smith

21.1	Some General Comments	349
21.2	The Lindberg Central Limit Theorem	350
21.3	Cramér's Theorem on the Normal Distribution	359
	References	361

CHAPTER 22 HIERARCHICAL PRIORS AND MIXTURE MODELS, WITH APPLICATION IN REGRESSION AND DENSITY ESTIMATION 363
M. West, P. Müller and M. D. Escobar

22.1	Introduction	363

22.2	Hierarchical Models with Mixture Priors	364
22.3	Posterior Computations	366
22.4	A Regression Example	372
22.5	A Multivariate Density Estimation Example	376
	Acknowledgements	385
	References	385

Index 387

Preface

This volume contains a collection of previously unpublished papers, which, on behalf of their authors and his many other friends and colleagues around the world, we offer as a tribute to Dennis Lindley, who celebrated his 70th birthday on 25 July, 1993.

As is clear from his extensive publication list and from Peter Armitage's biographical essay, Dennis Lindley's interests and involvement over the years have touched upon many *Aspects of Uncertainty*.

A unifying theme in all his work has been an insistence on basing developments on firm logical foundations and on thinking things through from first principles.

Senior colleagues recall him as one of the first of his generation of post-war academics in the UK to insist that Probability Theory be taught from a rigorous measure-theoretic perspective. Rethinking the foundations of the Theory of Queues led to the celebrated 'Lindley's equation'. And, although his fundamental contributions and commitment to Bayesian statistics and decision theory are universally recognized, it is less well known that he arrived at the Bayesian position as a result of seeking to establish a rigorous axiomatic justification for classical statistical procedures!

The papers in this volume reflect Dennis Lindley's seminal influence on many areas of research and on his students and colleagues. May his influence long continue!

P. R. Freeman
A. F. M. Smith
September, 1993
London

List of Contributors

Professor P. Armitage
2 Reading Road
Wallingford
Oxon
OX10 9DP

Dr R. E. Barlow
IEOR Department
University of California
4177 Etcheverry Hall
Berkeley
CA 94720
USA

Professor G. A. Barnard
Mill House
54 Hurst Green
Brightlingsea
Colchester
Essex
CO7 OEH

Professor D. J. Bartholomew
Department of Statistics
London School of Economics and
Political Science
Houghton Street
London
WC2A 2AE

Professor J. A. Bather
Mathematics Division
University of Sussex

Brighton
BH1 9QH

Professor J. M. Bernardo
Departamento de Estadistia
Universidad de Valencia
Facultad de Mathemáticas
46100–Burjassot
Valencia
SPAIN

Professor D. A. Berry
Institute of Statistics and Decision
   Sciences
Duke University
Durham
NC 27706
USA

Professor P. J. Brown
Department of Statistics and
   Computational Mathematics
The University
PO Box 147
Liverpool
L69 3BX

Professor R. V. Brown
2018 Lakebreeze Way
Reston
VA 22091–4021
USA

LIST OF CONTRIBUTORS

P. J. Browne
Mathematics Division
University of Sussex
Brighton
BH1 9QH

Professor S. Campodónico
Department of Operational
  Research and Stastics
George Washington University
707 22nd Street NW
Washington DC 20052
USA

Dr K. Chaloner
Department of Applied Statistics
University of Minnesota
352 Classroom Office Building
1994 Buford Avenue
St Paul
MN 55108
USA

Professor A. P. Dawid
Department of Statistical Science
University College London
Gower Street
London
WC1E 6BT

Dr J. J. Deely
Department of Mathematics
University of Canterbury
Christchurch
New Zealand

Dr M. D. Escobar
Department of Statistics
Baker Hall 232-212
Carnegie-Mellon University
Pittsburgh
PA 15213
USA

Professor P. R. Freeman
Ship Cottage
Cadgwith
Helston
Cornwall
TR12 7JX

Professor S. French
School of Computer Studies
The University
Leeds
LS2 9JT

Professor E. I. George
Department of Management Science
  and Information Systems
CBA 5.202
University of Texas at Austin
Austin
TX 78712-1175
USA

Professor M. Goldstein
Department of Mathematical
  Sciences
Durham University
Sciences Laboratories
South Road
Durham
DH1 3LE

Professor B. M. Hill
Department of Statistics
University of Michigan
419 S State
Ann Arbor
MI 48109-1027
USA

Dr J. S. J. Hsu
Department of Statistics and Applied
  Probability
University of California at Santa
  Barbara

Santa Barbara
CA 93106-3110
USA

Professor J. B. Kadane
Department of Statistics
Carnegie-Mellon University
Pittsburgh
PA 15213
USA

Dr F. R. Lad
Department of Mathematics
University of Canterbury
Christchurch
New Zealand

Dr H. Le
Department of Mathematics
University of Nottingham
University Park
Nottingham
NG7 2RD

Dr N. D. Le
British Columbia Cancer Agency
Biometry Section
600 W. 10th Avenue
Vancouver
British Columbia V5Z 4E6
Canada

Dr T. Leonard
Department of Statistics
University of Wisconsin-Madison
1210 West Dayton Street
Madison
WI 53706-1693
USA

Dr U. E. Makov
Department of Statistics
University of Haifa
Mount Carmel

31999 Haifa
Israel

Dr M. B. Mendel
IEOR Department
University of California
4177 Etcheverry Hall
Berkeley
CA 94720
USA

Dr P. Müller
Institute of Statistics and Sciences
Duke University
Durham
NC 27708-0251
USA

Professor A. O'Hagan
Department of Mathematics
University of Nottingham
University Park
Nottingham
NG7 2RD

Dr G. Parmigiani
Institute of Statistics and Decision
  Sciences
Duke University
Durham
NC 27706
USA

Professor N. D. Singpurwalla
Department of Operational Research
  and Statistics
George Washington University
707 22nd Street NW
Washington DC 20052
USA

Professor A. F. M. Smith
Department of Mathematics

*Imperial College of
 Science, Technology
 and Medicine
Huxley Building
180 Queen's Gate
London
SW7 2BZ*

Professor W. L. Smith
*302 Meadow Lane
Chapel Hill
NC 27514
USA*

Dr M. West
*Institute of Statistics and Sciences
Duke University
Durham
NC 27708–0251
USA*

Dr J. V. Zidek
*Department of Statistics
University of British Columbia
Vancouver
British Columbia V6T 1Z2
Canada*

CHAPTER 1

Dennis Lindley: the First 70 Years

Peter Armitage
Emeritus Professor of Applied Statistics, University of Oxford

Dennis Lindley's teenage ambition was to become an architect. It is interesting to speculate how the post-war history of architecture might have been affected if he had persisted in this choice of career (for one cannot imagine any profession being uninfluenced by his membership). Intellectually and aesthetically, his sympathies would have been with the modernist movement. But that movement formed the post-war establishment, and Dennis's radical temperament would surely have reacted against it. I suspect that that we should have seen a quite different form of post-modernist architecture from that which has emerged during the last few decades.

However, all that is pure conjecture. Architecture's loss was a gain for statistics. Dennis's conversion to mathematics owed a great deal to a remarkable mathematics schoolmaster, M. P. Meshenberg, who persuaded his parents that he had serious potential as a mathematician, and should remain at school to take the Cambridge entrance examination rather than leave early to join an architectural office. Mr Meshenberg gave Dennis personal tuition for the examination over a period of several months during the bombing of London, mostly in an air-raid shelter.

Dennis Victor Lindley was born on 25 July 1923. His family lived in Clapham, a southern suburb of London, and moved in 1930 to Surbiton, a south-western outer suburb. His father was a local building contractor. Dennis's secondary education took place at Tiffins', a well-known grammar school in Kingston upon Thames. His preparation for Cambridge involved additional cramming for Latin, a subject in which modest competence was

---

Aspects of Uncertainty edited by P. R. Freeman and A. F. M. Smith. © 1994 John Wiley & Sons Ltd.

required. Dennis attributes his success in obtaining an entrance scholarship less to his mathematical prowess than to the subject presented to candidates as the topic for a general essay: the English parish church. Architecture and mathematics had combined to ensure a sound start to his academic career.

During the war science students were normally allowed to attend a university for two years. The more able mathematics students at Cambridge skipped the first year (Tripos Part I), started with the second-year Preliminary Examination, and at the end of their second year took the final Tripos Part II. In normal times these students would occupy themselves in the third year with the Tripos Part III, roughly at Master's level. Dennis entered Trinity College, Cambridge in 1941, took Part II in 1943 and obtained a first-class degree (thereby becoming a wrangler). His studies were then interrupted, but he returned to take Part III in 1946–7, gaining a distinction.

At Trinity the senior mathematician in residence was A. S. Besicovitch, an analyst of international distinction, and a lovable personality who had an entirely individual view of the syntax and vocabulary of the English language. Besicovitch gave most of the one-to-one supervisions in pure mathematics, and Dennis recalls also with appreciation his applied mathematics tutorials with Frank Powell at Caius College. Of statistics there was very little. Maurice Bartlett and John Wishart were away on wartime duties. A course was given by J. Oscar Irwin, who had moved temporarily from London to Cambridge, and a number of mathematics students leaving in 1943 were strongly advised to take this course so as to equip themselves to do statistical work as their war service.

Dennis was appointed, with many of his contemporaries, to a unit of the Ministry of Supply, named SR17, concerned largely with quality control and sampling inspection. The background to the formation of this unit, and an outline of its activities, have been described by Barnard and Plackett (1985). Its director was John Womersley, an applied mathematician with experience in the cotton industry, and those appointed before the 1943 cohort arrived included George Barnard, Frank Anscombe, Robin Plackett and Denis Newman. The group worked in central London, near Baker Street station. The raids by manned planes had ceased by 1943, but they were succeeded in the following year by the V1 flying bombs and the V2 rockets. These had an irritant effect, but caused very little interruption of daytime work.

Dennis was assigned to a subgroup concerned mainly with methodological development, and occupied a room for much of the time with Arnold Baines, Stanley Collings, Jim Godwin, Christopher Winsten and myself. None of this group had had any extensive training in statistics (Irwin's course at Cambridge being too short and too theoretical to impart much of a feeling for applied work). There was, therefore, a good deal of mutual instruction, which extended beyond statistics to all aspects of the human condition. Arnold Baines in particular had an encyclopaedic knowledge matched to a liberal outlook, and would often be goaded in a friendly way by Dennis's left-wing and somewhat astringent views. Dennis had an extensive knowledge of Marxist theory, but his enthusiasm was entirely intellectual, untouched by considerations of party politics (which were in any case largely in abeyance during the war). It is tempting to see a connection here with his later espousal of another all-embracing system— Bayesian inference, not of course in the content of these two systems, but in their claims to be self-contained and comprehensive.

Most of the projects undertaken during this period by Dennis, and indeed by other members of SR17, were ephemeral. Perhaps the most durable contribution from the unit was the work of Plackett and Burman (1946) on fractional factorial designs, for which Dennis provided some results which formed the basis of one of his earliest papers [2].

George Barnard recalls an incident during SR17 days, when he attended meetings of a committee on anti-aircraft fuses.

At one of these the question came up of estimating the 'time to shot start' (T)—that is, the length of the interval between the firing of the gun and the initial movement of the shell up the gun barrel. For technical reasons the interval in question had to be obtained as the difference between time readings on two independent clocks. Dennis landed the job of calculating the 95% confidence interval for T, which turned out to lie entirely within the negative part of the real axis. I remember the dubious looks on the faces of the rest of the committee when I explained that it was a property of a 95% interval that, once in 20 times, it would fail to cover the true value, and that this had clearly occurred in the present case.

George suggests, perhaps only partly in jest, that Dennis's disenchantment with confidence intervals derives unconsciously from this early unfortunate experience.

Several members of SR17 were engaged in the development of sequential schemes for sampling inspection. The calculation of their operating characteristics required exact values of binomial coefficients with high indices, and the occupants of Dennis's room filled in idle moments by constructing a large Pascal's triangle which eventually filled most of one wall of the room. Some of the entries, which went up to about $n = 125$, had enormous numbers of digits and were quite tricky to calculate with the sort of electromechanical machines then available. Of the five occupants, only Dennis was deemed to have sufficiently neat handwriting to justify public exposure, and he was given the job of transcribing entries into the final table. This masterpiece of wallpaper design was made available to the compilers of the Royal Society's *Table of Binomial Coefficients*, but is now unfortunately lost.

During this period Dennis started research in linear algebra at Birkbeck College, London, under Paul Dienes, a Hungarian refugee. He may have had in mind the possibility of a future career in non-statistical mathematics, but the main stimulus came from his schoolmaster, Mr Meshenberg, who urged him to keep his mathematical faculties in good shape. Dennis eventually abandoned this line of work, although he does claim to have found necessary and sufficient conditions for a matrix to have a square root.

In 1945 SR17 was disbanded, although most of the members were required to continue their period of national service. Dennis moved, with some other colleagues, to join a new statistics section in the Mathematics Division of the National Physical Laboratory at Teddington, a London suburb not far from his home in Surbiton. The two senior statisticians were Edgar Fieller and Edward van Rest, both with pre-war experience of biological statistics. There was much excitement about the planning and construction of the electronic computer ACE, under the guidance of Alan Turing, who wandered around as a sort of free spirit, and was often to be seen in running shorts on the towpath of the Thames.

Dennis was able to complete some theoretical projects which had started to intrigue him in SR17. The most important of these was his work on regression and functional relationships, a topic proposed by Fieller. The resulting paper, published in 1947 [3], is a major contribution, and is written in an elegant if somewhat idiosyncratic style which he did not persist with in later papers. Another project initiated by Fieller concerned the case for and against using Sheppard's correction for the estimate of variance in a significance test as distinct from pure estimation. This was published later, in [5].

His final year at Cambridge, apart from its academic success, enabled him to make the acquaintance of some senior statisticians. Bartlett and Wishart were back, Harold Jeffreys gave somewhat opaque lectures on probability and R. A. Fisher had recently arrived as Professor of Genetics. Dennis and I first met Fisher in an unusual context. We had attended an inaugural meeting of a new Society for Psychical Research, the initiators of which proposed to conduct a supposedly decisive experiment to discriminate between telepathy and clairvoyance. The chairman asked whether there were any statisticians in the audience who would like to help. Dennis and I put our hands up, and were told to go and see one of the dons who was supposed to be an expert: the chairman thought his name was Fisher. We duly went along to Fisher's room one evening when we were generously plied with beer and given a very friendly tutorial on experimental design and the like. The proposed experiment never took place, but the encounter was memorable.

Fisher was customarily friendly and helpful to young people who offered no resistance to his views, but was liable to become strongly antagonistic towards those who were less compliant. He and Dennis had virtually no direct contact after that first meeting, even when Dennis was on the academic staff. Fisher's attitude to Dennis changed radically during the 1950s, following a review by Dennis [16] of his book *Statistical Methods and Scientific Inference* published in Fisher's favourite journal *Heredity*, and papers [15] and [17] criticizing Fisher's use of fiducial distributions in a Bayesian manner. The strength of Fisher's feelings can be judged by the disagreeable references to Dennis in Fisher's collected correspondence (Bennett 1990).

In 1947 Dennis married Joan Armitage, a mathematics graduate from London University, whom he had met on a walking tour of the English Lake District (Figure 1.1). This was the beginning of a stable and durable partnership. Their children, Janet, Rowan and Robert, were born in 1949, 1951 and 1956, respectively (Figure 1.2).

Dennis returned to the National Physical Laboratory for one year, but in 1948 he was appointed to the Statistical Laboratory at Cambridge where he stayed until 1960, first as demonstrator (the equivalent, for science subjects, of the newly created grade of assistant lecturer, with responsibilities weighted towards research rather than teaching), then as lecturer and later (after Wishart's early death) as Head of the Laboratory. This was a period of remarkable brilliance in Cambridge statistics. Wishart was joined by Frank Anscombe, Henry Daniels, David Cox and Dennis, and research students included Walter Smith (becoming a lecturer in 1957) and Ewan Page (later Vice-Chancellor of Reading University). Fisher was round the corner, with distinguished visitors such as C. R. Rao, but he distanced himself from the Statistical Laboratory and a closer contact was his younger colleague George Owen. And then there was Harold Jeffreys, whose contact with statistical organizations was minimal, and who had resolutely maintained his advocacy of inverse probability, in the face of a hostile statistical establishment, for over 20 years. Dennis's growing admiration for Jeffreys' contributions to statistical inference is expressed in [60], [87] and [112], and in the video-recording of an interview of Jeffreys by Dennis for the Royal Statistical Society.

It is remarkable and regrettable that during his appointments at Cambridge Dennis was not offered a fellowship at his old college, Trinity. At that time the commonality of interest between college and university, at both Cambridge and Oxford, was less widely accepted than is now the case, but even so the omission is striking, and perhaps is an indication of the relatively weak position then occupied by statistics in the academic hierarchy.

In an interview by Ray Harris (1983), Dennis recalled that his conversion to the Bayesian approach was gradual. He was deeply impressed by Jeffreys' writings, and uneasy about the traditional approaches to inference, both of Fisher and of Neyman and Pearson. His 1953 paper [10] on inference reveals an attempt to recast frequentist ideas in a semi-Bayesian framework, using a concept of 'minimum unlikelihood', but he insisted that he was an unrepentant frequentist. The crucial event was a one-year visit to the University of Chicago in 1954–5 at the invitation of Allen Wallis whom he had met in London. In Chicago he worked closely with

L. J. (Jimmie) Savage, whose influential book, *The Foundations of Statistics*, had been published in 1954. At that time Dennis's ambition was to redress what he regarded as the logical deficiencies of the Neyman–Pearson approach, but more with the idea of making it mathematically rigorous than of overthrowing it. Savage was at much the same point in his thinking. Together they moved towards an unambiguously Bayesian position. Dennis also visited Stanford before his return to Cambridge.

Walter Smith recalls various contacts with Dennis during their time together at Cambridge in the 1950s, which were 'simultaneously acts of kindness and crucial for my career'. In 1950–51 Dennis gave an advanced course on probability theory, making use of measure theory, with mimeographed notes, which Wally regards as 'the most important for me of all the lectures I ever attended as a graduate student'. Perhaps more importantly, Dennis encouraged Wally to branch away from his thesis topic concerned with nerve pulses in cockroaches, and to develop an interest and expertise in the theory of queues, in which Dennis was then writing his celebrated paper [8]. Wally also recalls Dennis's learning Russian at this time, and testing his skills by translating a paper by Kolmogorov. Dennis's concern for his colleagues' interests is illustrated by an incident during his period as Head of the Laboratory: he arranged for a skilful non-graduate computing assistant to be awarded degree status.

The well-known *Cambridge Elementary Statistical Tables*, published jointly with J. C. P. Miller in 1953, and in the second edition with W. F. Scott in 1985, could be seen, perhaps, as a last concession to frequentist statistics. They arose from a request from the Cambridge University Press, which had had to provide *ad hoc* tables annually for various examinations, and realized the advantage of having a standard set permanently available.

The wartime developments in statistics and operational research led to a gradual enhancement of these subjects in British universities during the following decades. In many institutions statistics edged its way in through joint honours degrees and statistics groups within mathematics departments, but the creation of chairs in statistics was a slow process. London led the way, with Egon Pearson in the established chair at University College, Roy Allen and Maurice Kendall appointed to chairs at the London School of Economics in 1944 and 1949 and George Barnard to a chair at Imperial College in 1954. Maurice Bartlett was appointed at Manchester in 1947 and Henry Daniels at Birmingham in 1957. Cambridge did not create a chair until 1962.

But where the rest of England hesitated, Wales stepped in. In 1960 Dennis was appointed Professor and Head of the Department of Statistics at the University College of Wales, Aberystwyth. The funding position here was relatively generous, for both buildings and academic posts, and Dennis received good support for his subject. Among his colleagues were Mervyn Stone, David Bartholomew, Ann Mitchell and Roger Miles.

Figure 1.1 *(above)* In the Lake District, 1946. Dennis Lindley (top left), Joan Armitage (later Lindley) (bottom right).

Figure 1.2 *(below)* Three generations of the Lindley family, 1989.

Figure 1.3 *(right)* At Berkeley, early 1980s.

Figure 1.4 Dennis and Joan at Minehead, 1992.

Figure 1.5 The archetypal progressive: Volvo, *Guardian* and green wellies.

By the time of his appointment at Aberystwyth Dennis was clearly the leader of the British Bayesian movement. Another enthusiast, I. J. Good, had departed for the United States, and the veteran Wilfred Perks confined his Bayesian activities to the theoretical fringes of the actuarial profession. Dennis was increasingly influential amongst the younger academic statisticians. He could be relied on to present a combative display at discussion meetings of the Royal Statistical Society (RSS), and gave the impression of having a personal mission to change the face of British statistics. This stance might have been faintly ridiculous in a lesser man, but no one doubted the sincerity of his views or the intellectual power of his arguments.

Dennis's debating strength had been clear from his earliest appearances at society meetings. Walter Smith recalls the first meeting he ever attended, the RSS symposium on stochastic processes in 1949. He was impressed by a person

who seemed, considering his manner, almost too young. He went for the giants with terrier-like intrepidity. 'What', he repeatedly demanded of an increasingly disconcerted speaker, in a tone that seemed to be approaching irritation, if not anger, 'What *is* a stochastic process?' As an innocent by stander I had to admit that on this topic the self-assertive young man never received a satisfactory answer from any quarter: it was then a few years before Doob's book would arrive in England.

During his period at Aberystwyth he had visiting appointments at the Harvard Business School, where he found like minds in Howard Raiffa and Robert Schlaifer, whose joint book *Applied Statistical Decision Theory* had recently been published, and at the University of Rome, at the invitation of Bruno de Finetti (whom he had met at the Berkeley symposium, and again later in London). In an interview by John Deely in 1990 at the University of Canterbury, New Zealand, Dennis was asked which of the statisticians he had known could be classified as geniuses. Having discounted Fisher as a genius whom he did *not* know, he nominated Jeffreys and de Finetti, and spoke particularly highly of Schlaifer.

In 1965 Dennis published the two-volume work *Introduction to Probability and Statistics (from a Bayesian Viewpoint)*. This has been extremely influential in explaining Bayesian methods in not too technical a way. The second volume, on inference, stresses the correspondence that often exists between Bayesian and frequentist methods. With an appropriately diffuse 'ignorance' prior, the posterior probability that a parameter is on a specified side of some hypothesized value is often the same as the tail-area probability in a frequentist significance test. Later in his career Dennis would regard this feature with less enthusiasm, but it had didactic value at a time when most readers were unsympathetic to the Bayesian approach.

In 1967 Maurice Bartlett vacated the chair of statistics at University College London (UCL) to move to Oxford, and Dennis was appointed as his successor. The department had been formed on the retiral of

Karl Pearson as Galton Professor of Eugenics in 1933, with R. A. Fisher appointed to the eugenics chair, E. S. Pearson to the readership (later chair) in statistics and J. B. S. Haldane to a chair in biometry. Jerzy Neyman had spent many years there before the war in partnership with Egon Pearson. The academic ancestry was, therefore, formidable, and, as Ray Harris remarks, the electors 'showed courage and tolerance—which Professor Lindley himself regards as necessary for progress in any science—in choosing such an accomplished and eloquent Bayesian to head the Department of Statistics'.

Egon Pearson remained as an honorary professor in the department for several years, although his long stint as editor of *Biometrika* had terminated in 1966. Pearson was a model of courtesy and tolerance, and never allowed his philosophical reservations about the Bayesian approach to mar his cordiality towards Dennis. Other colleagues in the department included Toby Lewis, Alan Hawkes, Adrian Smith, Mervyn Stone, Peter Freeman and Phil Dawid. Also at UCL, in the neighbouring Galton Laboratory, Cedric Smith had for some time been flying the Bayesian flag in statistical genetics. Contrary to some expectations, the syllabus at UCL did not change overnight. Dennis taught a course on Bayesian statistics, but if he had any ambition to revolutionize the whole teaching programme it would have foundered on the relative paucity and impracticability of the Bayesian methods on offer at that time.

José Bernardo, a research student of Dennis's at UCL, later became a close friend and colleague. He recalls admonishments from Dennis both for an undue liking for topological spaces and for misdemeanours such as driving a car on the UCL lawn and breaking the mainframe code to use more CPU time than he was entitled to.

During the appointment at UCL, Joan and Dennis lived in Hampstead Garden Suburb, some five miles from the centre of London. Dennis had a year's visit to the University of Iowa in 1974–5, as a distinguished visiting professor, but otherwise he remained academically anchored in Gower Street. In the mid-1970s the department was enlarged to absorb the previous Department of Computer Science, and Dennis headed the joint enterprise. Although he admits to an aversion to committee work, he had served as Secretary of the Royal Statistical Society's Research Section from 1956 to 1958, and as Chairman from 1963 to 1965. In 1968 he received the Society's Guy Medal in silver. He acted as Business Manager of *Biometrika* from 1969 to 1977, and has continued to serve as a trustee of that journal.

Dennis published two books in 1971: *Making Decisions*, which was translated into Spanish, Italian and German, and had a second edition in 1985; and *Bayesian Statistics: a Review*, a short SIAM monograph. Both these books were well received and have continued to be cited.

In 1977, at the age of 54, Dennis surprised the academic statistical world by announcing his retiral. He made clear his intention to continue research and other statistical activities, and attributed the move to an increasing distaste for academic administration. (He emphasizes that there was no disagreement with UCL, for which he has continued to express affection and admiration.) To those who lived through the pressures of the 1980s, stemming from the increasing financial stringency and governmental regulation, the 1970s seem in retrospect to have been comparatively idyllic, and I am sure that Dennis feels that he escaped just in time. Nevertheless, his exit from university statistics left a big gap, and must have seemed bewildering to his younger followers who might reasonably have expected to see him leading the development of a strong Bayesian school in Britain for the next 15 years or so. It would be quite wrong, though, to cast him as a 'lost leader', for he has remained a prominent figure during the subsequent period, both in Britain and abroad. But many will regret that his experience and enthusiasm have not been more directly available for the next generation of students and researchers.

Dennis immediately celebrated his liberation from the constraints of academic statistics by growing a beard. It is perhaps a Darwinian rather than a Fisherian beard: apart from the similarity of shape, Darwin had also started his beard in mid-career.

Dennis and Joan moved to Minehead, in Somerset, where they still live. Dennis accepted long-term honorary professorial posts at Aberystwyth and Warwick, a three-year appointment at Bath and two separate appointments as Erskine fellow at Canterbury, New Zealand. In this way he maintained contact with major statistics departments and has had an active programme of seminars in each. Until the late 1980s the general plan was to visit one or more overseas university during the winter months, returning to England during the climatically more hospitable summer, when Dennis and Joan could enjoy their garden and fruit orchard and explore the magnificent countryside within easy reach of their home. Dennis refers to his hobbies as statistics in the winter and gardening in the summer.

During his so-called 'retirement', Dennis has had visiting posts in Australia, Brazil, Canada, South Africa and the United States. In his many visits to the United States he has worked at the University of California, Berkeley (Figure 1.3) and Davis; the Business School, University of Chicago; the University of Wisconsin, Madison; Florida State University, Tallahassee; George Washington University, Purdue University and Duke University. He was the Wald lecturer for the Institute of Mathematical Statistics in 1988.

Dennis's research interests have been predominantly in the foundations of statistics and decision theory, and in the detailed implementation of Bayesian principles. His less frequent involvement with data analysis has, nevertheless, brought him into contact with a number of different fields of application. These interests have included medicine, reliability and educational statistics. (In the latter field he worked closely with the late

Melvin Novick; see [95].) But his attention has been drawn most strongly to the application of probability and statistics to legal problems, on which he has published prolifically since his 1975 paper [42]. His contributions in this area have been confined mainly to the theoretical aspects of the assessment of probabilities of innocence and guilt, and he has, for instance, avoided the trauma of appearance as an expert witness in specific cases.

The international Bayesian community has found a stimulating focus in the four-yearly meetings at the University of Valencia hosted by José Bernardo, the first of which was held in 1979; the fifth is planned for 1994. Dennis has played an active role in these conferences, as a speaker ([61], [83], [102] and [115]), as a programme committee member and joint editor of the series of conference proceedings, and latterly as President.

I have made no attempt in this memoir to give a comprehensive and critical survey of Dennis's publications. This would be a formidable task, best left to those who have worked in similar fields. Many insights into the influence of his work will be given in the later chapters of this book. In such a rich feast it is difficult to identify the choicest dishes. Nozer Singpurwalla recalls a conversation with Dennis during a breakfast stop on a long drive in 1992 between Washington, DC and Durham, North Carolina. Nozer asked Dennis which of his papers he considered to be his best, expecting him to choose his 1956 paper [13] on the information provided by an experiment:

Dennis stroked his beard several times and said that my question was a difficult one to answer. When I insisted that he produce an answer, he declared that his paper with Adrian Smith on the linear model [36] was his best, because it had the most impact. I asked Dennis to think harder and try again. Dennis by now was getting interested in this trivial pursuit, and after a few more strokes of his beard declared that the one on queuing, involving 'Lindley's equation' [8], might also qualify. This pastime continued for a few more iterations, during which Dennis suddenly switched roles by asking me which of my papers I considered the best. The answer for me was very easy: the paper with Dennis on adversarial testing [113]. Upon hearing this, Dennis at the top of his voice said 'Ah! now I see, is this what you wanted to hear me say?'

In the taped interview with John Deely, referred to earlier, Dennis selects two other papers as his favourites: the 1982 paper [72] on the 'inevitability of probability', and the 1984 paper [81] on a Bayesian lady tasting tea. (The latter arose from admiration of the writings of both Fisher and Neyman on the subject, and the posing of a challenge to himself to formulate a Bayesian solution.) The truth is, of course, that there are very many papers in the Lindley archives of which he can scarcely avoid a sense of pride, and his personal pleasure surface no doubt has an extremely flat maximum.

Many statisticians who have seen Dennis in action at conferences and other meetings will have come away with a misleading impression, as a result of his frequent and withering attacks on his colleagues' views. The distinction between animosity to individuals (which he does not exhibit) and hostility towards certain opinions (which he frequently exhibits) is vitally important. In the interview with John Deely, Dennis points out that the belligerence of which he is often accused is a necessary attribute for someone fighting the establishment, the upholders of which can afford to take a more relaxed and tolerant attitude. His stance is one of complete confidence in the validity of his case, and he has no sympathy with those, like myself, who advocate an eclectic approach in which Bayesian and non-Bayesian methods coexist in some degree of amity. In his view one approach is right and the other wrong, one is rational and the other not, and it is neither necessary nor desirable to admit compromise. He is somewhat surprised that the conversion of the statistical world, although under way, is so slow. He agrees, though, with de Finetti that the process will have been completed by 2020, and hopes to survive to see his prophecy fulfilled.

His uninhibited criticism can sometimes bruise the recipient unaware of the distinction between personal and philosophical disagreement. His friend Richard Barlow recalls his first meeting with Dennis, in London in 1966. Richard gave a paper on tolerance intervals. 'At the end of the talk, Dennis rose from the audience and said something like "Why are people still talking about tolerance intervals? My research student has taken care of this problem properly. Have you read his thesis?"'
Richard was left with a distinctly unfavourable view of Dennis, which persisted until they met again in Berkeley, after which they became firm friends. (Dennis was invited to the Operations Research Department at Berkeley, but never felt at home in the Statistics Department; he was led later to use the phrase 'Berkeley' statistics to indicate both an approach used by that department with which he found himself out of sympathy, and also the fact that Bishop Berkeley had been in dispute with Thomas Bayes.)

Even a brief acquaintance with Dennis shows that any disagreements on technical matters are overruled by his congeniality. He has throughout his career been a keen participant in the activities of the Royal Statistical Society, many members of which know him primarily as a highly sociable colleague rather than as the protagonist of a particular approach to the subject. Indeed, he is something of a *bon viveur*, having recently written an article about the statistics of wine tasting, which earned him an invitation to contribute to a forthcoming *Oxford Companion to Wine*.

Dennis must have more friends and acquaintances around the world than all but a few other statisticians. Some will see him as an inspiring leader, some as a formidable antagonist, others as an archetypal progressive (Figure 1.5, of which he is very fond, shows him exhibiting the features claimed in a social study to define a recognizable subset). All will see him

as a delightful and stimulating friend, and one of the most penetrating thinkers in the statistical world today. It is gratifying to find his enquiring mind as active as ever at the age of 70, and we offer our felicitations.
HAPPY BIRTHDAY, DENNIS!

ACKNOWLEDGEMENTS

I am grateful to Richard Barlow, George Barnard, José Bernardo, Nozer Singpurwalla and Walter Smith for sending me their recollections of Dennis Lindley, and to Richard Barlow for permission to reproduce Figures 1.3 and 1.4. I am indebted to Dennis for the pleasure of many conversations, for permission to reproduce Figures 1.2 and 1.5, and for the opportunity to see the taped interview by John Deely.

REFERENCES

[The numbered references to Dennis Lindley's publications refer to the bibliography below]

Barnard, G. A. and Plackett, R. L. (1985), Statistics in the United Kingdom, 1939–45, in A. C. Atkinson and S. E. Fienberg, eds, *A Celebration of Statistics*, New York, Springer-Verlag, pp. 31–55.

Bennett, J. H. (1990) (ed.). *Statistical Inference and Analysis: Selected Correspondence of R. A. Fisher*, Oxford, Clarendon Press.

Harris, R. R. (1983) Dennis Lindley, *The Professional Statistician*, **2** (4), 4–6.

Plackett, R. L. and Burman, J. P. (1946) The design of multifactorial Books experiments, *Biometrika*, **33**, 305–25.

BOOKS

Cambridge Elementary Statistical Tables (with J. C. P. Miller). Cambridge: University Press (1953), 35 pp. Second edition *New Cambridge Elementary Statistical Tables* (with W. F. Scott) (1985), 80 pp.

Introduction to Probability and Statistics (From a Bayesian Viewpoint), 2 vols, Cambridge: University Press (1965), 259, 292 pp. Japanese translation.

Making Decisions. John Wiley & Sons, Inc., New York (1971), 195 pp. Spanish, Italian and German translations. Second edition (1985), 207 pp.

Bayesian Statistics: A Review. SIAM, Philadelphia (1971), 83 pp.

BIBLIOGRAPHY

[1] Linear 'curves of best fit' and regression lines. *Nature*, 24 August 1946.
[2] On the solution of some equations in least squares. *Biometrika*, **33**, 326–7 (1946).

[3] Regression lines and the linear functional relationship. *J. Roy. Statist. Soc.* (suppl.), **9**, 218–44 (1947).
[4] Mortality during the 1st world war and during and after the 2nd world war (with E. R. Bransby). *Mon. Bull. Ministry of Health and Pub. Health Lab. Service*, 2–12 (1950).
[5] Grouping corrections and maximum likelihood equations. *Proc. Camb. Philos. Soc.*, **46**, 106–10 (1950).
[6] A regression problem. *Proc. Camb. Philos. Soc.*, **47**, 337–46 (1951).
[7] The mathematical theory of marshalling and queuing. *Oper. Res. Quarterly*, **3**, 4–8 (1952).
[8] The theory of queues with a single server. *Proc. Camb. Philos. Soc.*, **48**, 277–89 (1952).
[9] Estimation of a functional relationship. *Biometrika*, **40**, 47–9 (1953).
[10] Statistical inference. *J. Roy. Statist. Soc.*, **B15**, 30–76 (1953).
[11] The estimation of velocity distributions from counts. *Proc. Int. Congress Math.*, pp. 427–44 (1954).
[12] Application of biometric methods to problems of classification in ecology (with R. E. Hughes). *Nature*, **175**, 806 (1955).
[13] On a measure of the information provided by an experiment. *Ann. Math. Statist.*, **27**, 986–1005 (1956).
[14] Binomial sampling schemes and the concept of information. *Biometrika*, **44**, 179–186 (1957).
[15] A statistical paradox. *Biometrika*, **44**, 187–192 (1957).
[16] Review of '*Statistical Methods and Scientific Inference*' by R. A. Fisher, *Heredity*, **11**, 280–82 (1957).
[17] Fiducial distributions and Bayes theorem. *J. Roy. Statist. Soc.*, **B20**, 102–7 (1958).
[18] A survey of the foundations of statistics. *Appl. Statist.*, **7**, 186–98 (1958).
[19] Tables for making inferences about the variance of a normal distribution (with D. A. East, P. Hamilton). *Biometrika*, **47**, 433–7 (1960).
[20] The robustness of interval estimates. *Bull. Inst. Internat. Statist.*, **38**, 209–20 (1961).
[21] Dynamic programming and decision theory. *Appl. Statist.*, **10**, 39–51 (1961).
[22] The use of prior probability distributions in statistical inference and decision. *Proc. Fourth Berkeley Sympos. Math. Stat. Prob.*, **1**, 453–68 (1961).
[23] An experiment in the marking of an examination. *J. Roy. Statist. Soc.*, **A124**, 285–313 (1961).
[24] Review of '*Theory of Probability*' (3rd edition) by H. Jeffreys. *J. Amer. Statist. Ass.*, **57**, 922–4 (1962).
[25] A treatment of a simple decision problem using prior probabilities. *The Statistician*, **12**, 1–11 (1963).
[26] The Bayesian analysis of contingency tables. *Ann. Math. Statist.*, **35**, 1622–43 (1964).

[27] Sequential sampling: Two decision problems with linear losses for binomial and normal random variables (with B. N. Barnett). *Biometrika*, **52**, 507–32 (1965).
[28] A report on university staff and students in statistics. *J. Roy. Statist. Soc.*, **A129**, 467–470 (1966).
[29] Review of '*Mathematical Theory of Probability and Statistics*' by R. von Mises. *Ann. Math. Statist.*, **37**, 747–54 (1966).
[30] The choice of variables in multiple regression. *J. Roy. Statist. Soc.*, **B30**, 31–66 (1968).
[31] The Bayesian estimation of a linear functional relationship (with G. M. El-Sayyad). *J. Roy. Statist. Soc.*, **B30**, 190–202 (1968).
[32] Decision-making. *The Statistician*, **18**, 313–26 (1969).
[33] Bayesian least squares. *Bull. Inst. Internat. Statist.*, **43**, 152–3 (1969).
[34] A non-frequentist view of probability and statistics, in *The Teaching of Probability and Statistics* (ed. Lennart Rade), Almquist and Wicksell, pp. 209–22 (1970).
[35] The estimation of many parameters, in *Foundations of Statistical Inference* (eds. V. P. Godambe and D. A. Sprott), Toronto: Holt, Rinehart and Winston, pp. 435–55 (1971).
[36] Bayes estimates for the linear model (with A. F. M. Smith). *J. Roy. Statist. Soc.*, **B34**, 1–41 (1972).
[37] Bayesian statistics. *Bull. Inst. Math. App.*, **8**, 183–7 (1972).
[38] The future of statistics—a Bayesian 21st Century. *Supp. Adv. Appl. Prob.*, **7**, 106–15 (1975).
[39] Probability and medical diagnosis. *J. Roy. Coll. Phycns. Lond.*, **9**, 197–204 (1975).
[40] The role of utility in decision-making. *J. Roy. Coll. Phycns. Lond.*, **9**, 225–30 (1975).
[41] The effect of ethical design considerations on statistical analysis. *J. Roy. Statist. Soc.*, **C24**, 218–28 (1975).
[42] Probabilities and the law, in *Utility, Probability and Human Decision Making*, (ed. Wendt and Vick), Dordrecht: Reidel, pp. 223–32, (1975).
[43] Bayesian statistics, in *Foundations of Probability Theory, Statistical Theories of Science*, vol. II. (ed. Harper and Hooker), Dordrecht: D. Riedel, pp. 353–63 (1976).
[44] A class of utility functions. *Ann. Statistics*, **4**, 1–10 (1976).
[45] Inference for a Bernoulli process (a Bayesian view) (with L. D. Phillips). *Amer. Statistician*, **30**, 112–19 (1976).
[46] A Bayesian solution for two-way analysis of variance. *Coll. Math. Soc. Janos Bolyai*, **9**, 475–96 (1976).
[47] Costs and utilities, in *Decision Making and Medical Care—Can Information Science Help?* (eds. F. T. de Dombal and F. Gremy), Amsterdam: North-Holland, pp. 101–12 (1976).

[48] A survey of statistical thought. *Bull. Appl. Stat.*, **4**, (1977).
[49] The measurement and evaluation of risk. *Post Mag. and Insurance Monitor*, August 1977.
[50] A problem in forensic science. *Biometrika*, **64**, 207–13 (1977).
[51] The distinction between inference and decision. *Synethese*, **36**, 51–8 (1977).
[52] Probability and the law. *The Statistician*, **26**, 203–20 (1977).
[53] The concept of coherence in statistics, in *I Fondamenti dell' Inferenze Statistica*, Florence, pp. 179–207 (1978).
[53A] The Bayesian approach. *Scand. J. Statist.*, **5**, 1–26 (1978). [Reprint of 53.]
[54] The use of more realistic utility functions in educational measurement (with M. R. Novick). *J. Educ. Measurement*, **15**, 181–91 (1978).
[55] Fixed-state assessment of utility functions (with M. R. Novick). *J. Amer. Statist. Ass.*, **74**, 627–33 (1979).
[56] Analysis of life tables with grouping and withdrawals. *Biometrics*, **35**, 605–12 (1979).
[57] Statistics as a mathematical discipline. *New Zealand Statistician*, **14**, 15–20 (1979).
[58] On the reconciliation of probability assessments (with A. Tversky and R. V. Brown). *J. Roy. Statist. Soc.*, **A142**, 146–180 (1979).
[59] L. J. Savage—his work in probability and statistics. *Ann. Statistics*, **8**, 1–24 (1980).
[60] Jeffreys' contribution to modern statistical thought, in *Bayesian Analysis in Econometrics and Statistics* (ed. A. Zellner), Amsterdam: North-Holland, pp. 35–9 (1980).
[61] Approximate Bayesian methods, in *Bayesian Statistics* (eds. J. M. Bernardo *et al.*), Valencia: University of Valencia Press, pp. 223–45 (1980).
[62] The role of exchangeability in inference (with M. R. Novick). *Ann. Statistics*, **9**, 45–58 (1981).
[63] Improving judgment by reconciling incoherence (with R. V. Brown). *Theory and Decision*, **14**, 113–32 (1982).
[64] Bayes empirical Bayes (with J. J. Deely). *J. Amer. Statist. Ass.*, **76**, 833–41 (1981).
[65] The subjectivist view of decision-making. *European J. Oper. Research*, **9**, 213–22 (1982).
[66] Coherence, in *Encyclopedia of Statistical Sciences*, vol.2 (eds. S. Kotz and N. L. Johnson), New York: Wiley, pp. 29–31 (1982).
[67] Bruno de Finetti, in *Encyclopedia of Statistical Sciences*, Suppl. (eds. S. Kotz and N. L. Johnson), New York: Wiley, pp. 46–7 (1982).
[68] Leonard J. Savage, in *Encyclopedia of Statistical Sciences*, vol. 8 (eds. S. Kotz and N. L. Johnson), New York: Wiley, pp. 265–7 (1982).

[69] Bayesian inference, in *Encyclopedia of Statistical Sciences*, vol. 1 (eds. S. Kotz and N. L. Johnson), New York: Wiley, pp. 197–204 (1982).
[70] The improvement of probability judgments. *J. Roy. Statist. Soc.*, **A145**, 117–126 (1982).
[71] The Bayesian approach to statistics, in *Some Recent Advances in Statistics* (eds. J. Tiago de Oliveira and Benjamin Epstein), New York: Academic Press, pp. 65–87 (1982).
[72] Scoring rules and the inevitability of probability. *Inter. Statist. Rev.*, **50**, 1–26 (1982).
[73] Reply to Shafer: Lindley's paradox. *J. Amer. Statist. Ass.*, **77**, 334–6 (1982).
[74] The role of randomization in inference. *Philos. Sci. Ass.*, **2**, 431–46 (1982).
[75] Reconciliation of probability distributions. *Oper. Res.*, **31**, 866–80 (1983).
[76] Theory and practice of Bayesian statistics. *The Statistician*, **32**, 1–11 (1983).
[77] The problem of missing evidence (with Richard Eggleston). *Law Quarter Rev.*, **99**, 86–99 (1983).
[78] Lectures on Bayesian statistics. University of Sao Paulo (1983), 48 pp.
[79] Refereeing. *Math. Intelligencer*, **6**, 56–60 (1984).
[80] Prospects for the future: The next 50 years. *J. Roy. Statist. Soc.*, **147**, 359–67 (1984).
[81] A Bayesian lady tasting tea, in *Statistics: An Appraisal* (eds. H. A. and H. T. David), Iowa State Univ. Press, pp. 455–79 (1984).
[82] Bayesian estimation of the dispersion matrix of a multivariate normal distribution (with J. M. Dickey and S. J. Press). *Comms. in Statistics*, **14**, 1019–34 (1985).
[83] Reconciliation of discrete probability distributions, in *Bayesian Statistics 2* (eds. J. M. Bernardo *et al.*), Amsterdam: North-Holland, pp. 375–390 (1985).
[84] Reply to Shafer: Savage revisited. *Statist. Sci.*, **1**, 486–88 (1986).
[85] The reconciliation of decision analyses. *Oper. Res.*, **34**, 289–95 (1986).
[86] Reply to Diaconis and Freedman: consistency of Bayes estimates. *Ann. Statistics*, **14**, 60–3 (1986).
[87] On re-reading Jeffreys, in *Pacific Statistical Congress* (eds. Ivor S. Francis *et al.*), Amsterdam: North-Holland, pp. 35–46 (1986).
[88] Reply to Efron: Why isn't everyone a Bayesian? *Amer. Statistician*, **40**, 6–7 (1986).
[89] Plural analysis: multiple approaches to quantitative research (with R. V. Brown), *Theory and Decision*, **20**, 133–54 (1986).
[90] The relationship between the number of factors and size of an experiment, in *Bayesian Inference and Decision Techniques* (eds. P. Goel and A. Zellner), Amsterdam: North-Holland, pp. 459–70 (1986).

[91] Multivariate distributions for the life lengths of components of a system sharing a common environment (with N. D. Singpurwalla). *J. Appl. Prob.*, **23**, 418–31 (1986).
[92] Another look at an axiomatic approach to expert resolution. *Management Sci.*, **32**, 303–6 (1986).
[93] The technical disease. *Bull. Inst. Math. Statist.*, **15**, 233–34 (1986).
[94] Reliability (and fault tree) analysis using expert opinions (with N. D. Singpurwalla). *J. Amer. Statist. Ass.*, **81**, 87–90 (1986).
[95] Melvin R. Novick: his work in Bayesian statistics. *J. Educ. Statist.*, **12**, 21–26 (1987).
[96] Examples questioning the use of partial likelihood (with C. A. de B. Pereira). *The Statistician*, **36**, 15–20 (1987).
[97] Bernoulli pairs with invariant reversals: an example of partial likelihood, in *Foundations of Statistical Inference* (eds. I. B. MacNeill and G. J. Humphrey), Dordrecht: Reidel, pp. 39–50 (1987).
[98] Using expert advice on a skew judgmental distribution. *Oper. Res.*, **35**, 716–21 (1987).
[99] The probability approach to the treatment of uncertainty in artificial intelligence and expert systems. *Statist. Sci.*, **2**, 17–24 (1987).
[100] A tale of two wells. *Statist. Sci.*, **2**, 38–40 (1987).
[101] Rejoinder. *Statist. Sci.*, **2**, 42–3 (1987).
[102] Statistical inference concerning Hardy-Weinberg equilibrium, in *Bayesian Statistics 3* (eds. J. M. Bernardo *et al.*), Oxford: Oxford Univ. Press, pp. 307–326 (1988).
[103] The use of probability statements, in *Accelerated Life Testing and Experts' Opinions in Reliability* (eds. C. A. Clarotti and D. V. Lindley). Italian Phys. Soc., Amsterdam: North-Holland, pp. 25–57 (1988).
[104] Introduction, in *Accelerated Life Testing and Experts' Opinions in Reliability* (eds. C. A. Clarotti and D. V. Lindley), Italian Phys. Soc., Amsterdam: North-Holland, pp. xi–xviii (1988).
[105] The present position in Bayesian statistics. *Statist. Sci.*, **5**, 44–89 (1990).
[106] Good's work in probability, statistics and the philosophy of science. *J. Statist. Planning and Inference*, **25**, 211–23 (1990).
[107] A survey of George Barnard's statistical work, in *Bayesian and Likelihood Methods in Statistics and Econometrics* (eds. S. Geisser *et al.*), Amsterdam: North-Holland, pp. 13–31 (1990).
[108] Regression and correlation analysis, in *Time Series and Statistics. The New Palgrave* (ed. J. Eatwell *et al.*), London: Macmillan, pp. 237–43, (1990).
[109] Statistical inference, in *Time Series and Statistics. The New Palgrave* (eds. J. Eatwell *et al.*), London: Macmillan, pp. 285–93 (1990).
[110] Thomas Bayes, in *Utility and Probability. The New Palgrave* (eds. J. Eatwell *et al.*), London: Macmillan, pp. 10–11 (1990).

[111] Subjective probability, decision analysis and their legal consequences. *J. Roy. Statist. Soc.*, **A154**, 83–92 (1991).
[112] Sir Harold Jeffreys. *Chance*, **4** (2), 10–14 (1991).
[113] On the evidence needed to reach agreed action between adversaries, with applications to acceptance sampling (with N. D. Singpurwalla). *J. Amer. Statist. Assoc.*, **86**, 933–7 (1991).
[114] Prob, in *The Use of Statistics in Forensic Science* (eds. C. C. G. Aitken and D. A. Stoney), London: Ellis Horwood, pp. 27–50 (1991).
[115] Is our view of Bayesian statistics too narrow? in *Bayesian Statistics 4* (eds. J. M. Bernardo *et al.*), Oxford: Clarendon Press, pp. 1–16 (1992).
[116] The analysis of experimental data. *Teaching Statistics*, **15**, 22–5 (1993).
[117] On the presentation of evidence. *Math. Scientist*, **18**, 60–3 (1993).
[118] Adversarial life testing (with N. D. Singpurwalla). *J. Roy. Statist. Soc.*, **B55**, 837–47 (1993).

2 Reading Road
Wallingford
Oxon
OX10 9DP

CHAPTER 2

The Operational Bayesian Approach

Richard E. Barlow*, Max B. Mendel†
University of California, Berkeley

2.1 INTRODUCTION

The operational Bayesian approach is a radical restructuring of the traditional Bayesian approach to inductive inference. The key ingredients in this approach are an emphasis on (1) the operational meaning of questions posed, and (2) the related field of specialization within which the questions are posed. Deterministic and logical relationships basic to the field of application are used in part to develop appropriate probability models. This approach requires a deeper understanding of the questions posed than the traditional approach and is consequently less mechanical in its application.

P. A. W. Bridgman (1927) was perhaps first to popularize use of the term 'operational' in the context of physical phenomena. De Finetti (1937) was one of the first to consider this approach in the context of probability theory. Dawid (1982) followed de Finetti's point of view by deriving subjective probability models based on observable random quantities. Mendel (1989) used the same approach, but more closely followed Savage (1954) in placing emphasis on the decision problem. Mendel advocates the 'indifference principle' among acts as a means for constructing likelihood probability models.

One of the best ways to visualize a decision problem and its ramifications is to use an 'influence diagram', see Shachter (1986). We will make use of this technique to explain the operational approach to inductive analysis.

---

* Research partially supported by the US. Air Force Office of Scientific Research (F49620–93–1–001) grant and by the Army Research Office (DAAL03–91–G–0046) grant to the University of California at Berkeley.
† Research partially supported by the Army Research Office (DAAL03–91–G–0046) grant to the University of California at Berkeley.

Aspects of Uncertainty edited by P. R. Freeman and A. F. M. Smith. © 1994 John Wiley & Sons Ltd.

We begin such an analysis with a question or questions which require selection of an act or a decision. Like Lindley (1985, p. 4) we make no distinction between action and decision. The question or decision problem must contain terms which have operational meaning. That is, a bet based on the outcome related to our decision must have the property that it can be settled, at least in principle. Using the influence diagram technique we first draw a box corresponding to the mutually disjoint decisions which can be taken (Figure 2.1). Then we think about a value function relative to our problem. Value functions suggest 'parameters' of interest. Although

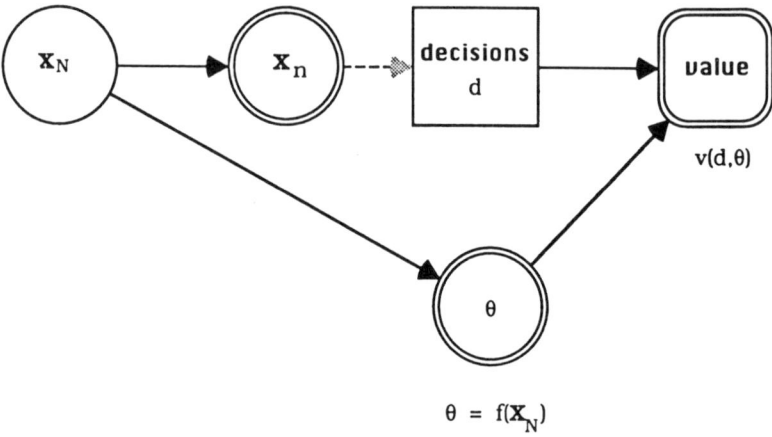

Figure 2.1 Influence diagram for a decision probelm.

in practice it may be difficult to specify a value function, thinking about one helps identify 'parameters'; i.e. functions of relevant but unknown observables. In general, value functions will depend on these parameters. Parameters may be additive functions of observables. Examples of relevant 'parameters' are energy, mass, force, money, etc. The value function and associated parameters together will determine the relevant sample space.

2.2 THE INDIFFERENCE PRINCIPLE

Consider a finite population of observable random quantities

$$\mathbf{X}_N = (X_1, X_2, \ldots, X_N)$$

which have been identified as relevant to a specified decision problem. We will be interested in 'parameters' considered as functions of $\mathbf{X}_N$ which are of the form

$$f(\mathbf{X}_N) = \sum_{i=1}^{N} \psi(X_i)$$

for an appropriate function ψ. We are interested in calculating the likelihood function $p(\mathbf{x}_n | f(\mathbf{X}_N) = N\theta)$ for $1 \leq n < N$.

In Figure 2.1, $\mathbf{X}_n = (X_1, X_2, \ldots, X_n)$ is a subset of $\mathbf{X}_N$ and, as Figure 2.1 indicates, is determined by $\mathbf{X}_N$. The question or problem of interest determines the decisions to be considered. A decision is taken based on the information, $\mathbf{x}_n$, available. If N is unbounded to our knowledge we may use an approximation to the likelihood based on letting $N \to \infty$.

In many physics and engineering problems the key 'parameter' of interest is energy; e.g. the kinetic energy of a particle in motion or the distortion energy related to strength of materials. Although energy can take many forms, it is conserved. This may provide the basis for the parameter of interest. We provide an example based on problems concerning the strength of materials in section 2.3. In other problems to be discussed later, the parameter of interest may be mass or force which can also be additive.

In problems concerning profit and expense, the parameter of interest may be money. Although our utility for money may not be linear, money itself is additive.

2.2.1 ℓ^q-isotropic distributions

If the joint distribution of $\mathbf{X}_N$ is absolutely continuous, the parameter of interest may correspond to the ℓ^q-norm of $\mathbf{X}_N$. For example, for a particle in motion with velocity vector

$$(\dot{X}_1, \dot{X}_2, \dot{X}_3)$$

the parameter of interest may be the kinetic energy while the data consists of the velocity in one coordinate direction. Since kinetic energy is proportional to the sum of squares of the coordinate velocities when the mass is known, the parameter of interest corresponds to the square of the ℓ^2-norm, namely

$$\sum_{i=1}^{3} \dot{X}_i^2 = 3\theta.$$

Considering the general case, suppose that our predictive probability function for $\mathbf{X}_N$ has the property that $p(\mathbf{x}_N) = p(\mathbf{y}_N)$ when $\mathbf{x}_N$ and $\mathbf{y}_N$ have the same ℓ^q-norm. We are in effect indifferent regarding vector outcomes having the same ℓ^q-norm. Even though our value function may depend on $\mathbf{X}_N$ only through the ℓ^q-norm, it does not follow that we would always be indifferent to vector outcomes having the same ℓ^q-norm. As Lindley has pointed out, 'My opinion about the data, were I to know the parameter, exists irrespective of the decision that might hinge on the data. To deny this would to fly in the face of scientific thinking which holds that pure science is meaningful.' In Barlow and Mendel (1992) we considered ageing as well as the ℓ^1-norm in deriving the likelihood.

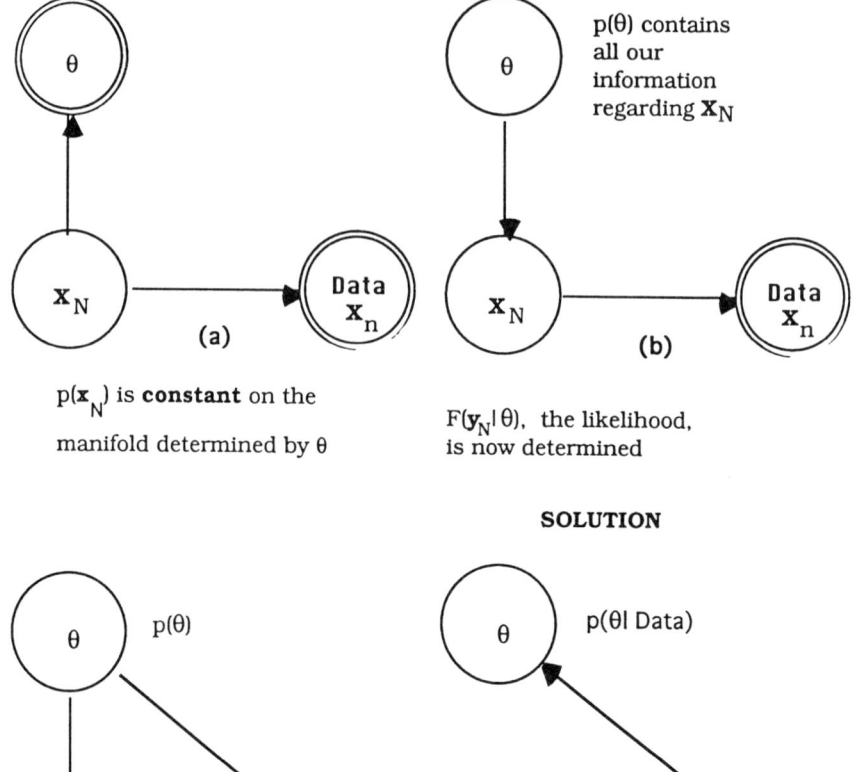

Figure 2.2 Application of the indifference principle.

The original inference problem where $N\theta$ is the qth power of the ℓ^q-norm of $\mathbf{X}_N$ is described in Figure 2.2(a). Reversing the arc from $\mathbf{X}_N$ to θ produces Figure 2.2. After reversal, both the θ-node and the $\mathbf{X}_N$-node are probabilistic. However, the distribution of $\mathbf{X}_N$ given θ is now singular, although $p(\mathbf{x}_n|\theta)$ exists for $1 \leq n < N$. In Mendel (1989), $p(\mathbf{x}_n|\theta)$ is derived. We have

$$p(\mathbf{x}_n|\theta) = \frac{\Gamma(N/q)}{\Gamma^n(1/q)\Gamma((N-n)/q)(N\theta^q)^{n/q}} \left(\frac{q}{2}\right)^n \left[1 - \frac{\sum_{i=1}^{n}|x_i|^q}{N\theta^q}\right]^{[(N-n)/q]-1}, \quad (1)$$

for $\sum_{i=1}^{n} |x_i|^q \leq N\theta^q$ and 0 elsewhere. For $N \to \infty$ and $q = 2$ we obtain the product of n independent normal densities where the parameter is the variance.

After reversing the arc from $\mathbf{X}_n$ to $\mathbf{X}_N$ in Figure 2.2(c), the $\mathbf{X}_N$ node is no longer informative relative to θ and is omitted in Figure 2.2(d).

2.3 INDIFFERENCE RELATIVE TO TRANSFORMED RANDOM QUANTITIES

Figure 2.2 described the steps in the solution of an inference problem where the parameter of interest is the qth power of the ℓ^q-norm, i.e. $\|\mathbf{X}_N\|^q = N\theta$ and $p(\theta)$ contains all our information regarding $\mathbf{X}_N$. This is a rather special case and most applications will depend on an indifference judgment relative to a specified transformation of $\mathbf{X}_N \mapsto \mathbf{W}_N$, where $\mathbf{X}_N$ and $\mathbf{W}_N$ need not be of the same dimension. Figure 2.3 is an appropriate diagram for at least three models.

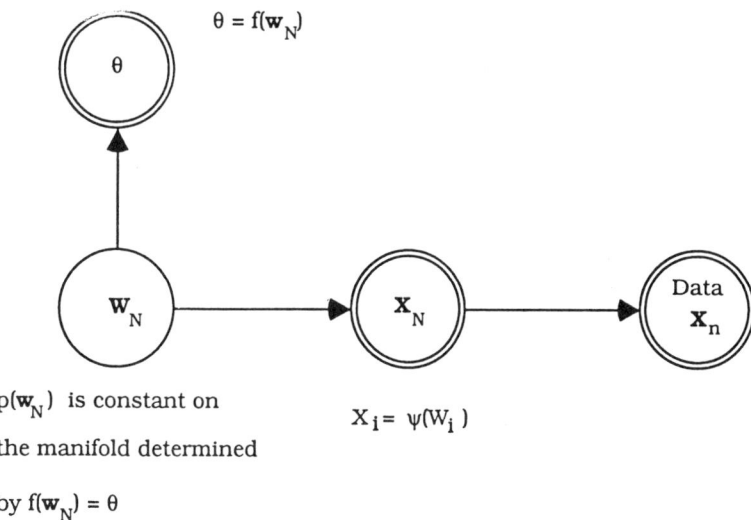

Figure 2.3 The indifference principle with respect to transformed random qualities.

Figure 2.3 describes:
1. The *generalized Weibull* model with $X_i = \psi^{-1}(W_i)$ and
$$f(\mathbf{W}_N) = \sum_{i=1}^{N} W_i = N\theta,$$
2. the *generalized gamma* model with $X_i = \psi^{-1}(W_i)$ and
$$f(\mathbf{W}_N) = \sum_{i=1}^{N} \psi^{-1}(W_i) = N\theta.$$

These models are derived in Barlow and Mendel (1992). A third model is the *Poissonian* model with $X_i = \sum_j W_{ij}$, where $\mathbf{W}_N$ is a 0–1 matrix describing the mappings from a set, S, with $j = 1, 2, \ldots, N\theta$ elements to $i = 1, 2, \ldots, N$ boxes or positions and

$$f(\mathbf{W}_N) = \sum_{j=1}^{N\theta} \sum_{i=1}^{N} W_{ij} = N\theta.$$

2.3.1 The generalized Weibull model

The generalized Weibull model occurs when ψ is bicontinuous, differentiable, increasing to infinity and $\psi(0) = 0$. In this case

$$p(\mathbf{w}_N) = p(\mathbf{w}'_N)$$

if $\sum_{i=1}^{N} w_i = \sum_{i=1}^{N} w'_i$ and the parameter of interest is $\sum_{i=1}^{N} w_i = N\theta$.

Lindquist (1992) makes use of this model relative to strength of materials. When the distortion energy u of a material reaches a threshold value the material will yield. The total elastic strain energy u_t absorbed by a volume of stressed material is the sum of a dilatational energy u_v, i.e. the elastic energy that causes a change in the material's volume, and the distortion energy, i.e. the elastic energy that results in shear deformations. Shear deformations are those which distort the material so that initial right angles no longer remain right angles.

Using this theory, yield predictions can be made for structural members under complex states of stress based only on the uniaxial or torsional stress test data. One simply calculates the distortion energy capacity for the material using the simple yield strength test results and uses this value to make yield predictions for members under more complicated states of stress. The formula used to calculate distortion energy for uniaxial stress is

$$u = \frac{\sigma^2}{6G} \qquad (2)$$

where σ is the uniaxial stress and G is the shear modulus (a material constant). Equation (2) follows from Hooke's law which is an empirical relationship between stress and strain.

If $p(\mathbf{u}_N) = p(\mathbf{u}'_N)$ when $\sum_{i=1}^{N} u_i = \sum_{i=1}^{N} u'_i$ and the parameter of interest is $\sum_{i=1}^{N} u_i = N\theta$, then

$$p(\mathbf{u}_n | \theta) = \frac{N-1}{N\theta} \cdots \frac{N-n}{N\theta} \left[1 - \frac{\sum_{i=1}^{n} u_i}{6GN\theta} \right]^{N-n-1}.$$

This is the ℓ^1-norm form of (1) for positive random quantities. By a change of variable we obtain the joint probability density for an exchangeable sample of items resulting in stress test data, $\boldsymbol{\sigma}_n = (\sigma_1, \sigma_2, \ldots, \sigma_n)$, namely

$$p(\boldsymbol{\sigma}_n | \theta) = \left(\prod_{i=1}^{n} \frac{2\sigma_i(N-1)}{6GN\theta} \right) \left[1 - \frac{\sum_{i=1}^{n} \sigma_i^2}{6GN\theta} \right]^{N-n-1} \quad (3)$$

When $N \to \infty$ this is the product of n Weibull densities with shape parameter 2.

If we had judged $p(\boldsymbol{\sigma}_n) = p(\boldsymbol{\sigma}'_n)$ when $\sum_{i=1}^{N} u_i = \sum_{i=1}^{N} u'_i$ then the likelihood for test stresses would have been based on the ℓ^2-norm. In this case the likelihood would not be given by (3). The first indifference assumption seems more reasonable to us unless additional information is available.

2.3.2 Representation of life distributions corresponding to ageing

The relationship between stress and lifetime is analogous in some ways to the relationship between stress and strain. Increasing tensile stress in a piece of material means increasing strain and in the case of lifetimes, increasing tensile stress tends to *decrease* lifetime. We have also observed in the case of certain lifetime experiments (Barlow et al. 1988) that compressing life lengths by subjecting units to a high stress seems to decrease predictability, i.e. the empirical distribution 'looks more exponential'. We next derive a representation for such life distributions after first reviewing an operational Bayesian concept of ageing.

Suppose that exchangeable units $\{1, 2, \ldots, N\}$ have survived to ages $\mathbf{y}_N = (y_1, y_2, \ldots, y_N)$. We say that their joint life distribution corresponds to ageing lifetimes if and only if for any two units, say i and j with $y_i < y_j$,

$$P(X_i > y_i + t | X_1 > y_1, \ldots, X_N > y_N)$$
$$\geq P(X_j > y_j + t | X_1 > y_1, \ldots, X_N > y_N) \quad (4)$$

for all $t > 0$. That is, the younger of the two units, namely unit i, should have the greater survival probability conditional on all units surviving to ages $\mathbf{y}_N$. This idea is related to the concept of majorization. We use the following ordering notation:

$$x_{[1]} > x_{[2]} > \ldots > x_{[N]}.$$

Definition 1 $\mathbf{y}_N$ majorizes $\mathbf{x}_N$, i.e., $\mathbf{x}_N \leq \mathbf{y}_N$ if and only if

1. $\sum_{i=1}^{j} x_{[i]} \leq \sum_{i=1}^{j} y_{[i]}, \quad j = 1, 2, \ldots, N;$

2. $\sum_{i=1}^{N} x_i = \sum_{i=1}^{N} y_i.$

See Marshall and Olkin (1979, pp. 7, 64).

Definition 2 A function f is *Schur-concave* if and only if
$$\mathbf{x} \preccurlyeq \mathbf{y} \Rightarrow f(\mathbf{x}) \geq f(\mathbf{y}).$$
A function f is *Schur-constant* if and only if
$$\mathbf{x} \preccurlyeq \mathbf{y} \Rightarrow f(\mathbf{x}) = f(\mathbf{y}).$$

Spizzichino (1992) has shown that (4) is true for all age vectors $\mathbf{y}_N$ and $t > 0$ if and only if
$$\bar{F}_N(\mathbf{y}_N) = P(X_1 > y_1, X_2 > y_2, \ldots, X_N > y_N)$$
is Schur-concave. This motivates the definition:

Definition 3 Units $\{1, \ldots, N\}$ age, or 'wear-out' with respect to age, in our opinion if and only if their joint survival distribution $\bar{F}_N(\mathbf{y}_N)$ is Schur-concave.

Starting with units whose lifetimes age in our opinion, suppose we shorten their lives by increasing their environmental stress. Let $\psi(X)$ be the transformed lifetime of a unit under increased stress. Suppose lifetimes are transformed coordinatewise as
$$(X_1, X_2, \ldots, X_N) \mapsto (\psi(X_1), \psi(X_2), \ldots, \psi(X_N)).$$
Now if ψ can be chosen in such a way that the joint transformed life distribution is Schur-constant, it then follows that

$$\bar{F}_N(\mathbf{y}_N) = \int_0^\infty \left[1 - \frac{\sum_{i=1}^n \psi(y_i)}{\eta}\right]_+^{N-1} d\Pi(\eta), \tag{5}$$

where Π is a probability distribution on $[0, \infty)$, $[a]_+ = a$ for $a > 0$ and $[a]_+ = 0$ for $a \leq 0$. $\bar{F}_N(\mathbf{y}_N)$ is Schur-concave if and only if ψ is convex. That is, the joint predictive life distribution can be represented as a mixture of generalized Weibull survival probabilities. This is proved in Barlow and Spizzichino (1992). The representation of the predictive distribution in (5) is independent of parameters other than ψ.

Specification of the transformation ψ would present the greatest challenge and would depend on particular circumstances as well as engineering knowledge.

2.4 CONCLUSIONS

The likelihood model used to analyse data is surely as subjective as any 'prior' probabilities. However, to make this model assessment we need to first concentrate on the question of interest and its relevant field of application. It has been suggested by Dev Basu that statistics departments will eventually disappear and that particular statistical models will be taught within particular fields of application. Whether or not this comes to pass, statistical inference must be more closely associated with particular fields of application in our view. This must occur if statistics is to be more than a cosmetic veneer attached to scientific papers.

REFERENCES

Barlow, R. E. and Spizzichino, F. (1993) 'Schur-concave survival functions and survival Analysis', *Journal of Computational and Applied Mathematics*, **40**.
Barlow, R. E. and Mendel, M. B. (1992) 'De Finetti-type representations for life distributions', *Journal of the American Statistical Association*, **87** (420), 1116–1122.
Barlow, R. E., Toland, R. H., and Freeman, T. (1988) 'A Bayesian analysis of the stress-rupture life of kevlar/epoxy spherical pressure vessels,' in *Accelerated Life Testing and Experts' Opinions in Reliability*, C. A. Clarotti and D. V. Lindley, eds. Bologna, Italy: Soc. Italiana di Fisica, pp. 203–36.
Bridgman, P. (1927) *The Logic of Modern Physics*, New York, Macmillan.
Dawid, A. P. (1982) 'Intersubjective statistical models' in *Exchangeability in Probability and Statistics*, G. Koch and F. Spizzichino, eds, Amsterdam, North-Holland.
De Finetti, B. (1937) 'Foresight: its logical laws, its subjective sources', *Annales de l'Institute Henri Poincaré*, **7**, 1–68. English translation in *Studies in Subjective Probability*, 2nd edn H.E. Kyburg, Jr., and H.E. Smokler, eds (1980), Huntington, New York, Robert E. Krieger, pp. 53–118.
Lindley, D. V. (1985) *Making Decisions* 2nd edn, New York, John Wiley & Sons.
Lindquist, E. (1992) 'Strength of materials and the Weibull distribution', UC Berkeley Technical Report ESRC 92-32, Engineering Systems Research Center. To appear in the journal *Probabilistic Engineering Mechanics*.
Marshall, A. W. and I. Olkin (1979) *Inequalities: Theory of Majorization and Its Applications*, New York, Academic Press.
Mendel, M. B. (1989) Development of Bayesian parametric theory with applications to control, MIT PhD Thesis.
Savage, L. J. (1954) *The Foundations of Statistics*, New York, John Wiley & Sons.

Shachter, R. (1986) 'Evaluating influence diagrams', *Operations Research*, **34**, 871–82.

Spizzichino, F. (1992) 'Reliability decision problems under conditions of aging' in *Bayesian Statistics 4*, J. M. Bernardo, J. O. Berger, A. P. Dawid, and A. F. M. Smith, eds, Oxford, Clarendon Press, pp 803–11.

IEOR Department
University of California
4177 Etcheverry Hall
Berkeley
CA 94720
USA

CHAPTER 3

Pivotal Inference Illustrated on the Darwin Maize Data

George A. Barnard
Mill House, Brightlingsea, Essex

3.1 THE DARWIN DATA

The following is offered to celebrate 50 years of enjoyable argument with Dennis Lindley.

Few data sets can have been more discussed in the statistical literature than the Darwin data quoted by R. A. Fisher in Chapter III of his *Design of Experiments* (*DoE*). This is natural because they represent the two commonest ways of expressing the result of one of the most important types of experiment—a 'comparative trial' of 'treated' versus 'untreated' experimental units. Darwin's data formed part of an 11-year study in which he compared the growth rates of cross-and self-fertilized plants of various species. They give the heights of matched pairs of maize plants. 'This plant is monoecious, and was selected for trial on this account, no other such plant having been experimented on. 'Further details of Darwin's discussion, with references to his book, are given as section 2 of *Data* (Andrews and Herzberg 1985).

It has been customary in the past to follow Fisher in basing his analyses on the assumption that the differences between heights of similar plants are normally distributed. This assumption can hardly be exactly true, but comfort was sometimes taken from studies showing that the distribution of Student's t, the criterion used by Fisher, is robust to departures from normality–especially so when the departures preserve symmetry. Fisher used the t criterion to assess the likely value of the typical difference δ,

---

Aspects of Uncertainty edited by P. R. Freeman and A. F. M. Smith. © 1994 John Wiley & Sons Ltd.

and also to assess the typical ratio θ, of heights of crossed versus selfed plants. It was left to Box and Tiao (1973) to point out that if the assumption of normality were false, Student's t might not be the proper criterion to use for the difference δ. They gave a Bayesian analysis for δ in which the possibility of non-normal densities was to some extent allowed for.

Heights of *Zea mays plants (in 1/8th inches)*
Crossed y_i; self-fertilized x_i

	y_i	x_i		y_i	x_i
Pot I	188	139	Pot II	176	160
	96	163		153	144
	168	160		172	149
Pot III	177	149	Pot IV	168	144
	163	122		177	102
	146	132		184	124
	173	144		96	144
	186	130			

The pivotal mode of approach starts from the premiss that our knowledge of distribution shape in cases such as this is approximate only. Thus, instead of treating the parameters as labelling a series of densities of known form, they are defined by the way they act on the observations in a pivotal quantity z_i, taken as $(y_i - x_i - δ)/σ$ for the difference in heights, and as $(θy_i - x_i)/σ$ for the ratio. Here σ is a scale factor, of no interest in itself, but necessary because we make assumptions about the shape, but not about the location or scale, of the distribution of z_i. The true value of the parameter of interest is then defined by the fact that it endows the pivotal in question with a distribution having some general property. In both of the present cases the property in question is that of symmetry about zero. The likelihood principle has no immediate application because the likelihood may not be well defined.

A pivotal quantity z is a Kolmogoroff random variable (KRV)[†], a function $z(x, θ)$ of observables x and parameters $θ$ whose density element $Kφ(z)dz$ is approximately known. We assume that given such a function

---

[†]In his classic *Grundlagen der Wahrscheinlichkeitsrechnung*, Kolmogoroff defined a (Kolmogoroff) random variable (KRV) as a real-valued measurable function on a probability space. An immediate consequence is that any measurable function of a KRV is itself a KRV. We extend the meaning of KRV to cover functions on probability spaces whose values are real-valued continuous functions of real variables; and retain the property that a continuous function of a KRV is itself a KRV. But as the late Morris de Groot was wont to point out, we may be able to make a reasonable assessment of our personal probability distribution of a real λ, but that does not imply that we can do the same for $λ^{100}$ or sin 100 λ. Such a λ would not be a KRV. Nor is the θ discussed here. But to call it a random variable (RV), as Fisher did, would not be inconsistent with frequent usage before 1950, when Kolmogoroff's text first became available in English.

z, coming to know the values x_0 of the observables will give us some information about the plausibility of various sets of values of θ. The nature of this information can vary, but in many cases it will be such as to be comprehensible to most persons. The basic idea is to find a known 1–1 continuous transformation T operating on z such that $T[z] = [A, B, C]$, where A does not involve θ, while B does not involve x, and only C involves both. We require that A be the maximal parameter-free function of z, in the sense that any parameter-free function of z must be a function of A. This requirement makes A unique, up to equivalence, because if A_1 and A_2 are both parameter-free functions of z, so is $[A_1, A_2]$. We then condition the joint density of $[B, C]$ on the observed value A_0 of A and it can turn out that the resulting conditional density can be more accurately known than than can the original density of z. For example, in the case of a single location parameter λ, when $z = (x - \lambda.1)$, we can set $T[z] = \bar{z} \cdot 1 + c$ subject to $1'c = 0$. We then have that $A = c$ while $C = (\bar{x} - \lambda)$. Provided ξ has a finite variance (and often even if it does not), unless n is excessively small, we could count ourselves unlucky if we went seriously wrong in taking $(\bar{x} - \lambda)\sqrt{n}$ to be nearly normal; and a look at c_0 will often warn us if we have been unlucky.

Again, we may be interested in a function $\lambda(\theta)$ and uninterested in its functional complement $v(\theta)$. It may turn out that $C = [C_1, C_2]$, where C_1 depends on λ_1, not on v, while C_2 depends on v, not on λ. Then after conditioning on A_0 the marginal density of C_1 may also be known more exactly than the original density of z; and we need only be satisfied that the value of v is irrelevant to the value of λ, and that nothing much, apart from the data is known of its value, to justify taking C_{10} and its marginal density as our data-provided source of information on λ.

In his role as a 'digester of data'—representing the essence of the raw data in a form more directly relevant to the questions clients ask—the statistician should make clear what he is assuming, and he should be ready to justify his assumptions or to explore alternatives. In the present case, it is known that the *differences* in heights of plants of this kind are *approximately* normally distributed. Some past discussions have assumed that the heights themselves are exactly normally distributed, though no one has thus far indicated what a maize plant of negative height would look like, or, indeed, how it might be found. Our discussion will assume that the 15 z_i can be taken as permutable (or exchangeable), so that their joint density $\varphi(z)$ can be taken to have the form:

$$\varphi(z) = \prod_i \xi(z_i). \qquad (1)$$

As Fisher pointed out, the absence of explicit randomization in Darwin's procedure casts doubt on the assumption of permutability. And as hinted in Andrews and Herzberg (1985)—and discussed below—alternative assumptions should be explored. But in the present we shall take (1) for granted, with $\xi(u)$ locally approximable by $K \exp -\frac{1}{2}u^2$, for some K.

3.2 ESTIMATING THE DIFFERENCE δ.

Writing $y - x = d$ we take the 15-vector $z = (d - \delta \cdot 1)/\sigma$ as our basic pivotal. Then we make a continuous 1–1 'location-scale' transformation T of z by putting

$$T[z] = s_z(t_z \cdot 1 + c) \quad \text{subject to} \quad 1'c = 0 \quad \text{and} \quad c'c = n(n-1) = 210. \tag{2}$$

Here s_z and t_z are scalars, while c is a 15-vector with endpoint restricted to a 13-dimensional sphere. A little algebra shows that, with the usual sample notation,

$$t_z = \bar{z}\sqrt{n}/s_z \quad \text{where} \quad \bar{z} = (\bar{d} - \delta)/\sigma \tag{3}$$

and

$$s_z^2 = \sum_i (z_i - \bar{z})^2/(n-1) = \sum_i \{(d_i - \bar{d})/\sigma^2(n-1) = s_d^2/\sigma^2 \tag{4}$$

while

$$c_i = (z_i - \bar{z})/s_z\sqrt{(n-1)} = (d_i - \bar{d})\sqrt{n}/s_d. \tag{5}$$

Taking s_z, t_z and the first 13 components of c as the new independent variables, the Jacobian of the transformation is of the form $\Delta(c)s_z^{n-1}$, where $\Delta(c)$ is a determinant depending on c but not on s_z nor on t_z. Thus the joint density of (s_z, t_z, c) is, from (1),

$$\Delta(c)s_z^{n-1}\prod_i \xi(s_z(t_z + c_i)) \tag{6}$$

and because c is parameter-free, while t_z still involves δ and s_z still involves σ, $T(z)$ is of the form (A, B, C) with $A = c$, $C = (t_z, s_z)$, and B is null. Here A is obviously maximal parameter free.

Now there is a logical principle which might be called the 'datelessness of knowledge' (DoK). It asserts that the inferences we can draw from knowing all the propositions P, Q, and $R, \ldots$ to be true are the same, no matter in what order we came to know $P, Q, R, \ldots$ to be true. Applying this to the present case, and denoting the observed values of the $y_i, x_i, \bar{y}, \bar{x} \ldots$ etc. by $y_{i0}, x_{i0}, \bar{y}_0, \bar{x}_0, c_0 \ldots$, we can imagine that we learn our data via a data processor corresponding to $T[\, . \,]$ which transforms the data in such a way that we learn (P) that $c = c_0$ before we learn (Q) the values of $\bar{d}$ and s_d. But when the observed c_0 are known, while t_z and s_z remain unknown, the relevant joint density of (t_z, s_z) is the conditional:

$$\chi(t_z, s_z | c_0) = K s_{z0}^{n-1} \prod_i \xi(s_z(t_z + c_{i0})) \qquad (7)$$

and if we agree to ignore s_z on the grounds that it does not involve the quantity of interest δ, and we also agree that the density we ascribe to s_z is unaffected by our knowledge of d_{z0}, then the relevant density of t_z is the marginal

$$\zeta(t_z | c_0) = K \int u^{n-1} \prod_i \xi(u(t_z + c_{i0})) \cdot du. \qquad (8)$$

The reader may object at this point that we might possess adscititious information (*OED*: adscititious='adopted from without') which could lead us to suppose that the value taken in our case by s_d/σ was not that of a random sample from its marginal distribution, but instead was rather on the high side, or on the low side. Such information might be expressible in terms of a prior density element $\pi(\sigma)d\sigma$. It was one of Jeffreys's major contributions to statistics (expounded for the first time in 1933 to his college mathematical society, when the present writer heard him) to observe that if, when s_d is unknown, s_z has density $\tau(s_z)$, say, and if σ has the *a priori* density element $\pi(\sigma)d\sigma$, then when $d = d_0$ is known, the probability density of $s_{z0} = s_{d0}/\sigma$ remains proportional to $\tau(s_{z0})$ if and only if $\pi(\sigma) = 1/\sigma$. In more modern terminology, $d\sigma/\sigma$ is a 'non-informative prior element' for σ. The general result is, that if a pivotal $p = (x - \theta)$ is linear in the parameter θ and has, for x unknown, density $\varphi(p)$, and if θ has prior probability element $\pi(\theta)d\theta$, then the posterior density of $p_0 = (x_0 - \theta)$ is proportional to $\varphi(p_0)$ iff $\pi(\theta)$ is constant. This remains true if 'approximately' is understood before 'linear', 'density', 'proportional' and 'constant'. No impropriety of distribution is required.

The present writer prefers to take the view that in carrying out our pivotal inference, with the given basic pivotal z, we are attempting to arrive at what the data say about δ—that is, what the present data say in the absence of adscititious information. If other information is available, expressible in the form of a prior density $\pi(\delta, \sigma)$ there is nothing to stop us extending the basic pivotal z to (z, δ, σ), with density

$$\prod_i \xi(z_i) \cdot \pi(\delta, \sigma)$$

and proceeding to extend $T[\,.\,]$ to a transformation $T^*[\,.\,]$ as before. When a prior specifying the density of all the unknown parameters is available, it will turn out that A will consist of all the observations, B will be presentable as (δ, σ), while C will be null. T^* will be a functional extension of T, and carrying out a pivotal inference without a prior will serve as a starting-point to which priors for some or all of the parameters can be added as information about them becomes available. In the absence of any

$\pi(\delta, \sigma)$, and assuming aproximate normality, we are entitled to say that given all the relevant data except $\bar{d}_0$ the density of $(\bar{d} - \delta)/9.746$ is as tabulated for t_{14}; and the observed $\bar{d}_0 = 20.933$. As Sprott (1978) showed, the presence in c of the two outliers corresponding to $d_i = -67$ and $d_i = -48$ signals that attention to the tails of the density of the d_i may be called for.

To make use of (8) we may note that the proposition $a < t_{z0} < b$ is logically equivalent to

$$\bar{d}_0 - bs_{b0}/\sqrt{n} < \delta < \bar{d}_0 - as_{d0}/\sqrt{n}. \tag{9}$$

In particular if, with Darwin. We are primarily interested to attach a probability p, based on the data, to the proposition that δ might be negative we have:

$$p = \int_t^\infty \zeta(v \mid c_0) dv$$

$$= \int_t^\infty \int u^{n-1} \prod_i \xi(u(v + c_{i0})) du \cdot dv \Big/ \int\int u^{n-1} \prod_i \xi(u(v + c_{i0})) du \cdot dv \tag{10}$$

where $t = \bar{d}_0 \sqrt{n}/s_{d0}$. If we assume ξ to be the normal density, this will give us the usual p-value.

3.3 LIMITS FOR THE RATIO Θ

Galton reviewed Darwin's data on six other kinds of plants besides the maize set we are discussing here. He remarked that in five cases of the seven the ratio selfed/crossed appeared to be within a few per cent of 0.8. In section 62.1 of *DoE* Fisher returns to the Darwin data to consider the question of estimating the *typical ratio* θ. The problem thus raised has already given rise to a considerable amount of discussion (Creasy 1954), and a pivotal inference approach sheds new light on it.

The problem is somewhat unusual in that the parameter of interest θ is defined by the fact that the density φ of the basic pivotal should possess certain features.

Using y and x as before, for the ratio in question to take the value θ means that the quantity $z_i = \theta y_i - x_i$ is, for each i, distributed symmetrically around zero. It was assumed by Galton that the heights of the plants could be taken as approximately normally distributed, subject only to the limitation that the heights are necessarily positive. And much of the later discussion by Fisher, Creasy, Fieller and others has made the same assumption. We shall make the much weaker assumption that if θ is the true value of the ratio, the observable values of z_i will generate a nearly normal distribution. That this must be centred on zero follows from the meaning of θ. Given a sequence of independent pairs of observations. (x_i, y_i), $i = 1, 2, \ldots, n, \ldots$, therefore, we adopt as our model that, for some σ in $0 < \sigma < \infty$, and some (true value) θ in $0 < \theta < \infty$,

$$z_i = (\theta y_i - x_i)/\sigma \quad i = 1, 2, \ldots, n \tag{1}$$

constitutes a sample of n independent observations from a density that is not too far from $N(0, 1)$. The assumption that this density is exactly $N(0, 1)$ is not open to the objection of 'negative heights' indicated above. For (1) to be exactly $N(0, 1)$ it is necessary and sufficient that, for all n,

$$\bar{z} = (\theta\bar{y} - \bar{x})\sqrt{n}/\sigma \tag{2}$$

should be very nearly $N(0, 1)$ independently of

$$s_z^2 = \sum_i (z_i - \bar{z})^2/(n - 1)$$
$$= \sum_i \{(\theta^2 y_i - \bar{y})^2 - 2\theta(y_i - \bar{y})(x_i - \bar{x}) + (x_i - \bar{x})^2\}/\sigma^2(n - 1) \tag{3}$$

which should be distributed as $\chi^2/(n - 1)$, and that these two are distributed independently of the vector c with components

$$c_i = (z_i - \bar{z})/s_z\sqrt{(n - 1)} \tag{4}$$

whose endpoint should be uniformly distributed on the $(n - 1)$-sphere with equation $\sum_i c_i = 0$, $\sum_i c_i^2 = 1$.

The position of the end-point c will vary with fixed θ and varying data, and with fixed data and varying θ. But for any given set of data very little can be inferred about the true value of θ from the position of any single c since all possible values are almost equally likely. Any information about θ carried by a given set of data will be carried by the quantities (2) and (3).

If something were known about the value of σ apart from the information provided by the data, each of $\bar{z}$ and s_z would carry information about θ. But if, as we assume, the only information bearing on σ comes from the data, to obtain information specifically about θ we must eliminate σ by dividing (2) by the square root of (3). The information about θ will then be carried by the observations together with the fact, deduced from our model, that

$$t_z(\theta) = \bar{z}\sqrt{n}/s_z = (\theta\bar{y} - \bar{x})\sqrt{n}/\sqrt{\{\theta^2 s_y^2 + 2\theta r s_x s_y + s_x^2\}}$$

has, for every $n > 1$, Student's distribution on $n - 1$ degrees of freedom.

Comparing two measurements such as we have here presupposes that both of them are positive. If we were to test whether the means of the y's and the x's were each positive we might use Student's

$$t_y = \bar{y}\sqrt{n}/s_y \quad \text{and} \quad t_x = \bar{x}\sqrt{n}/s_x,$$

so it is natural to express $t_n(\theta)$ in terms of these quantities so far as possible. Setting

$$m = s_x/s_y \quad \text{and} \quad r = \sum_i (x_i - \bar{x})(y_i - \bar{y})/(n-1)s_x s_y,$$

we find

$$t_z(\theta) = (\theta t_y - m t_x)/\sqrt{\{\theta^2 + 2mr\theta + m^2\}} = (\theta t_y - m t_x)/\sqrt{\{(m + r\theta)^2 + \theta^2(1 - r^2)\}}.$$

For the Darwin data $\bar{x} = 140.6$, $\bar{y} = 161.5333$, $s_x = 16.4141$, $s_y = 28.93555$, $n = 15$, $m = 0.56726$, $r = 0.33474$, so that $t_y = 21.62101$, $t_x = 33.17522$, and we find the following corresponding values of θ and of $t_{14}(\theta)$:

θ	0.70	0.75	0.80	0.85	0.90	0.95	1.00	1.05	1.10
t	−3.55	−2.41	−1.35	−0.38	0.53	1.37	2.15	2.88	3.56
Δt		1.14	1.06	0.97	0.91	0.84	0.78	0.73	0.68

As $\theta \to \infty$, $t_n \to t_y = 21.62$ while as $\theta \to 0$, $t_n \to -t_x = -33.18$, but the probability of t_{14} falling outside the range tabulated above, from -3.55 to $+3.56$, is only 0.0032, and the probability of its falling outside the range $(-33.18, 21.62)$ is less than 10^{-8}. With other sets of data, giving values of t_x and t_y smaller in magnitude than those occurring here, it will be seen that the tail probabilities correspond to the possibilities that the true means of the x- or of the y-observations are zero, giving θ-values of 0 and ∞ outside the permitted range $0 < \theta < \infty$. Thus the tail probabilities of Student's t associated with $t < -33.18$ and $t > 21.62$ can be interpreted as model-failure probabilities, negligible in the present case, but perhaps not negligible in other cases.

Adopting this interpretation, we see that over the plausible range we can write, as a simple approximation, close enough for most purposes:

$$t_{14} = (\theta - 0.87)/0.055$$

or, in 'estimate' form,

$$\hat{\theta} = 0.87 \pm 0.055 t_{14}.$$

This is interpretable as meaning that our knowledge of θ, assumed to exist, is approximately as if we had measured it, giving the value 0.87, with an instrument whose error distribution was that of $0.055 t_{14}$. This representation of the data would be useful to anyone desirous of checking Galton's conjecture, that the selfed/crossed ratio was about 0.8 over a range of different plant forms.

Noting that the probability is 95% that t_{14} lies between the limits ± 2.145, Fisher wrote (with α replacing our θ) 'the data contradict, at the 5 per cent level of significance, any statement of the form "The true average for cross-fertilization is the fraction θ of the true average for self-fertilization", whenever θ lies outside the limits 76.209 per cent and 99.98 per cent. The probability that θ lies between these limits is 95 per cent.'

Such language usage is perilous. The pivotal argument begins by supposing the basic z to be a KRV. In so far as we confine ourselves to functions of z, these also will be KRVs. But Fisher's statement was, at least at one time, taken by him to imply that θ—and hence any function of θ—was itself a random variable in Kolmogoroff's sense; but proceeding in this way in general leads to serious paradoxes. In pivotal inference only known continuous functions of the basic pivotal can be regarded as KRVs. This restriction seems to avoid all the paradoxes thus far associated with the Fsherian argument.

3.4 COMBINING SETS OF DATA

Let us now suppose that Pots I and IV in the Darwin data represent sets of experimental units not assumed to be exchangeable with those in pots II and III. If we knew more about the pots we might have grounds for suspecting that the crossed-selfed difference δ applying to pots I and IV might differ from that applying to pots II and III. Denoting the latter by δ', and setting $y_i - x_i = d_i$ for pots I and IV, and $y'_j - x'_j = d'_j$ for pots II and III, we would then have, with an obvious notation, a seven-rowed pivotal z with components $(d_i - \delta)/\sigma$ and an eight-rowed z' with components $(d'_j - \delta')/\sigma'$. We would wish to know whether there was evidence in the data that δ differed from δ'.

Treating z and z' as we treated the 15-rowed z in section 3.2, would lead us to expressions corresponding to (3):

$$t_z = \bar{z}\sqrt{7}/s_z = (\bar{d} - \delta)\sqrt{7}/s_d, \quad \text{and} \quad t_{z'} = \bar{z}'\sqrt{8}/s_{z'} = (\bar{d}' - \delta')\sqrt{8}/s_{d'},$$

with conditional densities given by expressions like (8). It would seem reasonable to regard t_z and $t_{z'}$ as independent, their joint density being obtained by multiplication. Then to examine the plausibility of various values of $(\delta - \delta')$ might perhaps be tempted to compute the distribution of

$$(s_d/\sqrt{7})t_z - (s_{d'}/\sqrt{8})t_{z'} = (d - d') - (\delta - \delta')$$

and use it as a pivotal giving information about $(\delta - \delta')$. But the DoK axiom serves to justify our pivotal transformations $T[\,.\,]$ only if we can, in principle, construct a data processor corresponding to $T[\,.\,]$ through which our raw data could be processed before we get to know it. But before we get to know the data we do not know the values of s_d and $s_{d'}$, so we cannot construct the necessary processor. To examine the plausibility of various values of $(\delta - \delta')$ we would need to find a known (differentiable) function $G(t_z, t_{z'})$ involving only the observables $\bar{d}$, $\bar{d}'$, s_d, $s_{d'}$, and $(\delta - \delta')$. But such a function G would have the property that $\partial G/\partial \delta = -\partial G \partial \delta'$ identically. But $(\partial G/\partial \delta) = \partial G/\partial t_z$, while $\partial G/\partial \delta' = \partial G/\partial t_{z'}$ and so if $\partial G/\partial \delta$ were identically equal to $-\partial G/\partial \delta'$, G would be a constant. See Barnard and Sprott (1983).

To make progress we must extend our basic pivotal. If we went so far as to specify a joint prior $\pi(\sigma, \sigma')$ we could set $s_d = \sigma s_z$ and $s_{d'} = \sigma' s_{z'}$, and then s_{d0} and $s_{d'0}$ would appear as constants in the extended A. We could then form the pivotal

$$G(t_z, t_{z'}, d_0, d'_0, (\delta - \delta')) = (s_{d0}/\sqrt{7})t_z - (s_{d'0}/\sqrt{8})t_{z'} = (\bar{d} - \bar{d}') - (\delta - \delta')$$

which would be of the form required to enable us to estimate the difference $\delta - \delta'$. If $\pi(\sigma, \sigma')$ corresponded to independent 'ignorance priors' proportional to $(1/\sigma)(1/\sigma')$, the densities of t_z and $t_{z'}$, would remain the same. But if we reparameterize the problem, setting $\bar{\sigma}^2$, the unknown variance of $\bar{d} - \bar{d}'$, equal to $(\sigma^2/7) + (\sigma'^2/8)$, and set $\lambda = (\sigma^2/7)/\bar{\sigma}^2$, the unknown fraction of this variance coming from $\bar{d}$, we find that the prior probability element $(d\sigma/\sigma)(d\sigma'/\sigma')$ becomes $(d\lambda/\lambda(1 - \lambda))(d\bar{\sigma}/\bar{\sigma})$ and the first factor diverges at both ends of its (0, 1) range. Adopting such a prior element amounts to asserting that for any $\varepsilon > 0$, the odds on λ lying outside the limits $(\varepsilon, 1 - \varepsilon)$ are infinite. Such an assertion is contradicted as soon as we observe the ratio $s_d/s_{d'}$ to be finite. The product $1/\sigma\sigma'$ is untenable as an approximate prior.

Further analysis shows that if we provide only a proper prior for λ, and are then prepared to ignore the value of $\bar{\sigma}$ as irrelevant, we can find a pivotal involving the observables, and λ, and $(\delta - \delta')$, of the required form. If ξ is taken to be normal, and the prior element for λ is taken to be of the form $\lambda^{\alpha}(1 - \lambda)^{\beta} \cdot d\lambda$ with α, β both > -1, it turns out to be possible to express the distribution of the resulting pivotal for $\delta - \delta'$ as a mixture of t-densities. See Duong and Shorrock (1992)—though their attempt to make the prior 'empirical' is open to question.

3.5 GENERAL LOGICAL CONSIDERATIONS

Unlike what has come to be referred to as the 'classical' approach to statistical inference, the thoroughgoing Bayesian position makes no sharp logical distinction between observables x and parameters θ. The pivotal approach *allows* for a sharp distinction to be drawn, without insisting on it. In experimental science a sharp logical distinction does exist, in that parameters θ are supposed to remain constant over repetitions of measurements on experimental units, while the measurements vary from one unit to an other. The point may also be made that it is a commonplace of measurements that they may be passed through a computer in such a way that what is recorded is $f(x)$, for some continuous f, rather than x

itself. A continuous function of an observation can be regarded as an observation. By contrast, a location parameter λ will specify a distribution of known shape and scale. But $\sin \lambda$ will not specify a distribution at all.

This is not the place to begin to write the book which could be written about the representation of measurements as real numbers. Every real measurement can be digitized, and functions of real measurements must therefore be thought of as digitized. We commonly avoid discussing awkward questions as to the possibility that a measurement g might be exactly represented as π, or as $\sqrt{2}$, by restricting ourselves to continuous functions. This is why, at the outset, we replaced Kolmogoroff's condition of measurability by the tighter condition of continuity.

The situation discussed in section 3.4, where we could make an estimation statement about $\delta - \delta'$ only if we were prepared to make an additional assumption about λ, is strictly comparable with that labelled by the term 'confounding' in the theory of experimental design. There we can estimate the sum of two effects, but in order to separate them we must assume, for instance, that one of them is small compared with the other. We could therefore call the λ of section 3.4 a 'confounded nuisance parameter'.

REFERENCES

Andrews, D. F. and Herzberg, A. (eds) (1985) *Data: A Collection of Problems From Many Fields*, New York, Springer-Verlag.

Barnard, G. A. and Sprott, D. A. (1983) The generalised problem of the Nile: robust confidence sets for parametric functions, *Annals of Statistics*, **11**, 104–113.

Box, G. E. P. and Tiao, G. C. (1973) *Bayesian Inference in Statistical Analysis*. Addison-Wesley Reading, Massachusetts

Creasy, M. A. (1954) Limits for the ratio of means, *J. Roy. Stat. Soc.*, **B16**, 186–194

Duong, Q. P. and Shorrock (1992) An empirical Bayes approach to the Behrens–Fisher problem, *Environmetrics*, **3** 183–192

Fisher, R. A. (1954) Contribution to a discussion of a paper on interval estimation by M. A. Creasy, *J. Roy. Stat. Soc.* **B16** 212–213

Sprott, D. A. (1978) Robustness and non-parametric procedures are not the only or the safe alternatives to normality, *Canadian J. Psych.*, **32** 180–185.

Mill House
54 Hurst Green
Brightlingsea
Colchester
Essex
CO7 OEH
UK

CHAPTER 4

Bayes' Theorem in Latent Variable Modelling

D. J. Bartholomew
London School of Economics and Political Science

4.1 SETTING THE SCENE

A latent variable is any quantity which, either in practice or in principle, cannot be directly observed. Such variables play an important part in modelling social phenomena and are the subject of much current interest both within and beyond the social sciences. The oldest examples are found in psychology and, especially, educational testing, where although such things as arithmetical ability or aptitude for languages cannot be directly measured these concepts are extremely useful when thinking scientifically about human abilities. Sociologists and political scientists are likewise interested in attitudes and intentions and economists deal in concepts like economic well-being or quality of life. Any empirical study which involves such quantities must necessarily introduce latent variables into the model on which it is based.

Latent variable modelling has reached its most developed form in latent path analysis which has been implemented in computer packages such as LISREL. In this field the interest centres on relationships between latent variables, and here the line of development merges with that stemming from modelling relationships between variables which are subject to measurement error.

It would be impossible to make any progress in latent variable modelling without there also being some variables which can be observed and which

---

Aspects of Uncertainty edited by P. R. Freeman and A. F. M. Smith. © 1994 John Wiley & Sons Ltd.

are related to the unobserved latent variables. Such variables are called manifest variables or, simply, indicators. A situation in which there are observable and unobservable variables is an obvious place for the deployment of Bayes' theorem since this will tell us what can be said about the latent variables after the indicators have been observed. Though this may seem obvious it is a course which has rarely been followed. This is, perhaps, partly due to the diverse disciplinary origins of much current methodology which have tended to obscure the essential unity of a wide class of models in present use. In some cases it also arises from the attempt to treat latent variables as parameters and thus to bring them into the ambit of standard frequentist inference.

The aim of this chapter is to pose the latent variable modelling problem within a framework which renders the use of Bayes' theorem entirely natural and to show that this offers practical and conceptual advantages. In particular it shows how to generalize continuous variable models to include categorical variables and how to resolve some long-standing disputes by avoiding the inappropriate conceptualizations which lie at their root. Although the line which we shall follow is Bayesian in the literal sense of making use of Bayes' theorem, it is not a full-scale Bayesian treatment. That would require us to regard all parameters which occur in the models as random variables with prior distributions expressing degrees of belief. In effect we have a two-tier model in which there are first of all unobserved random variables and then parameters which may or may not be regarded as random variables. Our approach neither requires nor excludes this further use of Bayes' theorem: it may therefore be best described as partially Bayesian.

4.2 A GENERAL FORMULATION

4.2.1 Notation

We shall denote latent variables by Greek letters and the manifest variables which relate to them by the corresponding italic letters. Thus ξ will represent a latent variable and x its indicators. We shall not distinguish notationally between scalars and vectors. In general, all random variables appearing may be thought of as vectors. It will be convenient to use notation appropriate to continuous variables in this section, but all results translate immediately into corresponding results for categorical, discrete or mixed variables. Probability (density) functions will be written generally as $f(\cdot)$ of $f(\cdot \mid \cdot)$ with no distinction between the name of a random variable and the values which it takes. These simplifications are justified since our aim is to focus on the conceptional framework and not to prove theorems.

4.2.2 A general latent variable framework

Let us suppose that we have a q-dimensional vector of latent variables, ξ, and that we have selected p indicators, x, which we believe to be related to ξ. Ideally we shall have $p \gg q$. In a random sample each individual will be characterized by (x, ξ) though only the x part will be known. The dimensionality of ξ may not be known, though in some of the most important applications it is supposed to be 1. In general terms the object of our analysis is to say something about ξ after x has been observed.

Since x and ξ are random variables they have a joint distribution and hence we can, in principle at least, find $f(\xi \mid x)$ which solves our problem. The joint distribution of x and ξ will be constructed from the prior (i.e. population) distribution of ξ, $f(\xi)$ and the conditional distribution of x given ξ, $f(x \mid \xi)$. Bayes' theorem then gives us what we require as

$$f(\xi \mid x) = f(\xi) f(x \mid \xi) / f(x). \tag{1}$$

In spite of its formal simplicity this result is not immediately useful. If we cannot observe ξ we cannot observe its population distribution either. Equally, if we cannot observe ξ we cannot fix its value to obtain empirical evidence about $f(x \mid \xi)$. All that we can observe directly is x and hence the only distribution about which we can make inferences directly is $f(x)$. This is not sufficient to determine $f(\xi \mid x)$ and our first step must be to see whether there is any further structure implicit in our formulation of the problem which will enable progress to be made.

If the x's are correlated with ξ, as they must be if they are to serve as indicators, then they will be correlated among themselves. Conversely, absence of correlation between the x's would imply no common influence. Hence, if on conditioning on ξ the x's are independent we may deduce that sufficient latent variables have been introduced to explain the correlation of the x's. If not, the dimensionality must be increased by one and the conditional independence tested again. The aim therefore is to find a q such that

$$f(x \mid \xi) = \prod_{i=1}^{p} f(x_i \mid \xi). \tag{2}$$

For any acceptable model, therefore, the marginal distribution of x must admit the representation

$$f(x) = \int f(\xi) \prod_{i=1}^{p} f(x_i \mid \xi) d\xi \tag{3}$$

for some q. The factorization of (2) is often misleadingly spoken of as an assumption as though it were introduced as a simplification to enable the analysis to proceed. It is better regarded as a definition of what it means to say that the interdependence of x is explained by ξ.

This still leaves us with the need to choose the form of $f(x_i | \xi)$ but the context of the practical problem will often provide some clues. Retrospectively, the suitability of any choice can be checked by comparing the empirical distribution of x with $f(x)$ and this is usually done when x is categorical.

A second source of confusion is the prior distribution $f(\xi)$ which is essentially arbitrary. This can be seen by noting that, any transformation $\xi \to \xi^*$ in (3), leaves $f(x)$ unchanged. (Only monotonic transformations of the elements will make practical sense.) There are thus infinitely many prior distributions which yield the same $f(x)$. This is tantamount to saying that there is no unique scale of measurement for ξ. This is just as it should be, because ξ is a construct which has no real existence. We are therefore at liberty to measure it on any scale that we find convenient. If, for example, we choose to use a scale which renders ξ standard normal this must be recognized as a convention and is not something on which there is or can be any empirical evidence.

It is common to find, in the literature, references to estimating the prior distribution in apparent contradiction to this result. To make the estimation of $f(\xi)$ possible it is necessary to specify the regression for x on ξ. For example, if $q = 1$ and $\xi \sim N(0, 1)$ and $x_i | \xi \sim N(\mu + \beta\xi, 1)$ then x will be multivariate normal. If we transform to $\xi^* = e^\xi$, ξ will be log normal and the regression of x_i on ξ^* will be $x_i = \mu + \beta \log \xi^*$, but the distribution of x will be unchanged. The requirement that the regression be linear in the latent variable and the fact that x is multivariate normal together then uniquely determine the form of the distribution of ξ. (This example is rather special in that the mean and variance of ξ are still arbitrary which means that one can still fix the origins and scale of measurement at will.)

Many of the apparent difficulties associated with the prior can be circumvented by the 'sufficiency principle' first introduced in Bartholomew (1984). This is more fully developed in Bartholomew (1987) but, in essence, it says that we should seek to select our conditional distribution such that there is a minimal sufficient statistic X for ξ, of the same dimension as ξ. The term 'sufficiency' here is used in the Bayesian sense of meaning that the posterior distribution of ξ, given x, depends on x only through q functions of x. These q functions, X, thus contain all the information in the x's about the latent variables. It turns out that the family of distributions which satisfies this requirement is the exponential family which, fortunately, is broad enough to cover most practical needs. In the most important special cases, including the normal and Bernoulli distributions, the X's, called components, are *linear* functions of the x's. These components often have considerable intuitive appeal in their own right and

BAYES' THEOREM IN LATENT VARIABLE MODELLING 45

they, or some function of them, may be used as proxies for the latent variables.

In this discussion we have implicitly taken any parameters as known but in practice they will have to be estimated. The likelihood function will depend on the form of the prior distribution through (3) and so, therefore, will the parameter estimates whether Bayesian or maximum likelihood. However, it may be shown that the dependence will be slight—a conclusion which is illustrated by the example in the following section.

4.3 THE NORMAL LINEAR MODEL

This is the most familiar latent variable model, usually known as the factor analysis model. We make three remarks about it from the perspective of the general treatment given in the last section.

The model is usually written

$$x = \mu + \Lambda \xi + e, \qquad (4)$$

where $\xi \sim N_q(0, 1)$, $e \sim N_p(0, \Psi)$ with ξ and e independent. Here I denotes the unit matrix and Ψ is a diagonal matrix of residual variances.

The conditional independence postulate, which is rarely remarked on for this model, is expressed by the independence of the e's. The correlations of the x's are induced by the presence of the common 'factors' ξ, and if these are sufficient to account for that correlation any further variation, represented here by the e's, must involve mutual independence.

It is also easy to see in this case why the form of the distribution of ξ is of little importance for the estimation of the parameters Λ, μ and Ψ. The sufficient statistics for the unknown parameters are the means of the x's and their covariance matrix. All of the standard methods of estimation involve minimizing some measure of the distance between the observed and expected covariance matrices. The likelihood function, assuming ξ to be normal, is one such measure, but in practice it is found that the estimates do not usually depend very strongly on what distance measure is taken. Different prior distributions would yield different measures, but in practice these yield similar estimates. Another way of looking at it is to say that the second moment properties of the model on which fitting depends are independent of the form of the distribution of ξ.

The greatest benefit from looking at the model in a Bayesian perspective is in relation to the factor scores problem. Having fitted the model it is sometimes desired to scale individuals; that is, to locate them in the latent space so that the position of one can be seen in relation to others. This is usually spoken of as estimating the value of ξ for the individual, but it is better thought of as a prediction problem since we are regarding ξ as a random variable. We have already seen that the components, derived using Bayesian considerations, provide just such a scaling. However, we can reach the same conclusion by approaching the problem more traditionally via the posterior distribution. A possible set of factor scores is then provided by $E(\xi\,|\,x)$ which, for the normal model, is given by $(I+\Lambda'\Psi^{-1}\Lambda)^{-1}\Lambda'\Psi^{-1}x$. These are, in fact, equivalent to the components which are given by $\Lambda'\Psi^{-1}x$. These scores, which can also be derived by other methods, have been criticized for being biased in that their conditional expectations, given ξ, are not equal to ξ. But it is far from clear why one should be interested in what happens in a long run of cases in which ξ is fixed. It seems more relevant to condition on what is known, namely x.

The difficulty of the traditional approach is highlighted when we consider what is usually referred to as the 'indeterminacy' of the factor scores which is held to be a serious drawback of the factor model. This arises from the attempt to solve (4) for ξ. There are $p+q$ unknowns, q ξ's and p e's, but only p equations; hence the ξ's are indeterminate. Much ingenuity has been expended on the attempt to render them determinate, but the Bayesian view of the matter shows that this is quite unnecessary. The values of the ξ's are quite properly uncertain, and our uncertainty about them is appropriately expressed by the posterior distribution of ξ. The ξ's are no more indeterminate than is any population parameter when all we have is a random sample. In both cases the indeterminacy is inherent in the problem and is handled by standard inferential procedures.

The factor scores problem, which is the subject of an enormous and controversial literature is thus seen to be no problem at all when properly formulated.

4.4 BINARY DATA

Binary responses are common in the social sciences where they often arise as answers to questions of the yes/no or agree/disagree kind. This too is a field fraught with confusion which here centres more on the appropriate choice of model than on the subsequent analysis. Much of the literature has been in the field of educational testing where children attempt a set of items which they get right or wrong. The latent variable in this case will usually be some ability which the items are designed to test.

It is instructive to approach the model selection problem within the framework set out above. We start with the conditional distribution of x_i given ξ. If we code the responses so that $x_i = 0$ or 1 the obvious choice is the Bernoulli distribution which we may write

$$f(x_i\,|\,\xi) = \{\pi_i(\xi)\}^{x_i}\{1-\pi_i(\xi)\}^{1-x_i} \tag{5}$$
$$= \{1-\pi_i(\xi)\}\exp\{x_i\,\text{logit} + \pi_i(\xi)\},$$

where $\pi_i(\xi)$ is the probability that an individual with ability ξ gets item i correct.

The sufficiency principle applies in this case and the general theory then shows that if we take

$$\text{logit } \pi_i(\xi) = \text{constant} + \alpha' \xi \qquad (6)$$

the sufficient statistics are $\alpha'x$ indicating that weighted sums of the binary indicator variables are the appropriate scores to use. An alternative model which is sometimes used takes the probit function in place of the logit. For practical purposes this is almost the same thing but it lacks the simplicity of the model given by (6).

Much work in this field has proceeded along an alternative track which, at first sight, is very appealing. The essential difference is to parameterize the latent variable by supposing that each sample member has their own 'ability' parameter. The simplest such model, associated with the name of Rasch (1960), may be written

$$\text{logit } \pi_{ij} = \gamma_i + \beta_j \qquad i = 1, 2, \ldots p; j = 1, 2, \ldots n, \qquad (7)$$

where γ_i is the 'item' parameter and β_j is the ability parameter; n is the sample size. If $x_{ij} = 1$ when the response of the jth individual to the ith item is positive and zero otherwise it is easily shown that the statistics $\sum_{j=1}^{n} x_{ij}$ ($i = 1, 2, \ldots p$) are sufficient for the parameters γ_i and $\sum_{i=1}^{p} x_{ij}$ ($i = 1, 2, \ldots n$) are sufficient for the β_j's. This accords with common sense and the common practice of taking, for example, the number correct as a measure of an individual's ability.

This model fits the traditional pattern of inference with a set of observations x with unknown parameters (γ, β) which can be estimated by standard methods. If a new examinee comes along with a score vector x then that candidate's ability can be estimated treating the item parameters estimated from the sample as if they were the true values. However, there are serious difficulties. In a typical case p can be as large as 50 or more and n can run into thousands, so the numerical problems of maximizing the likelihood are formidable. Furthermore, the estimates are not consistent because the number of parameters to be estimated is a linear function of the sample size. Further, if any individual gives the same response to every item the maximum likelihood estimator for the ability of that individual does not exist. Various devices have been used to get round these difficulties. One is to condition on the number of items right and use conditional maximum likelihood for the item parameters. Another is to introduce a distribution for the ability and use marginal maximum likelihood. This, of course, is identical with what we have proposed though in another guise, but it is usually regarded as unsatisfactory because of the arbitrariness of the choice of the distribution. Our analysis shows that this is not a drawback, but rather it gives formal expression to something which is inherent in the nature of the problem.

The Rasch model is thus best regarded as a special case of the logit model of (6) arising when the parameters, α, are all equal. The legitimacy of this assumption can be tested in the usual way and can be fitted using a simple modification of the basic program. The attractive sufficiency properties of the version in (7) are then preserved and the uncertainty about an individual's location on the ability scale is captured by the posterior distribution. The Rasch model of (7) is the subject of a vast and disputatious literature which would have been largely unnecessary if the latent ability had been treated as a random variable.

4.5 LINEAR STRUCTURAL RELATIONS MODELS

Path analysis was introduced by the biologist, Wright (1934) but it has now become a major tool of social research. A path model expresses the interdependence of a set of variables which are, usually, ordered in time. These variables might include such things as level of education, parental social class, level of earnings and so forth, and the aim of the analysis would be to explore the causal links between them. The pattern of relationships is often displayed by means of a path diagram on which the nodes represent variables and the directed arcs joining them the causal links. The relationships are assumed to be linear and regression methods are used to estimate the strengths of the links.

Since many variables of social interest are latent it was natural to generalize the path model to include relationships between latent variables. This requires the introduction of manifest variables through which the behaviour of the latent variables can be investigated. A typical formulation of this problem which has been implemented in the computer package LISREL is as follows.

The latent variables are divided into two classes. The exogenous variables, denoted by ξ, are assumed not to be causally influenced by any other variables within the system. The endogenous variables, denoted by η, are influenced by at least one other variable within the system. The causal model for the latent variables is then written

$$\eta = \Gamma\eta + B\xi + \zeta, \qquad (8)$$

where ζ is an independent error term. If the η's are temporally ordered so that any η depends only on those which precede it in the sequence the model is said to be recursive. However, the general formulation given in (8) allows for the reciprocal influence of one η on another.

Associated with ξ and η are sets of manifest variables x and y. The link between the two types of variable is called the measurement model and takes the form of two normal factor models. Thus

$$x = \Lambda_x \xi + e_x, \qquad y = \Lambda_y \eta + e_y. \qquad (9)$$

The aim of LISREL analysis is to estimate all the parameters of the model, but especially those in the causal part of the model (8). Methods of fitting, maximum likelihood or otherwise, depend on the data only through the covariance matrix of the variables x and y. For this reason the process is sometimes referred to as covariance structure analysis.

If we look at this type of model from the more general vantage point of section 4.2 the measurement part is given by the two conditional distributions $f(x \mid \xi)$ and $f(y \mid \eta)$. The joint distribution of x and y is

$$f(x, y) = \int f(\xi, \eta) f(x \mid \xi) f(y \mid \eta) d\xi d\eta \qquad (10)$$

and from this we can make the same deductions as in the earlier case. In particular that the prior distribution $f(\xi, \eta)$ is essentially arbitrary. This fact is of especial interest here because the prime purpose of the analysis concerns the structure of the prior. In the general framework the causal part of the model is embodied in $f(\eta \mid \xi)$. Any transformation of ξ and/or η will alter the form of the dependence of η on ξ. Uniqueness can only be ensured, as before, by assumptions like normality and linearity.

There are important questions of identifiability for the linear model of (8) and (9), but these are compounded when the more general framework is considered. Bartholomew (1992) gives examples to show how many different models involving different numbers of categorical or metrical ξ's can yield exactly the same covariance structure.

The sufficiency principle applies here also, but it relates only to the measurement part of the model. If, therefore, the distributions involved are from the exponential family there will be two sets of components, sufficient for ξ and η respectively, which can be used to locate individuals in the latent space. A rather more direct way of investigating the causal links would be to conduct a path analysis in which the components serve as proxies for the latent variables. This is also proposed in Bartholomew (1992) where, in the normal case, conditions are given for the correlations between the components to be close to those between the corresponding latent variables.

These conclusions, though modest, serve to emphasize that the LISREL model is a rather delicate tool which should be used with more circumspection than is often the case.

REFERENCES

Bartholomew, D. J. (1984) 'The foundations of factor analysis', *Biometrika* **71**, 221–232.

Bartholomew, D. J. (1987) *Latent Variable Models and Factor Analysis*, London, Griffin; and New York, OUP.

Bartholomew, D. J. (1992) 'Relationships between latent variables' invited paper, Seventh International Conference on Multivariate Analysis, New Delhi, Dec. 1992. To appear in *Sankhya*.

Rasch, G. (1960) *Probabilistic Models for Some Intelligence and Attainment Tests*, Copenhagen, Paedagogiste Institute.

Wright, S. (1934) 'The method of path coefficients', *Ann. Math. Statist.*, **5**, 161–215.

Department of Statistics
London School of Economics and Political Science
Houghton Street
London
WC2A 2AE
UK

CHAPTER 5

Bayesian Sampling Schemes for Auditors

J. A. Bather and P. J. Browne
University of Sussex

5.1 INTRODUCTION

It is well established for auditors to rely on random sampling in order to judge whether the records of a company give a true reflection of its financial position. Special statistical methods have been developed, such as Bayesian discovery sampling: see van Batenburg and Kriens (1989). The approach here is also Bayesian, but it is more directly related to a recent paper by Bather (1992) which describes a simple stochastic model based on a Markov process with two states. We shall investigate a generalization which allows greater flexibility in representing the underlying situation.

The new model has three pure states and two types of fault, major and minor. In the ideal state 0, all the accounts of the company are satisfactory and no faults can be discovered by inspecting items selected at random from them. However, a transition to state 1 might occur before the next annual audit. This represents a deterioration in the financial control of the firm in which a certain proportion of the items in its records contain minor faults. Any that are discovered by sampling must be corrected before the auditors can approve the accounts for that year. From state 1, random transitions can occur to either state 0 or state 2 before the next audit. The latter corresponds to a further deterioration, after which the proportion of minor faults has increased and major faults are also present. Minor faults can be corrected, but major faults are much more serious. The discovery of a major fault leads, in effect, to an immediate termination of the sequence of routine audits. We can imagine a detailed investigation of all the company records followed, perhaps, by bankruptcy and charges

---

Aspects of Uncertainty edited by P. R. Freeman and A. F. M. Smith. © 1994 John Wiley & Sons Ltd.

of fraudulent behaviour. One of the aims of our study will be to find sampling rules that detect the presence of any major faults as soon as possible after a transition to state 2 occurs.

From now on we shall be concerned with a single company and a sequence of external audits which continues until a major fault is discovered. The states of the company are generated by a Markov chain with the transition matrix

$$P = \begin{pmatrix} 1-p & p & 0 \\ q & 1-q-r & r \\ 0 & 0 & 1 \end{pmatrix}.$$

We suppose that an audit occurs each year and that the time taken by this is relatively short, so that the state of the company remains fixed during the sampling of items from the accounts in any particular audit. Transitions can only occur during the intervals between audits and they must correspond to positive entries in the matrix P. It will be assumed that the parameters p, q, r satisfy the following conditions:

$$p, q, r \in (0, 1), \quad q + r < 1. \tag{1}$$

Thus, states 0 and 1 are transient, but state 2 is absorbing.

If the company is in state 0, there are no faults to be discovered in the accounts. In state 1, there are some minor faults and we suppose that each of the items sampled independently during an audit will reveal a minor fault with probability α. The model ensures that, after a number of years, the company will fall into state 2 and this means that a proportion β of the accounts will have minor faults and another proportion γ will contain major faults. Thus, each item inspected can produce three possible outcomes: it is satisfactory (s) with probability $1 - \beta - \gamma$, it has a minor fault (m) with probability β and termination occurs, following discovery of a major fault, with probability γ. The parameters α, β, γ are assumed to satisfy the conditions:

$$\alpha, \beta, \gamma, \in (0, 1), \quad \alpha \leq \beta, \quad \beta + \gamma < 1. \tag{2}$$

We now introduce the costs associated with certain actions and events. The main aim of the chapter is to find sampling rules that minimize the expected total cost of all the actions taken by the auditors before the final termination occurs. Let c be the cost of sampling one item at any stage and let K be the cost of correcting a minor fault. We also suppose that a loss L is incurred by the auditors whenever they approve the accounts with the company in state 2. This loss represents the penalty for each year of delay in discovering the presence of major faults in the financial records of the company. The constants c, K and L are all positive and we may imagine that $c < K < L$, although these inequalities will not be needed in the analysis. Note that only the ratios between the last three parameters are important: for example, the sampling cost c could be treated as the unit of cost.

The state of the company cannot be observed directly by the auditors. Because of this, we must extend the set of states to allow any prior or posterior distribution on the pure states 0, 1 and 2. A convenient way of doing this is to use coordinates (x, y) to represent probabilities $(1 - x - y)$, x, y on states 0, 1, 2 respectively. The state space for our decision problem now consists of all points (x, y) in the triangle defined by $x, y \geqslant 0$, $x + y \leqslant 1$. There are two alternative actions for the decision-maker at any point and three types of transition that can occur in the state space. We shall investigate the minimum total expected cost $f(x, y)$, for any initial position (x, y), that can be attained by choosing an optimal sequence of actions.

Let us consider the basic actions and the transitions associated with them. They are

Action 1: wait,   Action 2: inspect one item.

The first action corresponds to approving the accounts in the current position (x, y) without further sampling. This leads to a delay until the next audit and a transition $(x, y) \to (x^{(w)}, y^{(w)})$, where the new point is obtained by applying the transition matrix P to the distribution vector $(1 - x - y, x, y)$. We find that

$$x^{(w)} = x(1 - p - q - r) + p(1 - y), \quad y^{(w)} = rx + y. \qquad (3)$$

Now consider action 2. This consists of choosing one item at random from a large set of accounts, which will be treated as infinite. In general, an audit is made up of independent repetitions of action 2 and the sample size is determined sequentially. When a single item is inspected, there are three possible results depending on whether a fault is discovered and, if so, what type it is. Let $(x, y) \to (x^{(s)}, y^{(s)})$ denote the transition that occurs if the item is satisfactory and let $(x, y) \to (x^{(m)}, y^{(m)})$, be the transition produced by a minor fault. As we have already noted, the process terminates when a major fault is discovered and there is no need to represent this explicitly in our notation. Given (x, y) and action 2, the probability of finding the item satisfactory is $1 - x\alpha - y(\beta + \gamma)$ and it is easy to check that the corresponding posterior distribution is represented by $(x^{(s)}, y^{(s)})$, where

$$x^{(s)} = \frac{x(1 - \alpha)}{1 - x\alpha - y(\beta + \gamma)}, \quad y^{(s)} = \frac{y(1 - \beta - \gamma)}{1 - x\alpha - y(\beta + \gamma)}. \qquad (4)$$

Similarly, the probability of finding a minor fault is $x\alpha + y\beta$ and this produces a transition to $(x^{(m)}, y^{(m)})$ where

$$x^{(m)} = \frac{x\alpha}{x\alpha + y\beta}, \quad y^{(m)} = \frac{y\beta}{x\alpha + y\beta}. \qquad (5)$$

Finally, action 2 at the point (x, y) leads to immediate termination with probability $y\gamma$. It is worth noting that action 2 is always immediate in its effect, whereas action 1 takes time.

There are costs associated with the possible transitions and, in the next section, we will consider minimizing expected costs over finite sequences of actions. In this way we can obtain approximations to the total expectation $f(x, y)$ at any point and study the dynamic programming equation satisfied by f. It will be shown that the function f is bounded and concave. This implies that f is continuous at every interior point of its triangular domain. In fact, it is also continuous on the boundary, but this property is more difficult to establish and the proof will be deferred until the final section of the paper, after the numerical illustration in section 5.3. The example given there relies on backward induction for computing the approximations to $f(x, y)$ and the expected duration $g(x, y)$ of the sequence of audits under the optimal policy. The policy determines a partition of the state space into subsets A_1 and A_2 where the actions 1 and 2 are optimal. An application of the optimal policy is also illustrated by a simulation of the path generated by a sequence of optimal actions.

5.2 DYNAMIC PROGRAMMING

Consider any initial point (x, y) in the state space, $x, y \geq 0$, $x + y \leq 1$. Let $f_n(x, y)$ be the minimum expected cost when at most n actions are allowed. If termination occurs at any stage, there are no further costs incurred. Thus $f_0(x, y) = 0$ and

$$f_1(x, y) = \min\{yL, c + (x\alpha + y\beta)K\},$$

where the two expressions on the right are the expected costs of choosing action 1 or action 2, respectively. In general, for $n \geq 1$, we can determine f_n from f_{n-1} by using the principle of optimality:

$$f_n(x, y) = \min \begin{bmatrix} yL + f_{n-1}(x^{(w)}, y^{(w)}), \\ c + (x\alpha + y\beta)\{K + f_{n-1}(x^{(m)}, y^{(m)})\} \\ + (1 - x\alpha - y(\beta + \gamma))f_{n-1}(x^{(s)}, y^{(s)}) \end{bmatrix}. \qquad (6)$$

Since all the costs are non-negative, it is clear that

$$f_n(x, y) \geq f_{n-1}(x, y) \qquad (7)$$

always holds. It will be shown that the sequence is bounded:

$$0 \leq f_n(x, y) \leq B, \qquad (8)$$

where B is a suitable constant. Hence, $f_n(x, y)$ increases to a limit $f(x, y) \leq B$ as $n \to \infty$ and f satisfies the dynamic programming equation obtained from (6):

$$f(x, y) = \min \begin{bmatrix} yL + f(x^{(w)}, y^{(w)}), \\ c + (x\alpha + y\beta)\{K + f(x^{(m)}, y^{(m)})\} \\ + (1 - x\alpha - y(\beta + \gamma))f(x^{(s)}, y^{(s)}) \end{bmatrix}. \qquad (9)$$

As we shall see, f_n is a concave function for each n and, hence, so is the limit function f. It follows that f is continuous at every interior point of its triangular domain: for example, see Chapter 3 of the book by Eggleston (1963). Later, in section 5.4, we will establish that f is also continuous on the boundary of the triangle, and it can then be verified that f is the unique continuous solution of equation (9).

We now turn to the proof of (8). Consider the policy of repeating the actions 1, 1, 2, 1, 1, 2, and so on. Let $h_j(x, y)$ be the expected cost of this policy over $3j$ actions. Then $f_n(x, y) \leq h_j(x, y)$ whenever $n \leq 3j$, so it will be enough to show that h_j is bounded uniformly in j. We define

$$B = \frac{c + K + 2L}{pr\gamma}.$$

The cost of three actions 1, 1 and 2 is at most $2L + c + K$, so $h_1(x, y) \leq B$ since $pr\gamma < 1$. Suppose that $h_{j-1}(x, y) \leq B$ for some $j \geq 2$ and all x, y. Now consider $h_j(x, y)$. The effect of applying action 1 twice is a transition to a new point (x^*, y^*), which can be found by using (3). In particular,

$$y^* = rx^{(w)} + y^{(w)} = rx(2 - p - q - r) + y(1 - pr) + pr \geq pr$$

for any x, y. The inductive hypothesis that h_{j-1} is bounded shows that the expected cost of taking action 2 at (x^*, y^*), followed by $j-1$ further repetitions of 1, 1, 2, is at most

$$c + (x^*\alpha + y^*\beta)\{K + B\} + (1 - x^*\alpha - y^*\beta - y^*\gamma)B \leq c + K + (1 - y^*\gamma)B.$$

Hence, we obtain

$$h_j(x, y) \leq 2L + c + K + (1 - y^*\gamma)B = B(pr\gamma + 1 - y^*\gamma).$$

Since $y^* \geq pr$, this means that $h_j(x, y) \leq B$ and the induction is complete.

The next step is to establish that the minimum expected cost functions f_n are concave. Consider a policy π for choosing a sequence of n actions. At each stage, there are two possible actions and four possible outcomes, including termination. The policy is a rule for choosing each action in the sequence, given all the previous actions and their results. There is no need to select further actions after termination, but we are concerned with of order 4^n possible sequences of actions and outcomes. The costs will depend on the states of the Markov chain that happen to coincide with the actions. In principle, we can evaluate the expected cost of any policy π over n stages by considering all possible paths formed by sequences of states, actions and outcomes. The three pure states are represented by the points (0, 0), (1, 0) and (0, 1). We can imagine evaluating the expected cost of π by conditioning on the true initial state of the Markov chain. Let $f_n^{(\pi)}(x, y))$ be the expection for a prior distribution represented by the point (x, y). Then we have

$$f_n^{(\pi)}(x, y) = (1 - x - y) f_n^{(\pi)}(0, 0) + x f_n^{(\pi)}(1, 0) + y f_n^{(\pi)}(0, 1).$$

The minimum expected cost $f_n(x, y)$ is obtained by choosing the particular policy π that is optimal at (x, y) in the finite set of all possible policies for n actions:

$$f_n(x, y) = \min_\pi f_n^{(\pi)}(x, y).$$

The functions $f_n^{(\pi)}$ are linear, so they are concave and it follows that f_n is concave. Hence, the limit function f is also concave and, as we remarked earlier, it must be continuous in the interior of its domain, at least.

The solution of the dynamic programming equation (9) determines an optimal sampling policy. The state space is partitioned into two disjoint subsets A_1 and A_2 according to which of the two actions is preferable:

$$A_1 = \{(x, y): f(x, y) = yL + f(x^{(w)}, y^{(w)})\},$$

$$A_2 = \{(x, y): f(x, y) < yL + f(x^{(w)}, y^{(w)})\}.$$

In general, there will be points (x, y) at which the choice of action is immaterial. Our definition of A_2 is based on the convention that action 2 is chosen only at points where there is a definite advantage over action 1.

Consider the policy of choosing action 1 or 2 according to whether the initial state (x_0, y_0) lies in A_1 or A_2 and then applying the same rule at subsequent states (x_t, y_t), $t = 1, 2, \ldots$, until termination occurs. We can also evaluate the expected time to termination for this policy, which corresponds to the number of repetitions of action 1. Let $g(x, y)$ denote this expectation, for an arbitrary initial state. Then

$$g(x, y) = 1 + g(x^{(w)}, y^{(w)}) \quad (x, y) \in A_1, \tag{10}$$
$$g(x, y) = (x\alpha + y\beta) g(x^{(m)}, y^{(m)}) + (1 - x\alpha - y(\beta + \gamma)) g(x^{(s)}, y^{(s)}),$$

when $(x, y) \in A_2$. However, a more convenient method of computing values of this function is to use a sequence of approximations g_n associated with the functions f_n obtained from (6). Let $g_0(x, y) = 0$ and, for $n \geqslant 1$, set

$$g_n(x, y) = 1 + g_{n-1}(x^{(w)}, y^{(w)}), \quad \text{if} \quad f_n(x, y) = yL + f_{n-1}(x^{(w)}, y^{(w)}) \tag{11}$$

$$g_n(x, y) = (x\alpha + y\beta) g_{n-1}(x^{(m)}, y^{(m)}) + (1 - x\alpha - y(\beta + \gamma)) g_{n-1}(x^{(s)}, y^{(s)})$$

otherwise. The numerical results will be illustrated in the next section, but we shall not attempt a detailed analysis of the method. Finally, in this section, let us note some further properties of the transitions given by equations (3)–(5). These properties indicate the directions in the state space taken by the random transitions generated by applying the optimal policy.

We recall that y is the probability that the company being audited is in state 2, which corresponds to the presence of major faults in the accounts. The transitions in the state space reflect this in a natural way. For example, (3) shows that $y^{(w)} = rx + y$ and so

$$y^{(w)} \geq y. \tag{12}$$

Action 1 always increases the value of y. By using the assumptions in (1) and (2) we can obtain properties of the other transitions. It is easily verified from (4) that

$$x^{(s)} + y^{(s)} \leq x + y, \quad y^{(s)} \leq y \tag{13}$$

The second inequality holds strictly, except when $y = 1$. Thus, y is always reduced by finding that an inspected item is satisfactory. In the case of discovering a minor fault, we obtain from (5) that

$$x^{(m)} + y^{(m)} = 1, \quad y^{(m)}/x^{(m)} \geq y/x, \tag{14}$$

where the last inequality requires that $x > 0$.

5.3 NUMERICAL ILLUSTRATION

The computer program used to evaluate the minimum expected cost function $f(x, y)$ is based on the value improvement method. The approximations $f_n(x, y)$ are obtained by applying relation (6) for $n = 1, 2, \ldots$ and the iterations continue until the increment between two successive approximations is small enough to neglect. The program splits up the ranges of x and y into steps of $1/100$. Thus, the triangular area of the state space is reduced to a total of 5151 points. The value improvement algorithm operates on this grid of points.

In theory the sequence of functions $\{f_n\}$ must converge to the solution f of the dynamic programming equation (9). However, some care is needed in order to interpolate values of the cost functions f_n at intermediate points within the grid mesh. The method adopted initially for interpolation fitted a hyperbolic paraboloid to the four corners of each grid square, but difficulties were experienced in ensuring that the approximations were concave everywhere. This interpolation problem was solved by splitting each square of the grid into a pair of triangles and by fitting a plane over each triangle. Various checks were carried out on the program. For example, it follows almost immediately from equation (9) that the value

$$f(0, 1) = (c + \beta K)/\gamma \tag{15}$$

and this provides a simple check on the numerical results.

The main purpose of the computer program is to evaluate the function f and this leads to an optimal policy by partitioning the state space into subsets A_1 and A_2 which determine the optimal choice of action at every point. The program also computes $g(x, y)$, the expected time to termination for any initial state (x, y), by using the iterations given in (11). The figures below illustrate the results in the case specified by the following parameters:

$$\begin{aligned} p &= 0.20 & q &= 0.10 & r &= 0.10, \\ \alpha &= 0.10 & \beta &= 0.20 & \gamma &= 0.10, \\ c &= 1.00 & K &= 3.00 & L &= 20.00. \end{aligned} \tag{16}$$

Figure 5.1 shows a plot of the function f and Figure 5.2 illustrates a path in the state space obtained by simulating the application of the optimal policy. The path consists of 50 transitions, starting at $(x_0, y_0) = (\frac{1}{3}, \frac{1}{3})$. It can be seen that the path never penetrates far below the boundary of the waiting region A_1, because action 1 always produces a transition upwards into A_2 during the waiting period before the next audit. Directions along the path can be followed by using the results given in (12), (13) and (14). Note that the parameter values in (16) were chosen for clarity of presentation: for

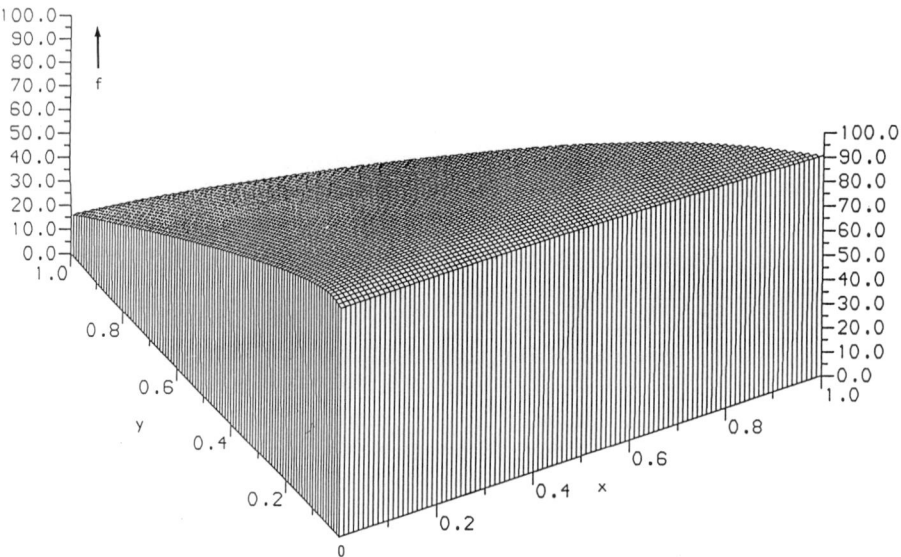

Figure 5.1 Plot of minimum expected cost $f(x, y)$: see equation (16).

example, much larger values of L would be more realistic. Figure 5.3 shows contours of the function g which indicate that its behaviour is linear.

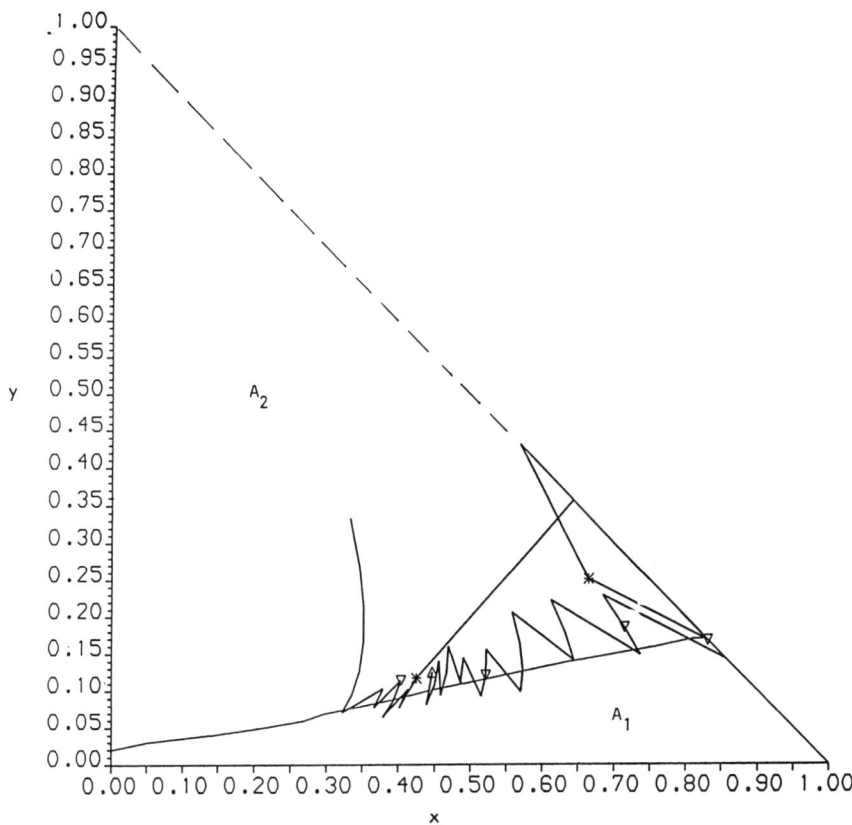

Figure 5.2 Path generated by 50 optimal actions starting at $(\frac{1}{3}, \frac{1}{3})$. A triangle marks the end of each set of 10 transitions and the stars show minor faults.

5.4 CONTINUITY AND UNIQUENESS

In section 5.2, it was shown that the minimum expected cost function f is bounded and concave. Hence, it is continuous at every interior point of the state space. Special arguments will be needed to show that f is also continuous on the boundary. A complete proof must cover many different cases, so we shall omit some of the details. The special techniques are decribed in Propositions 1 and 2. Once continuity has been extended to the whole of the triangular domain, it will be shown in Proposition 3 that f is the only continuous solution of the dynamic programming equation (9).

The state space consists of all points (x, y) such that $x, y \geq 0$, $x + y \leq 1$. Let $\{\xi, \eta\}$ be any point on the boundary of this triangle and

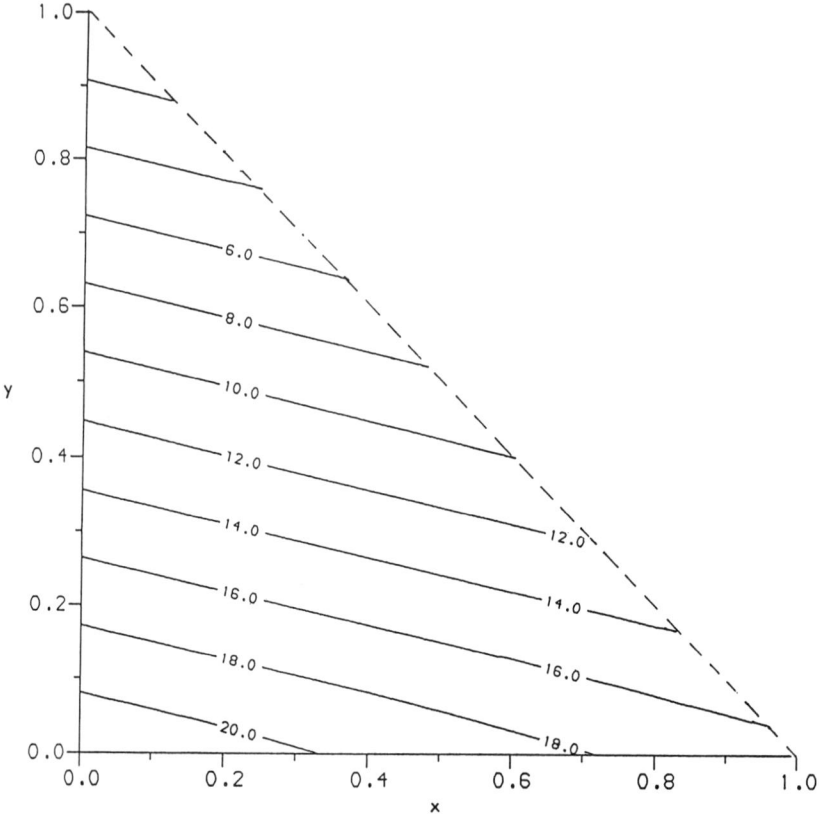

Figure 5.3 Contours of expected time to termination $g(x, y)$.

let $\{(x_j, y_j), j = 0, 1, \ldots\}$ be any sequence of states that converges to $\{\xi, \eta)$ as $j \to \infty$. Since f is a concave function, it cannot have an upwards jump at the boundary, so it is clear that

$$\liminf_{j \to \infty} f(x_j, y_j) \geq f\{\xi, \eta). \tag{17}$$

Continuity of f at the point $\{\xi, \eta)$ will follow if we can show that

$$\limsup_{j \to \infty} f(x_j, y_j) \leq f\{\xi, \eta) \tag{18}$$

for all such sequences.

Case (i) $(\xi, \eta) = (0, 1)$. This is an absorbing state and

$$(\xi^{(w)}, \eta^{(w)}) = (\xi^{(s)}, \eta^{(s)}) = (\xi^{(m)}, \eta^{(m)}) = (0, 1).$$

Equation (9) reduces to

$$f(0, 1) = \min\{L + f(0, 1), c + \beta K + (1 - \gamma)f(0, 1)\}.$$

It follows that $(0, 1) \in A_2$ and that $f(0, 1)$ is given by (15). We can also see that there is an advantage in sampling at neighbouring points. Consider a fixed policy π_τ which consists of the actions: 2, 2, ..., 2, 1, 2, 2, ..., 2, 1, 2, 2, ..., where action 2 is repeated τ times before action 1 in each cycle. The corresponding expected cost function $f^{(\tau)}(x, y)$ is linear and it can be shown that

$$f^{(\tau)}(0, 1) = \frac{c + \beta K}{\gamma} + \frac{(1 - \gamma)^\tau L}{1 - (1 - \gamma)^\tau}.$$

This can be made arbitrarily close to $f(0, 1)$ by choosing τ large enough. In general, $f(x, y) \leq f^{(\tau)}(x, y)$ and, since $f^{(\tau)}$ is continuous at $(0,1)$, this is enough to show that (18) holds for any sequence converging to that point. Thus, f is continuous at $(0, 1)$ and it is also clear that A_2 must contain a neighbourhood of this vertex.

The following result applies at boundary points of the triangle that lie in A_1.

Proposition 1 Let $(\xi, \eta) \in A_1$ be a boundary point. If $(\xi, \eta) \neq (0, 0)$, then $(\xi^{(w)}, \eta^{(w)})$ is an interior point and f is continuous at (ξ, η).

Proof Since $(0,1) \in A_2$, we may assume that $\eta < 1$. It is not difficult to check, using (3), that $(\xi^{(w)}, \eta^{(w)})$ is an interior point, except in the excluded case when $\xi = \eta = 0$. For any points (x_j, y_j), we have

$$f(x_j, y_j) \leq y_j L + f(x_j^{(w)}, y_j^{(w)})$$

by (9). Suppose that $(x_j, y_j) \to (\xi, \eta)$ as $j \to \infty$ and note that this implies $(x_j^{(w)}, y_j^{(w)}) \to (\xi^{(w)}, \eta^{(w)})$, where f is continuous. Since $(\xi, \eta) \in A_1$, we obtain

$$\limsup_{j \to \infty} f(x_j, y_j) \leq \eta L + f(\xi^{(w)}, \eta^{(w)}) = f(\xi, \eta),$$

so the inequality (18) holds and the proof is complete.

Case (ii) $(\xi, \eta) = (1, 0)$. At this point, equation (9) becomes

$$f(1, 0) = \min\{f(1 - q - r, r), c + \alpha K + f(1, 0)\}.$$

It follows that $(1, 0) \in A_1$, and Proposition 1 shows that f is continuous there. The continuity property also means that action 2 cannot be optimal near this vertex, so A_1 must contain a neighbourhood of $(1, 0)$.

Equation (9) shows that the remaining vertex $(0, 0)$ also lies in A_1, but $(\xi^{(w)}, \eta^{(w)}) = (p, 0)$, when $\xi = \eta = 0$, and this is another boundary point. We need a different approach in order to deal with this point.

Definition A boundary point (ξ, η) is a Δ point, for given $\Delta > 0$, if there is a sequence of points (x_j, y_j), converging to (ξ, η), such that
$$\limsup_{j \to \infty} f(x_j, y_j) \geq f(\xi, \eta) + \Delta.$$

Proposition 2 Let $(\xi, \eta) \in A_2$ be a boundary point and suppose it is a Δ point, for some $\Delta > 0$.

(a) If $\eta = 0$, then $(\xi^{(m)}, \eta^{(m)}) = (1, 0)$ and $(\xi^{(s)}, \eta^{(s)})$, with $\xi^{(s)} = (1 - \alpha)\xi(1 - \xi\alpha)^{-1}$, $\eta^{(s)} = 0$, is a $\Delta(1 - \xi\alpha)^{-1}$ point.
(b) If $\eta > 0$, then either $(\xi^{(m)}, \eta^{(m)})$ or $(\xi^{(s)}, \eta^{(s)})$ is a $\Delta(1 - \eta\gamma)^{-1}$ point.

Proof Let $(x_j, y_j) \to (\xi, \eta) \in A_2$ as $j \to \infty$. By equation (9)
$$f(x_j, y_j) \leq c + (x_j\alpha + y_j\beta)\{K + f(x_j^{(m)}, y_j^{(m)})\}$$
$$+ (1 - x_j\alpha - y_j(\beta + \gamma))f(x_j^{(s)}, y_j^{(s)}),$$
$$f(\xi, \eta) = c + (\xi\alpha + \eta\beta)\{K + f(\xi^{(m)}, \eta^{(m)})\}$$
$$+ (1 - \xi\alpha - \eta(\beta + \gamma))f(\xi^{(s)}, \eta^{(s)}).$$

If (ξ, η) is a Δ point, then the definition shows that
$$\Delta \leq (\xi\alpha + \eta\beta)\{\limsup f(x_j^{(m)}, y_j^{(m)}) - f(\xi^{(m)}, \eta^{(m)})\}$$
$$+ (1 - \xi\alpha - \eta(\beta + \gamma))\{\limsup f(x_j^{(s)}, y_j^{(s)}) - f(\xi^{(s)}, \eta^{(s)})\}.$$

(a) In the case $\eta = 0$, we have $(\xi^{(m)}, \eta^{(m)}) = (1, 0)$ and f is continuous there, so the first term on the right is zero. It follows that $(\xi^{(s)}, \eta^{(s)})$ is a $\Delta(1 - \xi\alpha)^{-1}$ point.
(b) Now suppose that $\eta > 0$. Then the result follows from the above inequality. Otherwise, neither $(\xi^{(m)}, \eta^{(m)})$ or $(\xi^{(s)}, \eta^{(s)})$ is a $\Delta(1 - \eta\gamma)^{-1}$ point and we have
$$\Delta < (\xi\alpha + \eta\beta)\Delta(1 - \eta\gamma)^{-1} + (1 - \xi\alpha - \eta(\beta + \gamma))\Delta(1 - \eta\gamma)^{-1} = \Delta,$$
which is a contradiction.

We are now in a position to complete the analysis. It will be enough to show that there are no Δ points on the boundary, which means that the inequality (18) always holds and f is continuous.

Case (iii) $(\xi, \eta) = (0, 0)$. As we remarked earlier, $(0, 0) \in A_1$ and $f(0, 0) = f(p, 0)$. Now suppose, for contradiction, that $(0, 0)$ is a Δ point for some $\Delta > 0$. There is a sequence of points $(x_j, y_j) \to (0, 0)$ as $j \to \infty$ and we have
$$\limsup_{j \to \infty} f(x_j, y_j) \geq f(0, 0) + \Delta.$$

By using (9), $f(x_j, y_j) \leq y_j L + f(x_j^{(w)}, y_j^{(w)})$, so
$$\limsup_{j \to \infty} f(x_j^{(w)}, y_j^{(w)}) \geq f(p, 0) + \Delta$$

and $(p, 0)$ is also a Δ point. In view of Proposition 1, this boundary point cannot lie in A_1, since $\Delta > 0$. We can now apply Proposition 2(a) repeatedly to obtain a sequence of points $(\xi_t, 0)$ with $\xi_0 = p$, all lying in A_2. We have $\xi_{t+1} = (1 - \alpha)\xi_t (1 - \xi_t\alpha)^{-1}$ for $t \geq 1$. The proposition also shows that $(\xi_t, 0)$ is a Δ_t point, where $\Delta_0 = \Delta$ and, in general, $\Delta_{t+1} = \Delta_t(1 - \xi_t\alpha)^{-1}$. It is easily verified that

$$\xi_t = \frac{p(1-\alpha)^t}{1 - p + p(1-\alpha)^t}, \quad \Delta_t = \frac{\Delta}{1 - p + p(1-\alpha)^t}.$$

Note that ξ_t converges to zero as $t \to \infty$ and Δ_t increases to the limit $\Delta(1 - p)^{-1}$. For each t, there is a sequence of points converging to $(\xi_t, 0)$ and we can select one point (x_{j_t}, y_{j_t}) from each sequence such that

$$f(x_{j_t}, y_{j_t}) \geq f(\xi_t, 0) + \Delta_t - 2^{-t},$$

say. This can be done in such a way that $(x_{j_t}, y_{j_t}) \to (0, 0)$ as $t \to \infty$. We can also use (17) with the last inequality:

$$\limsup_{t \to \infty} f(\xi_t, 0) \geq f(0, 0), \quad \limsup_{t \to \infty} f(x_{j_t}, y_{j_t}) \geq f(0, 0) + \Delta(1 - p)^{-1}.$$

We have shown that, if $(0, 0)$ is a Δ point, then it is also a $\Delta(1 - p)^{-1}$ point. But by repeating this argument, we can increase the value of $\Delta > 0$ without limit and this is a contradiction of the fact that f is a bounded function. Thus, $(0, 0)$ cannot be a Δ point and f must be continuous there.

We have now covered the most awkward case and the remaining boundary points are relatively easy to deal with.

Case (iv) Points $(\xi, 0)$, $0 < \xi < 1$. We have already considered the point $(p, 0)$ in case (iii). If $(\xi, 0)$ is a Δ point with $\Delta > 0$, we can construct a sequence $(\xi_t, 0)$, as before, such that $\xi_0 = \xi$, $\xi_t \to 0$ as $t \to \infty$ and the corresponding values of Δ_t form an increasing sequence with $\Delta_0 = \Delta$. This contradicts the fact that f is continuous at $(0, 0)$.

Case (v) Points $(0, \eta)$, $0 < \eta < 1$. In this case, we apply Propositions 1 and 2(b). If $(0, \eta)$ is a Δ point, $\Delta > 0$, then it must lie in A_2 and action 2 leads either to $(\xi^{(m)}, \eta^{(m)}) = (0, 1)$ or to $(\xi^{(s)}, \eta^{(s)})$ where $\xi^{(s)} = 0$, $\eta^{(s)} = (1 - \beta - \gamma) \eta (1 - \eta(\beta + \gamma))^{-1}$. But f is continuous at $(0, 1)$ so the other is a $\Delta(1 - \eta\gamma)^{-1}$ point. Repetition of this argument leads to a sequence of points $(0, \eta_t) \to (0, 0)$ with increasing Δ_t. Once again, this contradicts the continuity of f at the vertex $(0, 0)$.

Case (vi) Points (ξ, η), $\xi = 1 - \eta$, $0 < \eta < 1$. Suppose that $(1 - \eta, \eta)$ is a Δ point for some fixed $\Delta > 0$. We know that f is continuous at $(1, 0)$ and that neighbouring points lie in A_1, so there is a $\delta > 0$ such that every Δ point of the form $(1 - \eta, \eta)$ has $\eta \geq \delta$. Proposition 2(b) shows that, given

$\Delta > 0$ and a Δ point $(1 - \eta, \eta)$, we can find another point $(1 - \eta_1, \eta_1)$ with $\eta_1 \geq \delta$ which is a Δ_1 point and

$$\Delta_1 = \Delta(1 - \eta\gamma)^{-1} \geq \Delta(1 - \delta\gamma)^{-1}.$$

By repeating this argument we can find Δ_t points with $\Delta_t \geq \Delta(1 - \delta\gamma)^{-t}$ for $t = 1, 2, \ldots$. This contradicts the fact that f is bounded.

We have shown that f is continuous at every point of the state space. Finally, it will be shown that f is the only continuous function satisfying the dynamic programming equation (9).

Proposition 3 Let h be any continuous solution of (9). Then $h(x, y) = f(x, y)$, for all $x, y \geq 0$, $x + y \leq 1$.

Proof Let $M = \sup |f(x, y) - h(x, y)|$ and assume that $M > 0$. Since f and h are continuous functions, the supremum is attained at some point (x_0, y_0) in the state space. Suppose, for example, that $h(x_0, y_0) - f(x_0, y_0) = M$. If action 1 is optimal for f at this point and suboptimal for h, then equation (9) shows that

$$M \leq h(x_0^{(w)}, y_0^{(w)}) - f(x_0^{(w)}, y_0^{(w)}) \leq M$$

and the maximum is also attained at $(x_0^{(w)}, y_0^{(w)})$. On the other hand, action 2 may be optimal for f at (x_0, y_0) and, in this case, we obtain

$$M \leq (x_0\alpha + y_0\beta)\{h(x_0^{(m)}, y_0^{(m)}) - f(x_0^{(m)}, y_0^{(m)})\}$$
$$+ (1 - x_0\alpha - y_0\beta - y_0\gamma)\{h(x_0^{(s)}, y_0^{(s)}) - f(x_0^{(s)}, y_0^{(s)})\}.$$

The expression on the right cannot exceed $(1 - y_0\gamma)M$, so we must have $y_0 = 0$. Action 2 cannot be optimal at the point $(0, 0)$, so $x_0 > 0$ and it follows from the last inequality that the maximum difference is attained at both $(x_0^{(m)}, y_0^{(m)})$ and $(x_0^{(s)}, y_0^{(s)})$. In particular $(x_0^{(m)}, y_0^{(m)}) = (1, 0)$, since $y_0 = 0$ and action 1 is always optimal there so

$$h(1, 0) - f(1, 0) = h(1 - q - r, r) - f(1 - q - r, r) = M.$$

We have shown that, if the maximum of $h - f$ is attained at (x_0, y_0), then it must also be attained either at $(x_0^{(w)}, y_0^{(w)})$ or at the point $(1 - q - r, r)$. In either case we have a new point (x_1, y_1) at which $y_1 > 0$ and $h(x_1, y_1) - f(x_1, y_1) = M$. The argument can now be repeated and action 2 can be excluded since $y_1 > 0$. Hence M is attained at $(x_1^{(w)}, y_1^{(w)})$. In fact, there is a sequence of points defined by $(x_{t+1}, y_{t+1}) = (x_t^{(w)}, y_t^{(w)})$ for $t \geq 1$, such that $h(x_t, y_t) - f(x_t, y_t) = M$ always holds. However, it is a straightforward matter to verify that $(x_t, y_t) \to (0, 1)$ as $t \to \infty$ and, hence, both $h(x_t, y_t)$ and $f(x_t, y_t)$ converge to the same limit, given by (15). This contradicts our assumption that $M > 0$. Thus, $M = 0$ and $h(x, y) = f(x, y)$ at every point.

REFERENCES

Van Batenburg, P. C. and Kriens, J. (1989) 'Bayesian discovery sampling: a simple model for Bayesian inference in auditing'. *The Statistician*, **38**, 227–33.

Bather, J. A. (1992) 'A stochastic model for auditing', *J. Opl. Res. Soc*, **43**, 765–71.

Eggleston, H. G. (1963) *Convexity*. Cambridge, Cambridge University Press.

Mathematics Division
University of Sussex
Brighton
BN1 9QH
UK

CHAPTER 6

Optimizing Prediction with Hierarchical Models: Bayesian Clustering

José M. Bernardo
Universidad de Valencia, Spain,
Presidencia de la Generalidad Valenciana, Spain

6.1 INTRODUCTION

Dennis Lindley taught me that interesting problems often come from interesting applications. Furthermore, he has always championed the use of Bayesian analysis in practice, specially when this has social implications. Thus, when I was asked to prepare a chapter for a book in his honour, I thought it would be specially appropriate to describe some research which originated on a socially interesting area—politics—and may be used to broaden the use of one of the methodologies he pioneered, hierarchical linear models.

6.2 THE PREDICTION PROBLEM

Let Ω be a set of N elements, let $\mathbf{y}$ be a, possibly multivariate, *quantity of interest* which is defined for each of those elements, and suppose that we are interested in some, possibly multivariate, function

$$\mathbf{t} = \mathbf{t}(\mathbf{y}_1, \ldots, \mathbf{y}_N)$$

[†] José M. Bernardo is Professor of Statistics at the University of Valencia, and Adviser for Decision Analysis to the President of the State of Valencia.

Aspects of Uncertainty edited by P. R. Freeman and A. F. M. Smith. © 1994 John Wiley & Sons Ltd.

of the values of these vectors over Ω. Suppose, furthermore, that a vector **x** of covariates is also defined, that its values $\{\mathbf{x}_1, \ldots, \mathbf{x}_N\}$ are known for all the elements is Ω, and that a random *training sample*

$$\mathbf{z}_n = \{(\mathbf{x}_i, \mathbf{y}_i), i = 1, \ldots, n\},$$

which consists of n pairs of vectors (**x**, **y**), has been obtained. From a Bayesian viewpoint, we are interested in the predictive distribution

$$p(\mathbf{t} \mid \mathbf{z}_n, \mathbf{x}_{n+1}, \ldots, \mathbf{x}_N).$$

If the set Ω could be partitioned into a class $\mathbf{C} = \{C_i, i \in I\}$ of disjoint sets such that within each C_i the relationship between **y** and **x** could easily be modelled, it would be natural to use a hierarchical model of the general form

$$p(\mathbf{y}_j \mid \mathbf{x}_j, \boldsymbol{\theta}_{i[j]}) \quad \forall j \in C_i$$

$$p(\boldsymbol{\theta} \mid \boldsymbol{\varphi})$$

$$p(\boldsymbol{\varphi}), \tag{1}$$

where $i[j]$ identifies the class C_i to which the jth element belongs, $p(\mathbf{y} \mid \mathbf{x}, \boldsymbol{\theta}_i)$ is a conditional probability density, totally specified by $\boldsymbol{\theta}_i$, which models the stochastic relationship between **y** and **x** within C_i, $p(\boldsymbol{\theta} \mid \boldsymbol{\varphi})$ describes the possible interrelation between the behaviour of the different classes and $p(\boldsymbol{\varphi})$ specifies the prior information which is available about such interrelation.

Given a specific partition **C**, the desired predictive density $p(\mathbf{t} \mid \mathbf{z}_n, \mathbf{x}_{n+1}, \ldots, \mathbf{x}_N)$ may be computed by the following:

1. Deriving the posterior distribution of the $\boldsymbol{\theta}_i$'s,

$$p(\boldsymbol{\theta} \mid \mathbf{z}_n, \mathbf{C}) \propto \int \prod_{j=1}^{n} p(\mathbf{y}_j \mid \mathbf{x}_j, \boldsymbol{\theta}_{i[j]}) p(\boldsymbol{\theta} \mid \boldsymbol{\varphi}) p(\boldsymbol{\varphi}) d\boldsymbol{\varphi}; \tag{2}$$

2. Using this to obtain the conditional predictive distribution of the unknown *y*'s,

$$p(\mathbf{y}_{n+1}, \ldots, \mathbf{y}_N \mid \mathbf{x}_{n+1}, \ldots, \mathbf{x}_N, \mathbf{z}_n, \mathbf{C}) = \int \prod_{j=n+1}^{N} p(\mathbf{y}_j \mid \mathbf{x}_j, \boldsymbol{\theta}_{i[j]}) p(\boldsymbol{\theta} \mid \mathbf{z}_n) d\boldsymbol{\theta}; \tag{3}$$

3. Computing the desired predictive density

$$p(\mathbf{t} \mid \mathbf{z}_n, \mathbf{x}_{n+1}, \ldots, \mathbf{x}_N, \mathbf{C}) =$$
$$f[\mathbf{y}_1, \ldots, \mathbf{y}_n, p(\mathbf{y}_{n+1}, \ldots, \mathbf{y}_N \mid \mathbf{x}_{n+1}, \ldots, \mathbf{x}_N, \mathbf{z}_n)] \tag{4}$$

of the function of interest **t** as a well-defined probability transformation **f** of the joint predictive distribution of the unknown **y**'s, given the corresponding covariate values $\{x_{n+1}, \ldots, x_N\}$ and the known y values $\{y_1, \ldots, y_n\}$.

This solution is obviously dependent on the particular choice of the partition C. In this chapter, we consider the choice of C as a formal decision problem, propose a solution, which actually provides a new class of (Bayesian) clustering algorithms and succinctly describe the case study— Mexican State elections—which actually motivated this research.

6.3 THE DECISION PROBLEM

The choice of the partition C may be seen as a decision problem where the decision space is the class of the 2^N parts of Ω, and the relevant uncertain elements are the unknown value of the quantity of interest t, and the actual values of the training sample z_n. Hence, to complete the specification of the decision problem, we have to define a utility function $u[C, (t, z_n)]$ which measures, for each pair (t, z_n), the desirability of the particular partition C used to build a hierarchical model designed to provide inferences about the value of t, given the information provided by z_n.

Since, by assumption, we are only interested in predicting t given z_n, the utility function should only depend on the *reported* predictive distribution for t, say $q_t(\,.\,\mid z_n, C)$, and the actual value of t, i.e. should be of the form

$$u[C, (t, z_n)] = s[q_t(\,.\,\mid z_n, C), t]. \qquad (5)$$

The function s is known is the literature as a *score function*, and it is natural to assume that it should be *proper*, i.e., such that its expected value should be maximized if, and only if, the reported prediction *is* the predictive distribution $p_t(\,.\,\mid z_n, x_{n+1}, \ldots, x_N, C)$. Furthermore, in a pure inferential situation, one may want the utility of the prediction to depend only on the probability density it attaches to the true value of t. In this case (Bernardo 1979), the score function must be of the form

$$s[q_t(\,.\,\mid z_n, C), t] = A\log[p(t \mid z_n, x_{n+1}, \ldots, x_N, C)] + B \quad A > 0. \qquad (6)$$

Although, in our applications, we have always worked with this particular utility function, the algorithms we are about to describe may naturally be used with any utility function $u[C, (t, z_n)]$.

For a given utility function u and sample size n the optimal choice of C is obviously that which maximizes the expected utility

$$u^*[C \mid n] = \iint u[C, (t, z_n)] p(t, z_n) \mathrm{d}t\,\mathrm{d}z_n \qquad (7)$$

An analytic expression for $u^*[C \mid n]$ is hardly ever attainable. However, it is not difficult to obtain a numerical approximation. Indeed, using Monte

Carlo methods to approximate the outer integral, the value of $u^*[C\,|\,m]$, for $m < n$ may be expressed as

$$u^*[C\,|\,m] \approx \frac{1}{k}\sum_{l=1}^{k}\int u[C, z_{m(l)}, t)]p(t\,|\,z_{m(l)})\,dt, \qquad (8)$$

where $z_{m(l)}$ is one of k random subselections of size $m < n$ from z_n. This, in turn, may be approximated by

$$u^*[C\,|\,m] \approx \frac{1}{k}\sum_{l=1}^{k}\frac{1}{n_s}\sum_{j=1}^{n_s} u[C, z_{m(l)}, t_j)], \qquad (9)$$

where t_j is one of n_j simulations obtained, possibly by Gibbs sampling, from $p(t\,|\,z_{m(l)})$.

Equation (9) may be used to obtain an approximation to the expected utility of any given partition C. By construction, the optimal partition will agglomerate the elements of Ω in a form which is most efficient if one is to predict t given z_n. However, the practical determination of the optimal C is far from trivial.

6.4 THE CLUSTERING ALGORITHM

In practical situations, where N may be several thousands, an exhaustive search among all partitions C is obviously not feasible. However, the use of an agglomerative procedure to obtain a sensible initial solution, followed by an application of a simulated annealing search procedure, leads to practical solutions in a reasonable computing time.

In the aglomerative initial step, we start from the partition which consists of all the N elements as classes with a single element, and proceed to a systematic agglomeration until the expected utility is not increased by the process. The following is a pseudocode for this procedure:

C:= {all elements in Ω}
repeat
    for $i:= 1$ **to** N
        for $j:= i + 1$ **to** N
            begin
                $C^* := C \ominus (i, j), \quad \{C_i \to C_i \cup C_j)\}$
                if $u^*[C^*] > u^*[C]$ **then** C:= C^*
            end
until No_Change

The result of this algorithm may then be used as an initial solution for a simulated annealing procedure. Simulated annealing is an algorithm of random optimization which uses as a heuristic base the process of obtaining pure crystals (annealing), where the material is slowly cooled, giving time at each step for the atomic structure of the crystal to reach its lowest energy level at the current temperature. The method was described by Kirkpatrick *et al.* (1983) and has seen statistical applications, such as Lundy (1985) and Haines (1987). The algorithm is special in that, at each iteration, one may move with positive probability to solutions with lower values of the function to maximize, rather than directly jumping to the point with the highest value within the neighborhood, thus drastically reducing the chances of getting trapped in local maxima. The following is a pseudocode for this procedure:

> **get** Initial_Solution C_0, Initial_Temperature t_0, Initial_Distance d_0;
> $C := C_0$; $t := t_0$; $d := d_0$;
> **repeat**
> **while** (**not** d-Finished) **do**
> **begin**
>   **while** (**not** t-Optimized) **do**
>   **begin**
>   Choose_Random ($C_i \mid d$)
> $\delta := u^*[C_i] - u^*[C_0]$
>   **if** ($\delta \geqslant 0$) **then** $C := C_i$
>   **else if** ($\exp\{-\delta/t\} \leqslant$ Random) **then** $C := C_i$
>   **end**;
> $t := t/2$
> **end**;
> Reduce_Distance (d)
> **until** $d < \varepsilon$

In the annealing procedure, the distance among two partitions is defined as the number of different classes it contains.

6.5 AN APPLICATION TO ELECTION FORECASTING

Consider a situation where, on election night, one is requested to produce a sequence of forecasts of the final result based on incoming returns. Since the results of the past election are available for each polling station, each incoming result may be compared with the corresponding result in the past election in order to learn about the direction and magnitude of the swing for each party. Combining the results already known with a prediction of those yet to come, based on an estimation of the swings, in each of a set of appropriately chosen strata, one may hope to produce accurate forecasts of the final results.

In Bernardo and Girón (1992), a hierarchical prediction model for this problem was developed, using electoral districts within counties as a 'natural' partition for a three-stage hierarchy, and the results were successfully applied some weeks later to the Valencia State elections. One may wonder, however, whether the geographical clustering used in the definition of the hierarchical model is optimal for the stated prediction purposes.

With the notation of this chapter, a two-stage hierarchical model for this problem is defined by the set of equations

$$y_{j[i]} = x_{j[i]} + \theta_i + \epsilon_{0ij}, \quad j \in C_i \quad p(\epsilon_0 \mid \alpha_0), \quad E[\epsilon_0] = 0,$$

$$\theta_i = \varphi + \epsilon_{1i}, \quad i \in I, \quad p(\epsilon_1 \mid \alpha_1) \quad E[\epsilon_1] = 0, \quad \pi(\varphi, \alpha_0, \alpha_1), \quad (10)$$

where $y_j[i]$ is the vector which describes the results on the new election in polling station j which belongs to class C_i, $x_{j[i]}$ contains the corresponding results in the past election, the error distributions $p(\epsilon_0 \mid \alpha_0)$ and $p(\epsilon_1 \mid \alpha_1)$ have zero mean and are fully specified by the hyperparameters α_0 and α_1 and $\pi(\varphi, \alpha_0, \alpha_1)$ is the reference distribution (Berger and Bernardo, 1992) which corresponds to this model.

The function of interest is the probability vector which describes the final results of the new election, i.e.

$$t = \sum_{i \in I} \sum_{j \in C_i} \beta_{j[i]} y_{j[i]}, \quad (11)$$

where $\beta_{j[i]}$ is the (known) proportion of the population which lives in the poling station j of class C_i. The posterior distribution of t may be derived using the methods described above.

In this particular application, however, interest is essentially centred on a good estimate of t. Given some results from the new election, i.e. the training sample z_n, the value of t may be decomposed into its known and unknown parts, so that the expected value of the posterior distribution of t may be written as

$$E[t \mid z_n] = \sum_{i \in I} \sum_{j \in \text{Obs}} \beta_{j[i]} y_{j[i]} + \sum_{i \in I} \sum_{j \in \text{NoObs}} \beta_{j[i]} E[y_{j[i]} \mid z_n], \quad (12)$$

where

$$E[y_{j[i]} \mid z_n] = x_{j[i]} + \iint E[\theta_i \mid z_n, \alpha_0, \alpha_1] p(\alpha_0, \alpha_1 \mid z_n) d\alpha_0 d\alpha_1. \quad (13)$$

The conditional expectation within the double integral may be analytically found under different sets of conditions. In their seminal paper on hierarchical models, Lindley and Smith (1972) already provided the relevant expressions under normality, when y is univariate. Bernardo and Girón (1992) generalize this to multivariate models with error distributions which may be expressed as scales mixtures of normals; this includes heavy-tailed error distributions such as the matrix-variate Student t's. If an analytical expression for the conditional expectation $E[\theta_i \mid z_n, \alpha_0, \alpha_1]$ may be found,

OPTIMIZING PREDICTION WITH HIERARCHICAL MODELS 73

then an approximation to $E[y_{j[i]} \mid z_n]$ may be obtained by using Gibbs sampling to approximate the expectation integral.

In particular, when the error structure may be assumed to have the simple form

$$D^2[\epsilon_0] = \frac{1}{h_0} \Sigma, \quad D^2[\epsilon_1] = \frac{1}{h_1} \Sigma, \quad (14)$$

and the error distribution is expressible as a scale mixture of normals, then the conditional refrence reference distribution $\pi(\varphi, \mid h_0, h_1, \Sigma)$ is uniform and the first moments of the conditional posterior distribution of the θ_i's are given by

$$E[\theta_i \mid z_n, h_0, h_1, \Sigma] = \frac{n_i h_0 r_{.i} + h_1 r_{..}}{n_i h_0 + h_1} \quad (15)$$

$$D^2[\theta_i \mid z_n, h_0, h_1, \Sigma] = \frac{1}{n_i h_0 + h_1} \Sigma \quad (16)$$

where n_i is the number of polling stations in class C_i,

$$r_{.i} = \frac{1}{n_i} \sum_{j \in C_i} y_{j[i]} - x_{j[i]}, \quad i \in I$$

are the average sample swings within class C_i, and

$$r_{..} = \frac{1}{n} \sum_{j=1}^{n} y_j - x_j$$

is the overall average swing.

Since (14) are the rather natural assumptions of exchangeability within classes, and exchangeability among classes, and (15) remains valid for rather general error distributions, (12), (13) and Gibbs integration over (15) provide together a practical mechanism to implement the model described.

6.6 A CASE STUDY: STATE ELECTIONS IN MEXICO

On February 1993, I was invited by the Mexican authorities to observe their Hidalgo State elections, in order to report on the feasibility of implementing in Mexico the methods developed in Valencia. Although I was not supposed to do any specific analysis of this election, I could not resist the temptation of trying out some methods.

I had taken with me the code of the algorithm I use to select a set of constituencies which, when viewed as a whole, have historically produced, for each election, a result close to the global result. The procedure, which is another application of simulated annealing, is described in Bernardo (1992).

Using the results of the 1989 election in Hidalgo (which were the only available ones), I used that algorithm to select a set of 70 polling stations whose joint behaviour had been similar to that of the State as a whole, and suggested that the local authorities should send agents to those polling stations to report on the phone the corresponding returns as soon as they were counted. A number of practical problems reduced to 58 the total number of results which were available about two hours after the polling stations closed.

In the meantime, I was busy setting up a very simple forecasting model—with no hierarchies included—programmed in Pascal in a hurry on a resident Macintosh, to forecast the final results based on those early returns. This was in fact the particular case which corresponds to the model described in section 6.4, if the partition C is taken to have a single class, namely the whole Ω.

About 24 hours later, just before the farewell dinner, the provisional official results came in. Table 6.1, line-1 contains the official results, as percentages of valid votes of PAN (right wing), PRI (government party), PRD (left wing) and other parties. As is apparent from Table 6.1, line 2, my forecasts were not very good; the mean absolute error (displayed as the loss column in the table) was 3.28. Naturally, as soon as I was back in Valencia, I adapted the hierarchical software which I have been using here. The results (Table 6.1, line 3) were certainly far better, but did not quite meet the standards I was used to in Spain.

Table 6.1. Comparative methodological analysis. State of Hidalgo, 21 February 1993

	PAN	PRI	PRD	Others	Loss
Oficial results	8.30	80.56	5.56	5.56	
No hierarchies	5.5	76.8	9.3	8.4	3.28
Districts as clusters	6.4	80.6	7.7	5.3	1.09
Optimal clustering	8.23	80.32	6.18	5.27	0.31

On closer inspection, I discovered that the variances within the districts used as clusters in the hierarchical model were far higher than the corresponding variances in Spain. This prompted an investigation on the possible ways to reduce such variances and, naturally, this lead to the general procedures described in this chapter.

We used repeated random subselection of size 58 from the last election results in Hidalgo in order to obtain, using the algorithms described in section 6.3, the 1989 optimal partition of the polling stations. In practice, we made the exchangeability assumptions described by (14), assumed Cauchy error distributions, and chose a logarithmic scoring rule. We then used this partition to predict the 1993 election, using the two-stage hierarchical model described in section 6.4 and the 58 available polling station results. The results are shown in Table 6.1, line 4; it is obvious from them that the research effort did indeed have a practical effect in the Hidalgo data set.

6.7 DISCUSSION

Prediction with hierarchical models a very wide field. Although very often the clustering which defines the hierarchy has a natural definition, this is not necessarily optimal from a prediction point of view. If the main object of the model is prediction, it may be worth exploring alternative hierarchies, and the preceding methods provide a promising way to do this.

Moreover, there are other situations where the appropriate clustering is less than obvious. For instance, a model similar to that described here may be used to estimate the total personal income of a country, based on the covariates provided by the census and a training sample which consists of the personal incomes of a random sample of the population and their associated census covariates. The clustering which would be provided by the methods described here may have indeed an intrinsic sociological interest, which goes beyond the stated prediction problem.

Finally, the whole system may be seen as a proposal of a large class of well-defined clustering algorithms, where—as one would expect in any Bayesian solution—the objectives of the problem are precisely defined. These could be compared with the rather *ad hoc* standard clustering algorithms as explorative data analysis methods used to improve our understanding of complex multivariate data sets.

REFERENCES

Berger, J. O. and Bernardo, J. M. (1992) On the development of the reference prior method, *Bayesian Statistics 4* (J. M. Bernardo, J. O. Berger, A. P. Dawid and A. F. M. Smith, eds, Oxford, University Press, 35–60 (with discussion).

Bernardo, J. M. (1979) Expected information as expected utility. *Ann. Statist.* 7, 686–690.

Bernardo, J. M. (1992) Simulated annealing in Bayesian decision theory. *Computational Statistics* 1 Y. Dodge and J. Whittaker, eds. Heidelberg, Physica-Verlag, pp. 547–552.

Bernardo, J. M. and Girón, F. J. (1992) Robust sequential prediction form non-random samples: the election night forecasting case. *Bayesian Statistics 4*, (J. M. Bernardo, J. O. Berger, A. P. Dawid and A. F. M. Smith, eds), Oxford, University Press, pp. 61–77 (with discussion).

Kirkpatrick, S., Gelatt, C. D. and Vecchi, M. P. (1983) Optimization by simulated annealing, *Science*, **220**, 671–680.

Haines, L. M. (1987) The application of the annealing algorithm to the construction of exact optimal designs for linear regression models. *Technometrics* **29**, 439–447.

Lindley, D. V. and Smith, A. F. M. (1972) Bayes estimates for the linear model, *J. Roy. Statist. Soc*, **B34**, 1–41 (with discussion).

Lundy, M. (1985) Applications of the annealing algorithm to combinatorial problems in statistics, *Biometrika*, **72**, 191–198.

Departamento de Estadístia
Universidad de Valencia
Facultad de Matemáticas
4600 – Burjassot
Valenica
SPAIN

CHAPTER 7

Inference for a Covariance Matrix

Philip J. Brown[†] **Nhu D. Le**[‡] **and James V. Zidek**[§]
[†]*University of Liverpool*
[‡]*British Columbia Cancer Agency*
[§]*University of British Columbia*

7.1 INTRODUCTION

In this chapter we propose a generalization of the inverted Wishart (IW) distribution for Bayesian inference about a population covariance matrix Σ. This new partially conjugate prior distribution for Σ will be shown to have many of the desirable features of the IW distribution. At the same time it overcomes some of the shortcomings of the IW and other alternatives which have been suggested.

The importance of having a satisfactory prior distribution for Σ derives from the great importance of Σ itself. Statistical modelling often involves Σ as a central modelling feature. Sometimes it is suspected of having a special structure determined by latent variables. The path models of Wright (1934), the LISREL models of Jöreskog (1977) or the factor analysis model (see Spearman 1904) all exemplify the latent variable based approach. At other times, instead of being predetermined its structure merely determines how variables relate to one another with consequent predictive implications for a subset of the variables. Occasionally the presence of correlation is unsuspected and pernicious, as in the long-range dependence found in various measurement series (cf. Student 1927, Granger 1966 and Künsch et al. 1993).

Traditionally, models have explicitly or implicitly assumed normality of the variables. Spherically symmetric alternatives to the normal distribution

---

Aspects of Uncertainty edited by P. R. Freeman and A. F. M. Smith. © 1994 John Wiley & Sons Ltd.

are possible (cf Hampel et al. 1986, Ch. 5). We concentrate explicitly on the Gaussian distribution. Let $Y^{(i)}$, $i = 1, \ldots, n$ represent a random sample from a q-variate Gaussian population with known mean and $q \times q$ covariance matrix Σ. As noted in section 4 the assumption of a known mean may be relaxed. But as long as it is assumed known, we may without loss of generality take it to be the zero vector. With the $n \times q$ random matrix Y formed as $Y' = (Y^{(1)}, \ldots, Y^{(n)})$, we use the distributional notation of Dawid [6],

$$Y \mid \Sigma \sim \mathcal{N}(I, \Sigma). \qquad (1)$$

Here, given Σ, the rows are independent and within any row the covariance matrix is Σ. More generally, $\mathcal{N}(A, B)$ represents a random matrix; the covariance matrix of its ith and jth row vectors is $a_{ij}B$, that between its ith and jth column vectors, $b_{ij}A$. The usual non-Bayesian approach estimates the unknown Σ or $\Sigma(\theta)$ by the maximum likelihood estimator; here θ is an unknown vector of dimension less than the $m = \frac{1}{2}q(q+1)$ distinct elements of the symmetric matrix Σ. The LISREL, factor analysis and path models fall into such a prespecified structure. Typically computation of the estimates requires iterative methods.

Dempster (1972) provides a notable and perhaps somewhat neglected example. He sets to zero certain off-diagonal elements of the inverse of Σ. This corresponds to setting certain partial correlations to zero in the spirit of graphical models. The inverse elements of Σ are canonical parameters in the exponential family representation of the Gaussian distribution. Dempster's algorithm matches the moment parameters for a set of elements with their sample moments, while setting the canonical parameters to zero in the complementary set of parameters. We will have occasion to use a mixed canonical and moment parameter representation in our proposed prior distribution for Σ.

The Bayesian approach is natural if one wishes properly to assess the uncertainty of estimation and prediction. In developing a Bayesian approach to inference for Σ, Lindley (1978) (see also Dickey et al. 1985) has given careful consideration to the implications of assuming an IW prior distribution for Σ. In the next section we review the IW, emphasizing both its strengths and drawbacks as highlighted by Lindley. We go on to describe distinct alternatives to the IW before returning to a new natural conjugate approach. That approach emanates directly from and uses a conditional independence property of certain components of Σ implied by the IW. As we show in section 7.4, it is this independence property in conjunction with its conjugacy which makes the IW a tractable prior.

Recent rapid advances in the theory of Bayesian computation may make tractability seem unimportant. However, some applications like that made by Guttorp et al. (1993) require explicit representations for certain posterior expectations over Σ. These expectations yield the objective function required for combinatorial optimization in the design of monitoring networks.

Another point bears emphasis. However valuable the emerging Bayesian computational algorithms may be, there remains the fundamental requirement that priors be susceptible to implementation with realistic amounts of prior information. Care must be taken to assure reasonable simplicity of structure combined with interpretable hyperparameters; in this way 'identifiability' with respect to prior information is achieved. We believe that the prior proposed here meets the test implied by these considerations.

7.2 INVERTED WISHART PRIOR DISTRIBUTION

The natural conjugate (IW) prior distribution for Σ may be seen from the form of the density function from model (1), to be such that Σ^{-1} is Wishart. By

$$\Sigma \sim \mathcal{IW}(\delta; G), \qquad (2)$$

in the notation of Dawid (1981), we will mean Σ^{-1} has the Wishart distribution with degrees of freedom $v = \delta + q - 1$ and scale matrix G^{-1}. Note that the expectation of Σ exists for $\delta > 2$ when $E(\Sigma) = G/(\delta - 2)$. Using the shape parameter δ rather than the more obvious degrees of freedom has distinct advantages in terms of Bayesian manipulations and was suggested earlier by Lindley (1978).

Sometimes the hyperparameters δ and G can be specified by the investigator. If not a further layer of prior assumptions must be made. This may well yield $G = G(\theta)$ where θ involves a small number of parameters possessing, possibly, a further prior distribution. Alternatively, the coordinates θ may be estimated by plug-in empirical Bayes estimates for simplicity on the grounds that this yields a satisfactory approximation to the actual posterior distribution of Σ. Chen (1979) chooses an estimation-based approach. He uses the maximum marginal likelihood, thinking of Σ as unobserved missing data and using the EM algorithm of Dempster et al. (1979). Brown et al. (1992) develop a similar approach to estimating spatial covariance.

This IW layer provides the route for the typical Bayesian implementation of latent variable models like factor analysis. Rather than structuring Σ and putting a prior distribution on this structure, these approaches leave the $q \times q$ matrix Σ completely unstructured and impose the latent structure on G. The work of Chen (1979) exemplifies such approaches. He works quite generally with a variety of latent structures. Other examples are found in the work of Lindley (1978), which posits an intraclass structure for G; of Mäkeläinen and Brown (1988) in the context of regression prediction with an ordered sequence of regressors; of Brown and Mäkeläinen (1992) which develops an ARMA continuous Gaussian process structure; and of Brown et al. (1992) where a Kronecker product covariance structure is proposed. Pushing back the imposed structure to the hyperparametric level has a distinct advantage in making inference more robust to departures from the assumed structure. Such structure, if trusted at the primary, Σ, level may fail properly to assign uncertainty. However, difficulties in implementing more complex priors at a structured primary level may be diminished by advances in numerical integration.

The posterior distribution is readily deduced from model (1) and prior distribution (2):

$$\Sigma \mid \text{data} \sim \mathcal{IW}(\delta + n; G + Y'Y). \tag{3}$$

For $\delta + n > 2$, the posterior mean is

$$E(\Sigma \mid \text{data}) = (G + Y'Y)/(\delta + n - 2), \tag{4}$$
$$= (1 - w)E(\Sigma) + w(Y'Y/n)$$

provided $\delta > 2$, a weighted average of prior and data; the weight, w, is $n/(\delta + n - 2)$. This attractively simple linear form may be seen in a less benign light as unduly simplistic, reflecting the rather limited prior structure imposed by (2). Actually, the rather rich prior scale matrix G does have a full complement of $m = \frac{1}{2}q(q + 1)$ parameters. But the very limited structure for expressing uncertainty may be viewed with alarm. Just one hyperparameter, δ, is available to express this uncertainty.

It has to be admitted, however, that the inflexible structure of uncertainty in the IW prior just reflects its natural conjugate form; it is as if it arose from previous independent normal data. In a rough-and-ready way it may be adequate from some perspectives. It certainly produces straightforward analysis. And the conclusions from the posterior IW are often not unreasonable. Lindley (1978) has made this argument in the case of an exchangeable intraclass form of G.

One noteworthy feature of the IW in (2) leads us to a new family of tractable prior distributions. Suppose Σ is partitioned into

$$\begin{pmatrix} \Sigma_{11} & \Sigma_{12} \\ \Sigma_{21} & \Sigma_{22} \end{pmatrix}, \tag{5}$$

where Σ_{11}, Σ_{22} are $(q_1 \times q_1)$ and $(q_2 \times q_2)$ respectively, with $q_1 + q_2 = q$. Partition the data from model (1) conformably. Consider the one-to-one transformation of Σ to the parameters of the conditional distribution of $Y_1 \mid Y_2$, that is, $B = \Sigma_{22}^{-1} \Sigma_{21}$ and

$$\Gamma = \Sigma_{11.2} = \Sigma_{11} - \Sigma_{12} \Sigma_{22}^{-1} \Sigma_{21}$$

and the parameter of the marginal normal Σ_{22} of Y_2. This may be viewed as an exponential family mixed parameterization (c.f. Barndorff-Nielsen 1978). The transformation may be explicitly expressed through a decomposition we believe comes from work of Bartlett: $\Sigma = T\Delta T'$, where

INFERENCE FOR A COVARIANCE MATRIX

$$\Delta = \begin{pmatrix} \Gamma & 0 \\ 0 & \Sigma_{22} \end{pmatrix} \quad \text{and} \quad T = \begin{pmatrix} I & B' \\ 0 & I \end{pmatrix}.$$

With this transformation, Σ can be written as

$$\Sigma = \begin{pmatrix} \Gamma + B'\Sigma_{22}B & B'\Sigma_{22} \\ \Sigma_{22}B & \Sigma_{22} \end{pmatrix}. \tag{6}$$

We then have the following lemma.

Lemma 1 Suppose Σ is distributed as in (2), then after the transformation $\Sigma \to (B, \Gamma, \Sigma_{22})$,

1. Σ_{22} is independent of (B, Γ) and $\Sigma_{22} \sim \mathcal{IW}(\delta, G_{22})$;

2. $\Gamma \sim \mathcal{IW}(\delta + q_2, G_{11.2})$;

3. $B \mid \Gamma \sim G_{22}^{-1}G_{21} + \mathcal{N}(G_{22}^{-1}, \Gamma)$.

This result appears in Dawid (1988, Lemma 2) and is given in a slightly different guise in Dempster (1969, Theorem 13.4.2). But we believe it is attributable to Bartlett before that. Note in particular the independence in part (1) of the lemma.

Dawid (1988) has unearthed a further worrying aspect of the IW as a prior distribution, namely an implicit *determinism* in prediction. When q tends to infinity, the support of IW tends to a set of infinite dimensional covariance matrices; they correspond to an infinite dimensional normal distribution which permits exact prediction of Y_1, say, from the sequence $Y_2, Y_3,\ldots$. (Of course q is actually finite and the limiting result does not reveal which set of Y-variables can be used to accurately predict Y_1.) In Dawid (1992) he sees this determinism as symptomatic of the demise of the natural conjugate prior distribution in its failure adequately to come to terms with selection effects in statistical modelling.

However, despite the criticisms of the IW few alternatives have been offered. We examine some of these alternatives in the next section.

7.3 SOME ALTERNATIVE PRIOR DISTRIBUTIONS

The most immediate extension of the IW one might consider is a general scale modified rotatable distribution, following Dawid (1981). Recall that a square symmetric matrix Σ is rotatable if for any $q \times q$ orthogonal matrix Q, $Q'\Sigma Q$ has the same distribution as Σ. Thus $\mathcal{IW}(\delta, I)$ is rotatable. A scale modified form of this distribution, denoted $\Pi(G)$, involves pre- and post-multiplying by a square root of a square symmetric matrix, G. The IW is one particular example. But such examples are scarce. And it is hard to see, in terms of prior specification, that much has been gained with such a generalization.

The distribution of its eigenvalues, determines the distribution of such a rotatable matrix. Leonard and Hsu (1992) give a rich class of prior distributions whose members are not rotatable in general but do derive from eigenvalue distributions. They employ the matrix logarithmic transformation of Σ, the matrix logarithm being generated by the logarithm of the eigenvalues of Σ. The $m = \frac{1}{2}q(q + 1)$ parameters of the matrix logarithm of Σ, taken to be Gaussian, have a full complement of m means and $(m/2)(m + 1)$ covariances. These priors certainly have a plethora of hyperparameters and hence flexibility.

However, the proposed class is not without its problems. A primary difficulty lies in the lack of interpretability and hence intuitive meaning of logarithms of eigenvalues. An important facet of this problem would be marginalization invariance. Suppose you have constructed a $q \times q$ matrix logarithm prior distribution. Now you are given another variable and so need to extend your prior specification so as to preserve, marginally, your original specification. This seems to be a difficult task in general.

One final approach stemming from Mäkeläinen and Brown (1988) should perhaps be mentioned. They suppose: (i) $Y = Y_u + Y_*$; (ii) only the q-vector Y is observed; and (iii) Y_* follows a model (1) with prior distribution given by (2). Independently Y_u represents unexplainable error and might, for example, consist of q independent normal random variables with possibly unequal variances. This idea yields a way around the deterministic property of the IW. However, the resulting prior has not been developed, primarily because of the intractability it shares with most known alternatives to the IW. In contrast, our regression dissection of the IW in the following section yields a very tractable alternative.

7.4 GENERALIZED INVERTED WISHART (GIW)

Suppose one can partition the q-variates into two sets (and subsequently perhaps into further sets) in some natural way. This may require a permutation of coordinates. The partitioning may be possible because we can ascribe a special form to the covariance structure in one block, for example. We then split the $n \times q$ data matrix into Y_1, Y_2, with dimensions $n \times q_1$ and $n \times q_2$, respectively; here the rows of (Y_1, Y_2) have covariance structure given by (5). The parametric transformation to (B, Γ, Σ_{22}) preceding Lemma 1 arises quite naturally; B represents the 'slope' of the best linear predictor of Y_1 based on Y_2 while Γ is the residual covariance matrix of the resulting prediction errors. One now faces the favourable circumstance that elicitation of the prior turns on an assessment of the relationship of the observables rather than on the less intuitive object, Σ. With the IW prior distribution reorganized in the hierarchical structures seen in Lemma 1, the posterior distribution will also be of the same form and one can retransform back to the posterior distribution given by (3). But observe that in the (B, Γ, Σ_{22}) parameterization, one may specify the prior distribution somewhat more flexibly than the form prescribed for Lemma 1 while retaining a natural conjugate structure. This idea leads to the generalized IW distribution developed in this section.

Still in the case where the data vector consists of $k = 2$ blocks, consider the following prior structure with Σ_{22} independent of (B, Γ):

$$\Sigma_{22} \sim \mathscr{IW}(\delta_2; G_{22}) \tag{7}$$

$$\Gamma \sim \mathscr{IW}(\delta_1 + q_2; Q) \tag{8}$$

$$B \mid \Gamma \sim B_0 + \mathscr{N}(H, \Gamma). \tag{9}$$

This corresponds to the IW prior (2) if and only if

$$\delta_1 = \delta_2, \quad Q = G_{11.2}, \quad B_0 = G_{22}^{-1} G_{21}, \quad H = G_{22}^{-1}. \tag{10}$$

The above prior specification has $\frac{1}{2}q_2(q_2 + 1) + 1$ hyperparameters additional to the $\frac{1}{2}q(q + 1) + 1$ hyperparameters of the IW prior distribution (2). One may vary these extra hyperparameters freely so as to change the covariance structure from the simple structure determined by δ in the IW.

Note in this connection that the conditional normal model

$$Y_1 \mid Y_2 \sim B' Y_2 + \mathscr{N}(\Gamma, 1)$$

$$Y_2 \sim \mathscr{N}(\Sigma_{22}, 1)$$

and the prior distribution defined by (7), (8), (9) may be combined to show that

$$E(Y_1 Y_1') = \{B_0' G_{22} B_0 + [(\delta_2 + \tau - 2)/(\delta_2 + q_2 - 2)]Q\}/(\delta_2 - 2) \tag{11}$$

$$E(Y_1 Y_2') = B_0'\{G_{22}/(\delta_2 - 2)\} \tag{12}$$

$$E(Y_2 Y_2') = G_{22}/(\delta_2 - 2),$$

with $\tau = \text{trace}(G_{22}H)$. The mean in (12) is $G_{12}/(\delta_2 - 2)$ provided $B_0 = G_{22}^{-1} G_{21}$ as in (10). With this specification of B_0 the mean (11) is $G_{11}/(\delta_2 - 2)$ provided Q is proportional to $G_{11.2}$, with constant of proportionality

$$(\delta_1 + q_2 - 2)/(\delta_2 + \text{trace}(G_{22}H) - 2).$$

Thus δ_1 and the $q_2 \times q_2$ symmetric matrix H are completely free to vary while retaining the mean structure of (2).

Standard Bayesian (likelihood times prior) manipulations (see for example, Brown 1993 Appendix B.1) lead to the following posterior distribution, with Σ_{22} still distributed independently of (B, Γ):

$$\Sigma_{22} \sim \mathscr{IW}(\delta_2 + n; \hat{G}_{22}) \tag{13}$$

$$\Gamma \sim \mathcal{IW}(\delta_1 + q_2 + n; \hat{Q}) \tag{14}$$

$$B \mid \Gamma \sim \tilde{B} + \mathcal{N}(\hat{H}_0, \Gamma). \tag{15}$$

The various updated hyperparameter matrices required to define the posterior distribution stated above follow:

$$\hat{G}_{22} = G_{22} + Y_2'Y_2 \qquad \hat{H}_0 = (H^{-1} + Y_2'Y_2)^{-1}$$

$$\hat{Q} = Q + S + ([\hat{B} - B_0]'A_2[\hat{B} - B_0]) \qquad A_2 = [H + (Y_2'Y_2)^{-1}]^{-1}$$

$$S = (Y_1 - Y_2\hat{B})'(Y_1 - Y_2\hat{B}) \qquad \hat{B} = (Y_2'Y_2)^{-1}Y_2'Y_1$$

$$\tilde{B} = WB_0 + (I - W)\hat{B} \qquad W = (H^{-1} + Y_2'Y_2)^{-1}H^{-1}.$$

This posterior distribution induces the posterior distribution for Σ by the one-to-one reverse transformation $(\Sigma_{22}, B, \Gamma) \to \Sigma$ defined just before Lemma 1. This will not, of course, correspond to the IW posterior distribution except under the restricted hyperparametrization given by (10). However, such retransformation may be unnecessary, depending on the application. In fact some important properties of Σ and its relatives may be obtained directly from (13), (14) and (15).

Consider for example, the posterior expectation of the generalized variance. Since $|\Sigma| = |\Gamma^{-1}|^{-1}|\Sigma_{22}^{-1}|^{-1}$ and these latter two factors are distributed independently by the posterior, we obtain from the general theory of the Wishart distribution itself,

$$E[|\Sigma| \mid Y] = C|\hat{Q}||\hat{G}_{22}|,$$

where

$$C = (\delta_1 + q_2 + n - 2)^{-1} \ldots (\delta_1 + q_2 + n - q_1 + 1)^{-1}(\delta_2 + n - 2)^{-1}$$
$$\ldots (\delta_2 + n - q_2 + 1)^{-1}.$$

Only a slightly more complicated calculation is required to compute the posterior expectation of the Σ itself. First observe the following:

$$E[\Gamma \mid Y] = \hat{Q}(\delta_1 + q_2 + n - 2)^{-1}$$
$$E[\Sigma_{22} \mid Y] = \hat{G}_{22}(\delta_2 + n - 2)^{-1}$$
$$E[B \mid Y] = \tilde{B}.$$

The posterior independence of B and Σ_{22} implies

$$E[B'\Sigma_{22}B \mid Y] = \tilde{B}'E[\Sigma_{22} \mid Y]\tilde{B} + E[\Gamma \mid Y]\text{trace}\{E[\Sigma_{22} \mid Y]\hat{H}_0^{-1}\}.$$

These results yield $E[\Sigma \mid Y]$ through the corresponding expectations of the components of its partition in (6). More precisely

$$E[\Sigma_{11} \mid Y] = E[\Gamma \mid Y][1 + \text{trace}\, E[\Sigma_{22} \mid Y]\hat{H}_0^{-1}] + \tilde{B}' E[\Sigma_{22} \mid Y]\tilde{B},$$

$$E[\Sigma_{12} \mid Y] = \tilde{B}' E[\Sigma_{22} \mid Y], \quad E[\Sigma_{21} \mid Y] = E[\Sigma_{22} \mid Y]\tilde{B}.$$

The role of $E[\Sigma_{22} \mid Y]$ in the posterior expectation has been highlighted to pave the way to a recursive generalization of these results.

The foregoing results can be used recursively to the general case of $k \geq 2$ blocks. Partition the $n \times q$ data matrix into $Y = (Y_1, \ldots, Y_k)$ with Y_i having dimension q_i, $i = 1, \ldots, k$ where $q_1 + \ldots + q_k = q$. Suppose Σ is partitioned conformably, it being the covariance matrix of each of the independent rows of Y. For convenience, let $Y_{(i)} = (Y_i, \ldots, Y_k)$ and have dimension $n \times q_{(i)}$ where $q_{(i)} = q_i + \ldots + q_k$. The covariance matrix of $Y_{(i)}$ will be denoted by $\Sigma_{(ii)}$ and recursively partitioned by

$$\Sigma_{(ii)} = \begin{pmatrix} \Sigma_{ii} & \Sigma_{i(i+1)} \\ \Sigma_{(i+1)i} & \Sigma_{([i+1][i+1])} \end{pmatrix}. \tag{16}$$

The role of B and Γ above will be played successively by B_i and Γ_i with

$$B_i = [\Sigma_{(i+1][i+1])}]^{-1} \Sigma_{(i+1)i}$$

and

$$\Gamma_i = \Sigma_{ii} - \Sigma_{i(i+1)} [\Sigma_{([i+1][i+1])}]^{-1} \Sigma_{(i+1)i},$$

for $i = 1, \ldots, (k-1)$.

It is readily shown that the likelihood function can be expressed as a product, $L_1 \ldots L_{(k-1)} L_k$; here L_i derives from the conditional sampling density of Y_i given $Y_{(i+1)}$ and Σ, $i = 1, \ldots, (k-1)$ and L_k from that of Y_k. However, L_i depends on Σ only through B_i and Γ_i.

The prior distribution for Σ, to be conjugate, must factorize conformably with that likelihood. Our proposed generalization of the IW, the generalized inverted Wishart (GIW) is obtained by imposing the following extensions of the distributional assumptions in (7), (8) and (9):

Here Σ_{kk} is independent of the $\{(B_i, \Gamma_i), i = 1, \ldots, (k-1)\}$ and the latter are independent of each other:

$$\Sigma_{kk} \sim \mathscr{IW}(\delta_k; G_{kk}) \tag{17}$$

$$\Gamma_i \sim \mathscr{IW}(\delta_i + q_{(i)}; Q_i) \tag{18}$$

$$B_i \mid \Gamma_i \sim B_{0i} + \mathscr{N}(H_i, \Gamma_i) \tag{19}$$

and for all i.

Adopting the GIW prior distribution entails the factorization of the posterior. This in turn makes the earlier theory for the case $k = 2$ directly applicable. Thus *a posteriori*, Σ_{kk} remains distributed independently of the $\{(B_i, \Gamma_i)\}$ and

$$\Sigma_{kk} \sim \mathcal{IW}(\delta_k + n; \hat{G}_{kk}) \tag{20}$$

$$\Gamma_i \sim \mathcal{IW}(\delta_i + q_{(i+1)} + n; \hat{Q}) \tag{21}$$

$$B_i \mid \Gamma_i \sim \tilde{B}_i + \mathcal{N}(\hat{H}_{0i}, \Gamma_i). \tag{22}$$

Here

$\hat{G}_{kk} = G_{kk} + Y'_{(k+1)} Y_{(k+1)}$ $\qquad \hat{H}_{0i} = (H_i^{-1} + Y'_{(i+1)} Y_{(i+1)})^{-1}$

$\hat{Q}_i = Q_i + S_i + [\hat{B}_i - B_{0i}]' A_i [\hat{B}_i - B_{0i}]$ $\qquad A_i = [H_i + (Y'_{(i+1)} Y_{(i+1)})^{-1}]^{-1}$

$S_i = (Y_i - Y_{(i+1)} \hat{B}_i)'(Y_i - Y_{(i+1)} \hat{B}_i)$ $\qquad \hat{B}_i = (Y'_{(i+1)} Y_{(i+1)})^{-1} Y'_{(i+1)} Y_i$

$W_i = (H_i^{-1} + Y'_{(i+1)} Y_{(i+1)})^{-1} H_i^{-1}$ $\qquad \tilde{B}_i = W_i B_{0i} + (I - W_i) \hat{B}_i$

As a consequence of conjugacy of the prior, the posterior remains in the GIW family of distributions.

The recursive model for Σ derives from the relation $\Sigma_{(ii)} = T_i \Delta_i T'_i$ for $i = 1, \ldots, (k-1)$, where

$$\Delta_i = \begin{pmatrix} \Gamma_i & 0 \\ 0 & \Sigma_{([i+1][i+1])} \end{pmatrix} \quad \text{and} \quad T_i = \begin{pmatrix} I & B'_i \\ 0 & I \end{pmatrix}.$$

Thus $\Sigma = \Sigma_{(11)} = T_1 \Delta_1 T'_1$; $\Sigma_{(22)}$ now plays the role played earlier in this section by Σ_{22}, B_1 that of B and Γ_1 that of Γ. In turn, Δ_1 through $\Sigma_{(22)}$ can be expressed in terms of T_2 and Δ_2, and so on recursively. This recursive relationship captures a fundamental property of the GIW distribution and enables us to extend results for the case $k = 2$ to $k \geq 2$. In particular, the posterior expectation of Σ can be found explicitly by repeated application of the results obtained above for $k = 2$.

We will not attempt to elaborate further, the properties of the GIW. Instead we address in the next section, the issue of how the GIW can be implemented.

7.5 IMPLEMENTING THE GIW PRIOR

The results presented above assume a known mean. Le and Zidek (1992) show how this assumption can be avoided within the Bayesian framework. They require only that the mean depends linearly on observable covariates with coefficients which need not be known. And the responses as well as covariates may be functions of a scalar parameter, like time for example. However, the same covariates must be measured for each response even when they are allowed to depend on such a parameter. Moreover, the mean's covariate model must be an additive component of the response model. The work of Le and Zidek (1992) allows the responses to be 'deseasonalized', as it were, within the Bayesian framework. Then Σ would represent the covariance matrix of the residual (deseasonalized) responses. Le and Zidek (1992) also show that only a (the same) subvector of responses need be observed on each sampling occasion. The resulting analysis can be carried out within the framework of this chapter. For brevity, the details of the extension of the current work, though important in practice, will not be presented here (see Le and Zidek, 1992 for details).

We turn now to the specification of the hyperparametric covariance structure in the Gaussian framework of the last section. For each block $i = 1, \ldots, (k-1)$, δ_i (a scalar), B_{0i} (a $q_{(i+1)} \times q_i$ matrix), Q_i (a symmetric $q_i \times q_i$ matrix) and H_i (a symmetric $q_{(i+1)} \times q_{(i+1)}$) must all be specified along with δ_k. This entails the specification of

$$k + \sum_{i=1}^{(k-1)} [q_i(q_i+1)/2 + q_{(i+1)}(q_{(i+1)}+1)/2] + \sum_{i=1}^{(k-1)} q_i q_{(i+1)}$$

distinct parameters. This is a much larger number of hyperparameters than are required for the IW prior: $1 + q(q+1)/2$. So the task of specifying the GIW is more difficult than that of specifying the IW. But a substantial pay-off in flexibility is achieved. How to fully exploit this increased flexibility is a subject for future investigation.

The additional span of the family of GIW distributions derives from the varying numbers of degrees of freedom for its submodules. The second added source of flexibility is found in the $\{H_i\}$, which determine the covariance structure for the $\{B_i\}$.

We would just note in passing that the work of Zidek and Weerahandi (1992) suggests an avenue to the specification of the covariance structure for the $\{B_i\}$. Briefly, the linear model is often regarded as having been obtained by ignoring higher order terms from a Taylor expansion of the response function. In particular, slope coefficients may be regarded as the first derivatives of the response function evaluated at some point, central to the domain of the model. It follows that the covariance of two such slopes is just a second-order mixed derivative of the autocovariance function of the random response function. The overall character of that autocovariance function expresses qualitative features of the response process. For example, the order of differentiability of the process is reflected in the order of differentiability of the autocovariance function. And the degree of perceived autocorrelation in the response process is reflected in the rate at which the autocovariance function declines to zero. In practice, a simple parametric model may well approximate that function quite adequately. In that case, the covariance structure for the $\{B_i\}$ could readily be prescribed using just the parameters of the autocovariance function.

As noted in section 7.1 the primary deficiency of the IW lies not in the lack of hyperparameters in G. Primarily the GIW gives us a freedom denied by the IW to vary the $\{\delta_i\}$'s. In fact, we expect that it will usually be sufficient to take

$$H_i = G^{-1}_{(i+1)(i+1)}, \quad Q_i = G_{ii \cdot (i+1)} \quad \text{and} \quad B_{0i} = G^{-1}_{(i+1)(i+1)} G_{(i+1)i}$$

for $i = 1, \ldots, k - 1$, thus leaving only the δ_i's to be specified. Even then our approach may offer something novel, namely a new way of eliciting prior knowledge. Instead of specifying G as a whole, 'you' tackle smaller (level 1 independent) pieces. In a preliminary qualitative step you organize the observables in a hierarchical (block) structure. It might be natural to put the blocks of Y's thought most predictable at the upper reaches of the hierarchy with i near 1. This would make the least predictable block correspond to $i = k - 1$. In any case, for each block separately, you need only specify the hyperparameters for that block without regard to the other blocks. And the $\{B_{0i}\}$ as well as the $\{Q_i\}$, at least, are quite intuitive, being, respectively, the 'slopes' and residual covariance matrices for the best linear fits of successively the $\{Y_i\}$ on the $\{Y_{(i+1)}\}$. If you deem Y_i to be well explained by $Y_{(i+1)}$ then $G_{ii \cdot (i+1)}$ would be 'small'. And δ_i would reflect your confidence in your assessment. With the qualitative ordering proposed above, we would expect the $\{\delta_i\}$ to decline with increasing i.

To obtain an insight which may be helpful in specifying the hyperparameters, consider the case where

$$B_{0i} = G^{-1}_{(i+1)(i+1)} G_{(i+1)i} \quad H_i = G^{-1}_{(i+1)(i+1)}$$

and

$$Q_i = G_{ii \cdot (i+1)} \quad \text{for } i = 1, \ldots, k - 1.$$

Set $G = \alpha I + \beta \phi \phi'$ where ϕ is a $1 \times q$ vector of constants and $\beta > -\alpha \|\phi\|^{-2}$ to ensure that G is positive definite. Using a subscript notation consistent with that used in defining the GIW, we find

$$H_i = G^{-1}_{(i+1)(i+1)} = \alpha^{-1}[I_{(i+1)(i+1)} - \beta(\alpha + \beta\|\phi_{(i+1)}\|^2)^{-1})\phi_{(i+1)} \phi'_{(i+1)}]$$

$$Q_i = G_{ii.(i+1)} = \alpha I_{ii} + \alpha\beta(\alpha + \beta\|\phi_{(i+1)}\|^2)^{-1}\phi_i\phi_i'$$

and

$$B_{0i} = G_{(i+1)(i+1)}^{-1} G_{(i+1)i} = \beta(\alpha + \beta\|\phi_{(i+1)}\|^2)^{-1}\phi_{(i+1)}\phi_i'$$

for $i = 1, \ldots, k - 1$.

Since $\|\phi_{(i+1)}\|^2$ is increasing as $i \to 1$, we find in agreement with intuition that $G_{ii.(i+1)}$ decreases as $i \to 1$. The decrease is largest for i near $k - 1$. But the rate of decrease diminishes as i approaches 1. Again in agreement with intuition, the degree of reduction in uncertainty about Y_i afforded by regression on $Y_{(i+1)}$ is governed by β; no reduction is achieved when $\beta = 0$. The intraclass correlation model obtains, by setting the coordinates of the ϕ to 1, $\alpha = \sigma^2(1 - \rho)$ and $\beta = \sigma^2\rho$. In this case when $\rho \to 1$ or $\rho \to 0$ the benefit of regressing on $Y_{(i+1)}$ declines to insignificance. The first of these two results derives from the fact that at $\rho = 1$ the $\{Y_i\}$ are identical; nothing is gained through regression on $Y_{(i+1)}$.

The specific model suggested above seems too restrictive to be of general value. But its qualitative features may well obtain in more general situations. Clearly, further study is required.

Even with any added simplicity conferred by our stepwise approach to eliciting prior information about Σ, the task may remain quite difficult. As noted in section 7.2, it will usually be necessary to carry on to an additional stage, even if that involves using a vague prior to obtain an approximate posterior. Alternatively, a type II (i.e. marginal) maximum likelihood estimation may be adopted in the same spirit once the hyperparameter matrices have been specified as a model of lower dimensional second-stage hyperparameters.

The type II maximum likelihood estimates can be obtained directly with the marginal distributions of Y conditional on the hyperparameters. With an IW prior for Σ given the hyperparameters G and δ as in (2), the marginal distribution of Y is a multivariate Student t distribution with scale parameter G and degrees of freedom depending on δ. Moreover, with a GIW prior given in (17-19) for k blocks, the marginal distribution of Y can be presented as a sequence of conditional distributions. Here, as a generalization of the results derived in Le and Zidek (1992), Y_k has a multivariate Student t distribution with scale parameter G_{kk} and degrees of freedom depending on δ_k, and $Y_i \mid Y_{(i+1)}$ for $i = 1, \ldots, k - 1$, also has a multivariate Student t distribution with scale parameter depending on Q_i and $G_{([i+1, i+1])}$ and degrees of freedom depending on δ_i.

Another approach for obtaining the maximum likelihood estimates is to use the method proposed by Chen (1979) as mentioned in section 7.2. Here the posterior expectations of Σ_{ii} for $i = 1, \ldots, k$ can be obtained through the posterior distributions given in (20-22).

7.6 CONCLUDING REMARKS

Our proposed generalization of inverted Wishart (IW) prior distribution holds promise. By incorporating into it one of the latter's fundamental features (expressed in Lemma 1), the new distribution retains much of the tractability of its predecessor, an important consideration in many applications where results must be given in a reasonably explicit form. (Its tractability is demonstrated in section 7.4 by the calculation of moments of the posterior distribution.) At the same time our generalized IW (GIW) can express a much wider range of prior opinion, most notably in allowing variation in the degrees of freedom (uncertainty) assessed in the various subcomponents of the GIW distribution.

A by-product of our generalization of the IW is a seemingly natural approach to the elicitation of prior opinion for a covariance matrix. Elicitation is through a sequence of independent regression submodules with interest focusing on the 'slope' coefficients and residual covariance matrices of these regressions.

The price of the enhanced flexibility offered by the GIW is the need to specify more hyperparameters than obtain in the IW or, more generally, the need for a more complex second stage in a hierarchical prior model. Further work remains in that direction as well as in determining other properties of the GIW.

The idea of decomposing the joint distribution of a set of random variables into a running sequence of conditional distributions has been used in connection with the multinomial distribution. It leads to a sequence of binomial random variables. If the running sequence of binomial probabilities are assigned beta prior distributions then a generalised Dirchlet results, see Connor and Mosimann (1969) and Kokolakis (1983). The idea would seem to extend to general graphical models as described in Lauritzen and Spiegelhalter (1988).

ACKNOWLEDGEMENTS

We are grateful to Professors Philip Dawid and Timo Mäkeläinen for helpful comments on this chapter.

REFERENCES

Barndorff-Nielsen, O. (1978) *Information and Exponential Families in Statistical theory.*, Chichester, John Wiley & sons.

Brown, P. J. and Mäkeläinen, T. (1992) Regression, sequenced measurements and coherent calibration, in *Bayesian Statistics 4*, J. M. Bernardo, J. Berger, A. P. Dawid and A. F. M. Smith, eds, Oxford, Clarendon Press, pp. 97–108.

Brown, P. J., Le, N. D. and Zidek, J. V. (1992) *Multivariate Spatial Interpolation with Kronecker Covariance Structures.* Technical report, Department of Statistics, University of British Columbia. Vancouver.

Brown, P. J., (1993) *Measurement, Regression and Calibration*, Oxford, Clarendon Press.

Chen, C.-F. (1979) Bayesian inference for a normal dispersion matrix and its application to stochastic multiple regression analysis, *Journal of the Royal Statistical Society* **B41**, 235–48.

Connor, R. J. and Mosimann, J. E. (1969) Concepts of independence for proportions with a generalisation of the Dirichlet, *Journal of the American Statistical Association*, **64**, 194–206.

Dawid, A. P. (1981) Some matrix-variate distribution theory: notational considerations and a Bayesian application, *Biometrika*, **68**, 265–74.

Dawid, A. P. (1988) The infinite regress and its conjugate analysis, in *Bayesian Statistics 3*, J. M. Bernardo, M. H. DeGroot, D. V. Lindley, and A. F. M. Smith, eds, Oxford, Clarendon Press, pp. 95–110.

Dawid, A. P. (1992) Selection paradoxes of Bayesian inference, *Proceedings of the International Symposium on Multivariate Analysis and its Applications*, Hong Kong Baptist College.

Dempster, A. P., Laird, N. M. and Rubin, D. B. (1977) Maximum likelihood from incomplete data via the EM algorithm (with Discussion), *Royal Statistical Society* **B39**, 1–38.

Dempster, A. P. (1969) *Elements of Continuous Multivariate Analysis*, Reading Massachusetts, Addison Wesley.

Dempster, A. P. (1972) Covariance selection, *Biometrics*, **28**, 157–76.

Dickey, J. M., Lindley, D. V. and Press, S. J. (1985) Bayesian estimation of the dispersion matrix of a multivariate normal distribution, *Communications in Statistics: Theory and Methods* **14**, 1019–34.

Granger, C. W. J. (1966). The typical spectral shape of an economic variable. *Econometrika*, **34**, 150–61.

Guttorp, P., Le, N. D., Sampson, P. D. and Zidek, J. V. (1993) Using entropy in the redesign of an environmental monitoring network, in *Multivariate Environmental Statistics*, G. P. Patil and C. R. Rao eds, New York, North-Holland/Elsevier Science, pp. 175 –202.

Hampel, F. R., Ronchetti, P. J., Rousseeuw, P. J. and Stahel, W. A. (1986) *Robust Statistics: The Approach Based on Influence Functions*, New York, John Wiley & Sons.

Jöreskog, K. G. (1977) *Structural Equation Models in the Social Sciences: Specification, Estimation and Testing*, North-Holland, pp. 265–86.

Kokolakis, G.E. (1983) A new look at the problem of classification with binary data, *Statistician*, **32**, 144–152.

Künsch, H., Beran, J. and Hampel, F. (1993) Contrasts under long range correlations, *Annals of Statistics*, **21**, 943–964.

Lauritzen, S. L. and Spiegelhalter, D. J. (1988) Local computations with probabilities on graphical structures and their application to expert systems (with discussion), *Journal of the Royal Statistical Society*, **B50**, 157–224.

Le, N. D. and Zidek, J. V. (1992) Interpolation with uncertain spatial covariance: a Bayesian alternative to Kriging; *Journal of Multivariate Analysis*, **43**, 351–74.

Leonard, T. and Hsu, J. S. J. (1992) Bayesian inference for a covariance matrix. *Annals of Statistics*, **20**, 1669–1696.

Lindley, D. V. (1978) The Bayesian approach (with discussion), *Scandinavian Journal of Statistics*, **5**, 1–26.

Mäkeläinen, T. and Brown, P. J. (1988) Coherent priors for ordered regressions, in *Bayesian Statistics 3*, J. M. Bernardo, M. H. DeGroot, D. V. Lindley, and A. F. M. Smith, eds, Oxford, Clarendon Press, pp. 677–96.

Spearman, C. (1904). The proof and measurement of association between two things, *American Journal of Psychology* **15**, 72, 202.

Student (1927) Errors of routine analysis, *Biometrika*, **19**, 151–64.

Wright, S. (1934) The method of path coefficients, *Annals of Mathematical Statistics* **5**, 161–215.

Zidek, J. V. and Weerahandi, S. (1992) Bayesian predictive inference for samples from smooth processes, in *Bayesian Statistics 4*, J. M. Bernardo, J. O. Berger, A. P. Dawid and A. F. M. Smith, eds, pp. 547–563. Oxford, Clarendon Press.

Department of Statistics and
Computational Mathematics
The University
PO Box 147
Liverpool
L69 3BX
UK

British Columbia Cancer Agency
Biometry Section
600 W. 10th Avenue

Vancouver
British Columbia V5Z 4E6
Canada

Department of Statistics
University of British Columbia
6356 Agricultural Road
Vancouver
British Columbia V6T 1Z2
Canada

CHAPTER 8

The Role of Statistical Theory in Decision Aiding: Measuring Decision Effectiveness in the Light of Outcomes

Rex V. Brown
*George Mason University and
Decision Science Associates Inc.*

Unlike other chapters in this book, no doubt, this chapter does not propose a contribution to statistical theory *per se*. As a decision analyst and researcher (not a statistician), I will suggest how the Bayesian community might contribute—as Dennis Lindley already has in a major way—to the development of operational decision methodology. In particular, I will address a major current problem—how to evaluate a decision after the fact.

A tentative method, summarized as follows, is presented for predicting a decision's impact on a decider's utility, conditional on evolving evidence available by hindsight to an evaluator. Different stages of decision-making are to be evaluated: the action, the choice process or any intervention in that process (such as training or a decision aid). Decision effectiveness is interpreted as predicted impact on decider's utility. Utility scale can be interpolation between lifelong disaster and bliss, but it may be more operational to measure differences on that absolute scale relevant to comparisons of interest. The decider need not even be human provided utility can be given some meaning. Evaluations are based on whatever the

---

Aspects of Uncertainty edited by P. R. Freeman and A. F. M. Smith. © 1994 John Wiley & Sons Ltd.

evaluator has learned. Diagnostic information may include observed measures or elicited responses to the aftermath of action.

A modelling strategy is based on a causal sequence of stages: intervention, process, action, unfolding outcomes and ultimate utility. Predicted impact of decision on utility can be conditioned on any given stage. Though any one conditioned assessment may be suspect, several can be combined, and several empirical measures can be taken to characterize any stage.

8.1 SOME PAST DEVELOPMENTS IN PRESCRIPTIVE STATISTICS

8.1.1 Prescriptive decision research

People and logic problems of decision aiding

Prescriptive decision research is dedicated to helping deciders to make their decisions. It draws on normative theory—which addresses what a logically coherent decider might do—and descriptive theory—which addresses what a fallible real decider can actually do (Brown 1989). Prescriptive decision research is the normative tempered by the descriptive, though it has a province of its own.

As a decision analyst, I am concerned with developing decision-aiding methods and using them to help clients in the 'real world'. A successful decision aid needs to pass two tests: it must call for inputs people can provide and produce outputs people will use—the descriptive component; and the outputs must be consistent with the inputs—the normative component. Both must have sound theoretical underpinnings, and their research agendas interact. Brown and Vari (1992) have proposed descriptive research

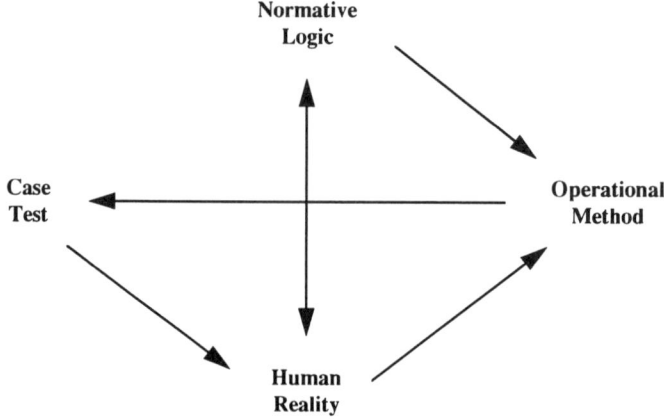

Figure 8.1 The decision aid development cycle.

to ensure that inputs and outputs are psychologically and institutionally realistic. This chapter considers what statistical theory is needed to ensure that the logical linkage between them is sound.

The decision aid development cycle

Figure 8.1 shows the main elements and influences in the typical process for developing decision-aiding technology. It is similar to the familiar principle in engineering design 'build–test–build-test' (where physics and mathematics play the descriptive and normative roles respectively). For example, a normative paradigm (such as maximizing personal expected utility) suggests a particular operational method (such as rolling back a decision tree and updating its probabilities by Bayes's rule). This is tried out on a test case (such as a choice of business investment) and lessons are learned about human reality (such as: lay people cannot provide meaningful probabilities in this form). Together with intuition and behavioural research this directly prompts a change in the method (such as elicitation procedure). It also prompts research on an alternative paradigm (such as fuzzy sets), which gives rise to further methodological innovation. And so on.

A fruitful collaboration

No normative decision theorist has done more than Dennis Lindley for prescriptive decision research, and derivatively for the success of applied decision analysis. I have been singularly fortunate to have had his unparalleled expertise, imagination and enthusiasm to draw on, when my work developing operational decision-aiding methodology called for some critical normative contribution. Our collaboration, though intermittent, has spanned nearly 30 years.

I will try to convey the nature of that productive interaction and what it has to say more generally about how statistical decision theory can feed and discipline applied decision 'technology' through some account of our joint work. This evolved in response to a consulting need to monitor the effectiveness of the aids we used. In section 8.2.1 I will discuss a methodological approach which grew out of this earlier work, on which further analytic infusion is needed from the Lindleys of this world, if useful methodology anchored in rigorous logic is to emerge.

8.1.2 An illustrative case: plural evaluation methodology[†]

In the mid-1970s the experience of decision-aiding consultants suggested that the conventional practice of 'singular' decision analysis—i.e. basing a

---

[†] This section is adapted in part from Brown (1989).

recommendation on a single model or analysis of the problem—often conflicts with common sense. However, the alternative of *plural* analysis lacked a theoretically respectable basis, or much in the way of defensible procedures for designing a plural strategy or merging the results. Decision theory at this time provided a test of *coherence* between alternative single analyses (e.g. between aggregated judgement and a disaggregated model), but not much more. This observation propelled us into an interleaved, largely opportunistic, programme of research and application, in the build–test–build–test mould, extending over two decades. It involved mathematical statisticians, pre-eminently Dennis Lindley, joined more recently by John Pratt, Stephen Watson, Jacob Ulvila and Kathy Laskey. The behavioural side was supported by Amos Tversky, Detlof von Winterfeldt and Marvin Cohen (Brown and Lindley 1978, Brown et al. 1984).

Early ad hoc research

The initial normative phase Our starting-point was normative and adaptive: to see if any variants of existing statistical theory addressed the problem. Lindley and I developed an adaptation of Bayesian updating and pre-posterior analysis (Brown and Lindley 1986). Psychologist Amos Tversky joined our project to ensure that the procedures developed would be psychologically sound.

We illustrated the Bayesian updating approach on a specific (but hypothetical) inference problem—assessing the probability of a binary event with alternative probability models. It seemed to work, and showed that, at least in this simple case, plural analysis enhanced the quality of the target judgement (Lindley et al. 1979). Lindley and others developed the normative basis further by adapting the general Bayesian updating paradigm to a variety of special cases, including plural utility evaluations and plural option selection (Freeling 1986; Lindley 1982, 1983, 1985, 1987a, 1987b; Laskey 1990).

Real-world testing: a submarine combat aid Consultant colleagues and I had been developing concepts for aids to help submarine commanders to make operational decisions in combat (Bromage et al. 1983) and target ranging in particular, i.e. estimating distance from a target, preparatory to firing a torpedo. We developed a computerized 'range pooling aid', for combining multiple estimates of range (Cohen and Brown 1980). It incorporated a normal Bayesian updating paradigm, developed by Lindley, which incorporated judgements of accuracy and interdependence about the several range estimation techniques the US Navy had been using.

However, when we tried it out on navy personnel, using plausible input data for a hypothetical engagement, we found that, when the multiple input estimates were widely divergent, it produced as output a pooled estimate with much smaller variance than experienced professionals judged reasonable. The normal model failed to account for the fact that divergent estimates should lead us to downgrade the accuracy originally ascribed to input estimates.

Lindley developed an alternative model based on the t-distribution, which corrected inappropriate assumptions and produced plausible results (Lindley 1985; Ulvila and Brown 1982). Two approaches to validating the aid, and its underlying principles, were attempted. The first was a statistical outcome-based test. It compared the accuracy of range estimates actually made by a submarine commander during a documented field exercise with the accuracy of estimates which the aid would *have* produced, based on the same available data. This exercise suggested the aid would have yielded significant improvement—of the order of 20% reduction in average error (Bromage *et al.* 1983).

The exercise was, however, essentially normative, based on untested (if plausible) ideal behavioural and institutional assumptions about how the aid would be used. To achieve descriptive realism, the next validation exercise was to replicate, as accurately as a shore-based laboratory would allow, the circumstances of actual application of the aid. The subjects were real submarine commanders and they were led through a real historical combat scenario from a fleet exercise. They were presented with visual and numerical data that they would expect to have, and with realistic timing, in a real engagement.

Removed as this experiment was from the traditional sophomore-in-a-contrived-situation, it still fell substantially short of adequate realism, i.e. simulating a real war, using a fully developed decision aid, in a stressful environment. However, it provided insights into what an aid of this type required of a user, what value it could give to him in return and what promising directions for improvement would be (Bromage *et al.* 1983). Further development of the range pooling aid and adoption by the fleet will now be the Navy's responsibility.

A structured research project

By the early 1980s we were routinely doing plural evaluation and finding it to be generally sound practice. However, we still did not have a prescriptive (i.e. useful) analytic methodology to help us do it *right*—and do it defensibly. Our design of plural research still came largely from intuitive practitioner tinkering. The National Science Foundation funded us to develop prescriptive methodology to support that design.

The first step was to evaluate, from a prescriptive point of view, the then-current state of the art and we found that virtually all the published work—ours and others'—was still largely normative. (A significant exception was pooling plural forecasts (Winkler and Makridakis 1983), where an operational methodology was emerging.) Most approaches were too burdensome to use on an applied decision-aiding project, especially at the design phase. It is not worth spending half the available budget to help spend the remaining half!

In particular, our own plural design and reconciliation efforts had been based on normative Bayesian updating and preposterior paradigms. Our attempts to use them ran into serious problems deriving from descriptive reality. Producing a likelihood function involved difficult second-order probability judgements, which few humans (including statisticians) could supply.

This led us to develop 'feedback adjustment', a more intuition-intensive approach to capture the same notions, but in a more cognitively accessible way (Brown and Lindley 1986; Brown *et al.* 1984). Suppose a subject S, has analysed a judgemental task (whether of inference, evaluation or choice), using two or more decision analytic models. If S were coherent, these would produce the same judgement as output, but in general they do not. The plural pooling task is to reconcile the several models so that their outputs do coincide.

A Bayesian paradigm would call for a prior on the components of each model and a likelihood function that reflects the conditional probabilities of S's component assessments, given their 'true' values. This is impracticably difficult in most situations. With feedback adjustment S manipulates all the input assessments at will, observing the degree of inconsistency in the resulting plural outputs until those outputs coincide, in which case coherence will have been achieved. With luck, he will properly, but informally, take into account the same higher-order judgements he would have needed for a formal Bayesian solution; for example, about how relatively 'firm' his input assessments are and any firmness dependence between them (joint likelihood function).

An experimental psychologist, Detlof von Winterfeldt, investigated how subjects do in fact pool and reconcile plural estimates *with and without* analytic support (such as feedback adjustment), and with what success. He adopted the traditional psychological experimental mode, i.e. using students as subjects and giving them hypothetical tasks (von Winterfeldt *et al.* 1992). These experiments also cast light on implicit plural analysis models which humans use, which may lead to improved formal aids.

A major prescriptive impediment to the normative Bayesian paradigm is the difficulty most subjects have in supplying statistical measures of their 'confidence' in the plural approaches—in particular, variances and covariances of second-order probability distributions. Anthony Freeling has suggested that those judgements about contending plural approaches might be represented to the subject as 'the amount of information' each contains, and shown graphically by circles, whose relative areas correspond to variances. 'Overlapping information' could be represented by intersecting areas and correspond to covariance. Alternative plural mixes of approaches could be compared according to the area of the union of the circles, as elicited.

Whether such intersecting circles can adequately provide the basis for such an aid depends on both descriptive and normative issues. Subjects' implicit interpretations of the intersecting areas is a descriptive issue amenable, with some ingenuity, to experimentation. Mathematical statistician John Pratt is exploring the normative question: what formal interpretation needs to be put on them, in order for conclusions from the output to be logically valid. (A complementary descriptive issue is how good people are at comparing the size of differently shaped areas).

The general effect of our prescriptive research programme on plural evaluation is that we have some promising and partially developed analytic aids and clearly developed lines for further prescriptive research, consistent with the basic 'build–test–build–test' paradigm.

8.2 STATISTICAL DEVELOPMENT NEEDED ON EVALUATING DECISION EFFECTIVENESS

8.2.1 Problem background

Evaluating decisions in the light of outcomes

Evaluating the effectiveness of previous actions, choice processes or interventions on the basis of their observed aftermaths is a major, and distinctively prescriptive, research problem.

The starter insight, as with plural evaluation above, was the observation that common sense (mine and others) conflicted with prevailing academic orthodoxy and was worth checking out analytically. In this case, the orthodoxy, permeating Decision Analysis 101, was that 'Good outcomes do not mean good decisions' (and there is a sense in which that is unexceptionable). Subtler leaders in the field conceded that the problem was meaningful and important, but intractable and hopeless to attack. The common sense was that, as we watch our fellows go about their business, we form impressions that some make better decisions than others, and that these impressions have some basis in fact. Pet lovers even feel they have observational grounds for arguing that cats are smarter than dogs. Their practical intelligence presumably means they take actions that, in some sense, serve their interests better.

Monitoring the effectiveness of decision training or aids is key to improving them. That effectiveness can be interpreted as the desirability, by hindsight, of the outcomes of actions it generates. Though this is never known with certainty, especially compared with 'the path not taken', the aftermath of past actions does provide diagnostic evidence.

Different types of data can be gathered, say, by observation (e.g. decision reversals), interrogation (e.g. expressed regrets) or canvassing expert opinion (e.g. in the given field). Developing a methodology for obtaining and analysing such data is daunting: the outcome of an action not taken can rarely be confidently established; the complete outcome of the action that is taken may be long delayed; how desirable the outcome is may be problematic; a difference in the outcome of options is usually confounded by factors other than the quality of the choice. The methodology has two separable parts: (1) an analytic framework which specifies the questions to ask and the use to be made of the answers; (2) techniques for answering the questions, through data gathering and analysis.

Psychologist Vince Campbell and I have proposed an approach to assessing the effectiveness of others' decisions, with application to career planning, including measuring instruments for an evaluator to use (Brown and Campbell, 1993; Baron and Brown 1991). We have expressed the argument qualitatively, in the informal language of a practising decision analyst. However, it raises theoretical and technical issues which need to be addressed by specialized statistical experts before any operational methodology can be considered normatively well grounded and algorithmically implementable. Some of these issues are noted in brackets [ ] throughout the chapter.

The research task

We wish to enable an evaluator, E, to make (and substantiate) evaluations about decider, D, of the following type of 'object being evaluated' (OBE):

1. A past action compared to its alternatives ('It was fortunate for the Israelis that they destroyed the Iraqi reactor in 1982; it has since become fairly clear that the international community would not have stopped its being put to military use.')
2. A class of actions or a decision process ('Judging from that and similar episodes, Israeli decision-making cannot be much improved upon in such cases.')
3. A difference between deciders or processes ('A concentration in decision analysis will help MBAs make career decisions worth on average $10 000 a year more, judging by a survey of recent graduates.')

As used here, a *choice* is 'good' if the action chosen has outcomes more desirable for D than alternative actions. A decision *process* or class is good if it produces better choices on average than contending processes. This is an external interpretation, only incidentally related to internal phenomena like what D knows and how well D uses his knowledge. In this sense, good outcomes *do* mean good decisions (contrary to decision analytic orthodoxy). However, this decision effectiveness can only be inferred with uncertainty by E (not by D) from available evidence. Much can be learned from the unfolding aftermath of actions—especially if choices are repeated.

State of the art

Decision scientists have noted that, to advance prescriptive (contrasted with descriptive and normative) decision science, the most pressing (if most difficult) research need is to develop a general methodology for evaluating decision effectiveness (Brown 1989; Brown and Vari 1992); but there have been few major attempts. Von Winterfeldt and Edwards (1986) propose a concept of 'relative expected loss' as a measure of the 'non-intuitiveness' of a decision-aiding procedure, which Politser (1991) has implemented in an experiment to evaluate 'gains' from medical decision analysis. However, non-intuitiveness of choice is not necessarily a good thing (Brown 1991). Intuition may be sounder than the aid we replace it by.

Studies by decision scientists of the effectiveness of education and other decision process interventions, almost all domain-specific, have typically focused on appraising the decision process directly, through self-report or observation. Evaluations focus on whether treated subjects possess target skills (e.g. 'Can D generate a list of goals for a given decision problem?'); or apply an approved procedure (e.g. 'Do verbal protocols show that D spontaneously articulates his goals?') to a greater extent than do control subjects. The relationship of skills and procedures to the achievement of D's goals is usually left implicit.

Measures have included self-reports of competence (Adams and Feehrer 1991), performance on contrived experiments with 'school solutions' (von Winterfeldt and Edwards 1986), expert evaluation of hypothetical nurse decisions and observed impact on 'undesirable' behaviour, such as experimental smoking.

Beyth-Marom *et al.* (1991) review of a group of studies relating to decision skills courses. Each study involves a single measure, with differing advantages and disadvantages, but none stands on its own as a firm basis for measuring the quality of real-life decisions. Watson and Brown (1975, 1978) document whether decision consulting improved clients' decisions, with impressionistic case studies and they propose a normative paradigm for evaluating decision analysis plans in advance. Nakdimon (1989) reviews the soundness of the 1982 Israeli bombing attack in detail, but in purely political journalistic terms.

8.2.2 Problem formulation

Phenomena addressed

The primary unit of enquiry is the individual *action*, whose selection constitutes a *choice*. We may be interested in the effectiveness of an action in itself[†];

[†] I will not deal here explicitly with generalizations of discrete action, such as complex strategies or incremental commitments (Brown 1987).

or of a *class* of comparable actions generated by a decision *process* common to more than one D (reflecting, say, the use of an aid); or of *decider*, *D*, himself, and the succession of choices D makes (reflecting, say, a training intervention). The focus is on the *aftermath* of action, i.e. the succession of observable events that lead from a chosen action to an ultimate *outcome* for D and to his satisfaction with the aftermath. Since he might have been equally satisfied with the aftermath of alternative actions, we are primarily concerned with an action's *consequence*, i.e. the difference the action made to D's interests.

Interpretation of decision effectiveness

The measure of evaluation may be qualitative, ordinal or quantitative, depending on the purpose of the exercise. A measure of decision effectiveness should be readily interpreted so it can be checked judgementally for plausibility, and sufficiently universal as to permit comparisons across deciders, at least within a particular domain. It should call for feasibly generated and analysed data and judgement.

For present purposes, I interpret the effectiveness of an action as E's expectation of the consequence to D of that action, conditional on E's knowledge at the time of the evaluation. The consequence is interpreted as the difference in D's utility attributable to the action. (Is utility, e.g. *a la* von Neumann–Morgenstern, sufficiently unambiguous for this purpose?)

'Attributable' means the difference between the total utility D will enjoy if he takes the action and if he does not. Accordingly, E will be addressing the probability distribution on such alternative actions. [Is this a satisfactory interpretation? Does it imply that if D will take a slight variant of the action OBE then the effectiveness of that action is low? Sounds unreasonable, but what would be more reasonable? Treating clusters of similar actions together?]

The effectiveness of a decision class or process (possibly D himself) is given by the mean of E's probability distribution of the consequences of the actions it comprises or generates. The effectiveness of an intervention is straightforward: E's expectation of the difference in effectiveness between the process before and after.

E's evaluation of a decision process can be broken into two parts:

1. Predicting the consequence of a given past action;
2. Predicting the consequence for a class of such actions (having evaluated one or more actions in the class).

Evaluating a process intervention (like the use of a decision aid) amounts to comparing the predicted consequences of two processes.

Evaluating a past action

Figure 8.2 shows a general schema for evaluating the evolving impact on outcome utility of an action, say a career move by D guided by a decision aid. The two curves represent E's comparison of aftermath outcomes to

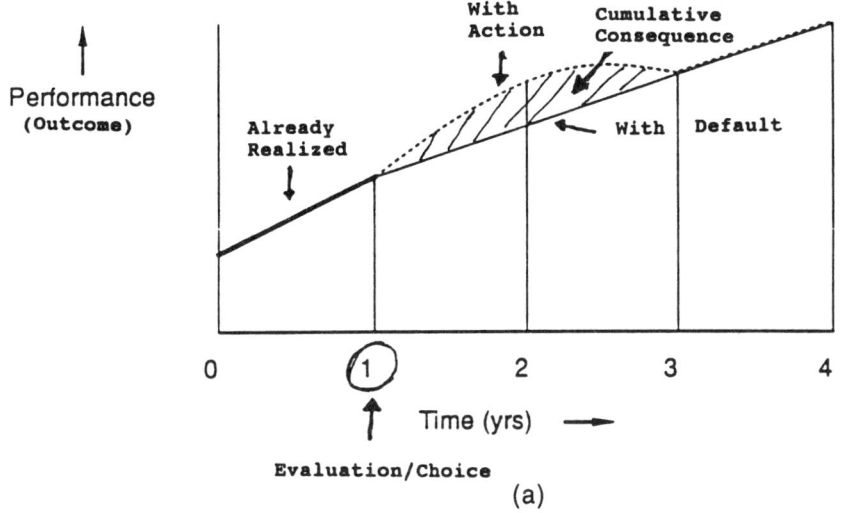

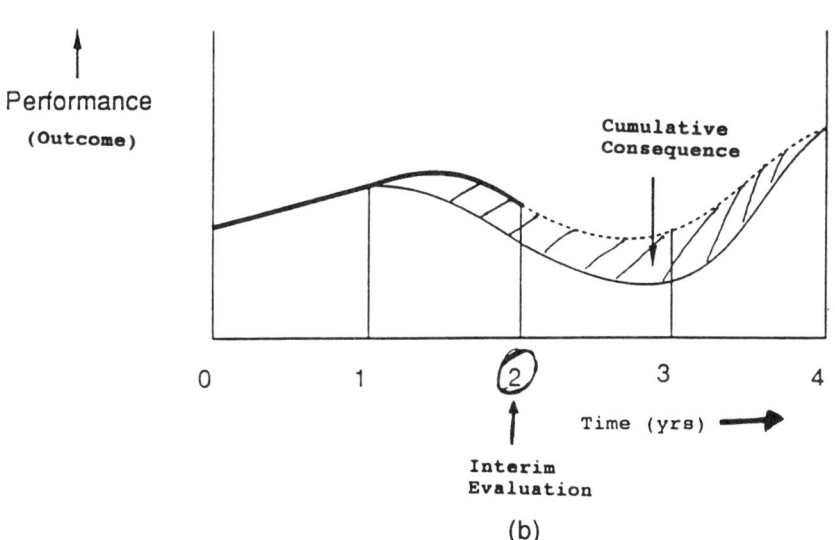

Figure 8.2 Evaluating the consequence of an action. (a) Evaluated before choice; (b) assessed some time after choice

D with and without the move. The cumulative difference, i.e. the area between the two curves, represents E's evaluation, with developing information, of differential consequence, i.e. impact on D's outcome utility. (The actual but never fully known consequence could be represented in a similar figure.)

Figure 8.2(a) shows a hypothetical evaluation conducted before the move. E has judged D's career outcome to be improving up to the time of the move (perhaps due to some unrelated development, such as raising the minimum wage). E's pre-choice judgement was that this trend would continue in a straight line over the next three years, and that the move would lead to still greater improvement for two more years, after which it would have no continuing effect. This is basically the judgement addressed by regular decision analysis. Our concern is how to update that judgement later. At all times E is basically making an assessment of the total area between the curves, based on data.

Figure 8.2(b) shows an updated evaluation a year later, after some career developments. In E's updated judgement, career performance has actually declined in the intervening year, but that it is attributed to something other than the move (perhaps a recession). Without the move, E judges performance would have been even worse. Moreover, in the light of that year of experience, E now judges that the impact will persist over three years, not two (perhaps because he has seen evidence that career decisions have longer-term effects). He is thus now persuaded that the total impact of the move will be even more positive than he first thought (reflected by the larger shaded area in Figure 8.2(b)).

[Under what circumstances is this representation rigorous, as opposed to merely suggestive? What interpretation does the vertical outcome scale need to have for the consequence area to be a meaningful measure of action consequence? Discounted for utility timing? Critical assumption of additive year-by-year utilities? Expectations acceptable as equivalent substitute for uncertain impacts?]

Utility scale

The analytic issue most critical to developing an operational methodology may be to specify the scale on which D's utility, and therefore the effectiveness of D's decisions, is to be measured. It needs to permit unambiguous interpretation, and sufficiently general in nature that meaningful comparisons can be made over time, between individuals and even across decision domains.

If the attribute is readily measurable, scale definition is straightforward (e.g. money in business situations). Where consequences are complex and intangible (e.g. prestige, peace of mind), we have a familiar, if by no means resolved, problem in decision analysis. However, in regular decision analysis, if the object is only to compare actions, the scale may need to do no more than indicate preference. Moreover, there may an obvious default action to use as a base, as in a go/no go choice.

I tend to favour a holistic approach interpolating utilities between well-defined anchor scenarios (Raiffa 1968).[†] For actions, these can be *arbitrary* good and bad outcomes (e.g. convenient examples of disastrous and glorious careers). Alternatively utilities can be related to a *plausible* range of aftermaths, i.e. how well or badly D might 'plausibly' have done. [A rigorous interpretation of 'plausible'?] Another scale might be based on 'ideal' decision processes defined by potentially accessible information, generalizing the concept of ideal impersonal probability (Brown 1993).

Zero might be the status quo. If we are evaluating candidate enhancements, the interpolated utility is then basically percentage of room for improvement (Brown 1989). ('Unaided students typically make career choices about a third of the way between the poorest prevailing and ideal decision-making practice. The aid improves this to about half-way, on average.')

The anchor points can be absolute, i.e. conceivable good and bad complete careers; or relative to the specific object of evaluation; e.g. the maximum difference an action might make, *given developments extraneous to this decision*. Relative scales, addressing the difference between alternative actions, may be preferred to absolute scales, which may be dominated by considerations unrelated to a given choice.

If the purpose of decision evaluation is simply to establish whether one decision process is better than another (say due to some intervention) or is improving, in a limited domain (say a given person's career), the scale need only be applicable to that domain (and for example only address consequences of interest to a single D whose career decisions are affected). The broader the findings sought, the more universal the scale needs to be. The scale may be different for different stages in the causal model: for example based on absolute outcomes for decision processes and on relative consequences for actions. [Is it logically feasible to concatenate them, or are they incompatible?]

Once utility is defined, the task is to predict the impact on it of whatever early level in the causal decision chain we wish to evaluate, say a specific action, decision process or intervention.

8.2.3 Strategy for modelling decision effectiveness

Our basic approach is to combine plural evaluation with multistage causal models of decision effectiveness within a decision theory paradigm. It builds

---

[†] This contrasts with a scale requiring a vector of attribute scores to be combined mathematically as in multi-attribute utility analysis (Keeney and Raiffa 1976). However, this is not an issue specific to measuring decision effectiveness.

on insights from a decision course evaluation by Laskey and Campbell (1991).

A multistage causal model of decision effectiveness.

Our approach uses a causal, or at least sequential, chain linking D or his decision process to outcome utility, via decision procedure, action and evolving events. Figure 8.3 (from Brown and Campbell 1993) illustrates the sequence for the evaluation of a decision skills programme.

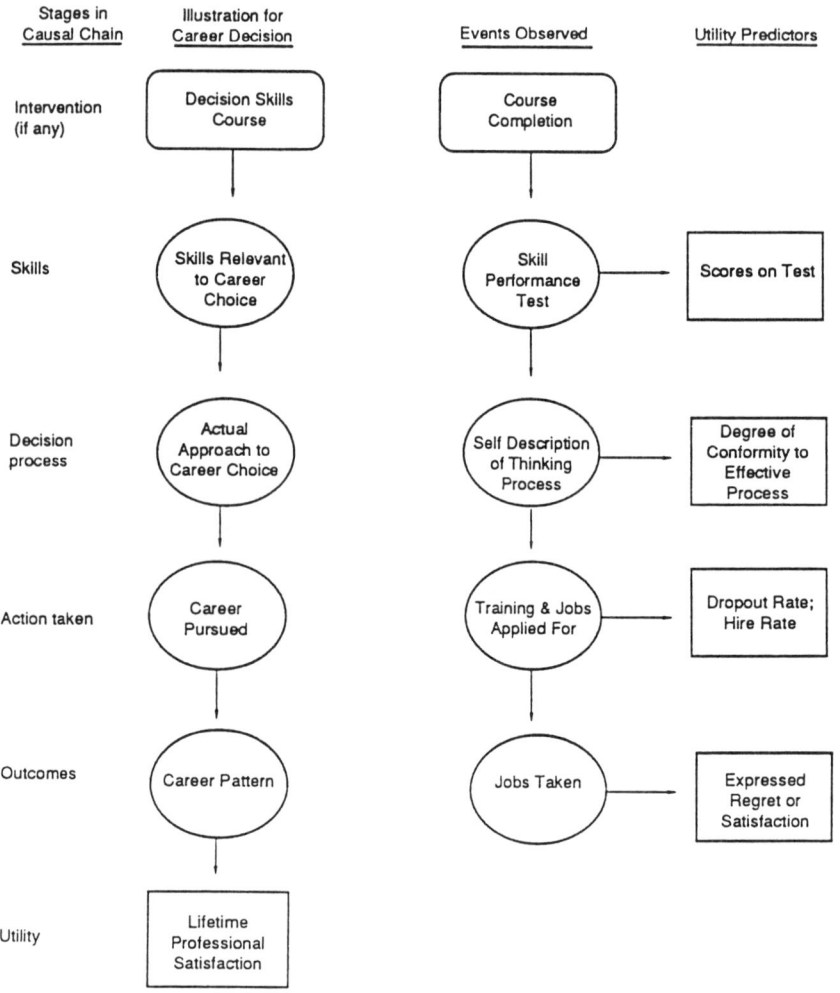

Figure 8.3 Sequential prediction of decision quality.

ROLE OF STATISTICAL THEORY IN DECISION AIDING 107

The column on the left considers a causal sequence of six stages.

1. An intervention (if any) that may affect D's decision procedure.
2. Decision-making skills at D's disposal (the intervention may be targeted at acquiring these skills or enhancing their application).
3. The decision procedure actually applied by the decision-maker (which may or may not make use of these skills).
4. The action taken.
5. The outcomes, which may be further decomposed into different stages and dimensions of aftermath.
6. The ultimate net utility experienced by the decision-maker.

Loosely speaking, the anticipated consequence of an OBE is derived from a hierarchy of anticipations down the chain. For example, the valuation of a student's training is the anticipated valuation of actions it produces, each of which is the anticipated valuation of the early outcomes the action gives rise to, and so on to ultimate utility.

(This is a simplified form of influence diagram. What is the significance of relaxing the simplification, by considering more complex dependencies, such as multiple conditioning?)

The second column of Figure 8.3 illustrates the different stages in the causal chain for a career decision. The third column shows the observable events that might be selected to evaluate each stage. The fourth column illustrates particular measures or indicators that could be quantitatively analysed in predicting utility.

Each stage on its own can be the basis of an independent evaluation, using conditioned assessment (see below). Several such stage models can be combined in an integrated evaluation, building on the 'plural evaluation' work discussed in section 8.1.2.

There are two phases in the evaluation of a decision process (or alternative OBEs at any later stage in the chain, such as actions): determining the effect of the alternative at some later stage and evaluating the effect (in terms of impact on utility). For the earlier stages, the effect is easier to measure, but harder to evaluate.

Developing and analysing evidence The component assessments called for in the causal chain would use whatever evidence is developed at the time of evaluation. One or more measures of the effect can be taken, such as elicited regret, observed reversal, expert appraisal (see next section). The measures need to be translated into an assessment of the probable effect itself and its impact on outcome utility.

Assessments for the interesting cases will be inevitably highly tentative, even as aftermath history accumulates. They may consist of little more than opportunistic data judgementally interpreted. A more experimental programme would have to be unrealistically large to be at all definitive.

Analytic strategy for assessing utility impact of a given action

Any OBE can be evaluated with conditioned assessment keyed to an intermediate stage variable. [Math representation, including 'meta' conditioning on evolving evidence?] For example, if the OBE is a decision process, then actions it produces (or their outcomes) could be the intermediate stage variable. In this context, conditioned assessment would involve probabilistically predicting contingent actions, and the utility impact of each action.

Let us consider more closely the latter, since the effectiveness of an action may be of interest in its own right. E's evaluation of D's action at the time of choice we interpret as E's expected consequence, in terms of D's utility, conditional on E's knowledge at the time of the evaluation. In Bayesian terms, it is E's prior expectation of the action's consequence, updated by later information.

We make no presumption about how D makes his choice, or what may be in his mind. D might not even be human, say, an organization, a stock trading programme or an animal. [How might utility be defined in such cases?]

Initially, i.e. before any aftermath has had time to unfold, the evaluation of action is the classical decision analysis case: expected utility conditional on option and on knowledge at time of choice. The novel circumstance is re-evaluating that determination by hindsight in the light of later evidence (data). [Math formulation?]

A critical issue (among many) is what *difference* the action makes. Assessing the consequence of an action involves two uncertainties: interpreting the default actions corresponding to anchor points on the scale, and locating the action on the scale. However, since consequence is interpreted as utility, whose expectation is also (logically) utility, only expected utility need be assessed. For any default action which is not obvious (as it would be in a go/no go decision) the assessment of outcome utility, and hence of the (differential) consequence of the action evaluated, will be conjectural.

Example with simple monetary measure To avoid having the difficult utility scale issue confusing other analytic issues, we can uncouple them initially by addressing examples with simple scales (e.g. chess, where the scale is win, draw or lose) or at least only address those components of outcome where scale is not an issue (e.g. the money dimension of a career choice). We can progressively relax unrealistic simplifications in more complex problems.

Suppose E is evaluating the effectiveness of D's action in investing his $50K savings, on Uncle Bert's advice, in an environmental stock in 1988. The measure of utility is expected dollars accumulated by 1995, when D will realize the proceeds. Let us consider two hypothetical evaluations: one made just prior to purchase and one four years later.

1. 1988 evaluation: stock looks like a good choice; E expects it to double money by 1995; i.e. $100K. Based on discussions with D at the time, E is pretty sure D would otherwise have left it in real estate, with a more modest expectation of $80K. Therefore, E's expected consequence of D's action: + $20K.
2. 1992 interim evaluation: stock price has dropped 40%. However, E predicts the election of an environmental president, and expects D to get back the $50K he originally paid. Real estate has generally dropped 20%, and E has no reason to think D's would do differently. But E expects Washington real estate to be up 30% over 1988 by 1995 and D's property should therefore sell at around $65K. E's new expected consequence of investment action: − $15K.

Note that, for now, we are measuring consequence in relation to the 'path not taken' which, as noted earlier, may not prove the most appropriate scale.

A more complex example Right after college, in 1958, D went into market research. Otherwise, he would have gone to graduate school. Would E consider that D did the right thing at the time? What more could E say after a few/many years? The test is how much satisfaction D will have gained from his career, and other parts of his life that this choice could affect. Utility is measured on a 0–100 scale from as bad to as good 'as it might plausibly get' given his situation at the time of choice in 1958. Zero is typified by a pessimistic scenario where he leads a life of drudgery in a dead-end low paying job, which stultifies his creativity, his wife and peers dismiss him as a loser, etc. 100 is typified by a knighthood and income in the top 1%.

In 1958 E evaluated the market research choice as 30 utiles on this scale, compared with 35 for the probable alternative, which E would accordingly have recommended to D. That is, E expected a poorish − 5 utiles as the consequence of the actions. Six years later D has changed careers and got a university job. Market research had not been much fun and his peers all seem to have got further ahead. E's outcome evaluation is updated from 30 to 40 on the same scale. However, E judges D would otherwise probably have had a PhD by now and be positioned for a more illustrious academic career. The default utility moves from 35 to 50. That is, E now values action as an even worse (− 10 'utiles')

Thirty years later D's career is turning out better than E had expected. D had taken advantage of consulting experience he got in market research and reached a richer material and intellectual life that E would expect from an initial graduate school action. On same scale D is now 15 utiles ahead of the path not taken. So it looks like he made a good choice after all.

[How to represent and validate these qualitative case analyses in rigorous mathematical form? Do they conform to the previous qualitative exposition?]

Evaluation of a decision process

Evaluation of the decision *process* (e.g. of D and his decision procedure) is the *expectation* of action effectiveness, defined above, as always given E's current knowledge. Its interpretation is comparable to that for action, but an additional layer of complexity is introduced, due to uncertainty about which actions may follow from any given decision process (Brown 1991).

Example of evaluating a class of interventions Let us extend the earlier investment action case to the task of evaluating the practice of D taking Uncle Bert's advice on investment *generally*. E's 1992 evaluation that Bert's stock pick in 1988 will have lost money for D does not necessarily mean E expects that acting on Bert's advice generally has negative consequences for D. However, E has one negative data point now, which makes him slightly less confident in Uncle Bert than he had been in 1988. When E tracks how D's cousins whom Bert also advises have fared, and finds they have all done well, his confidence is more than restored (though he knows they may just be smarter than D).

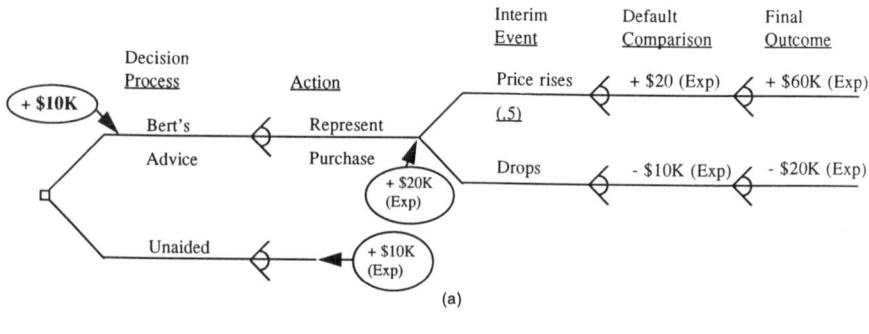

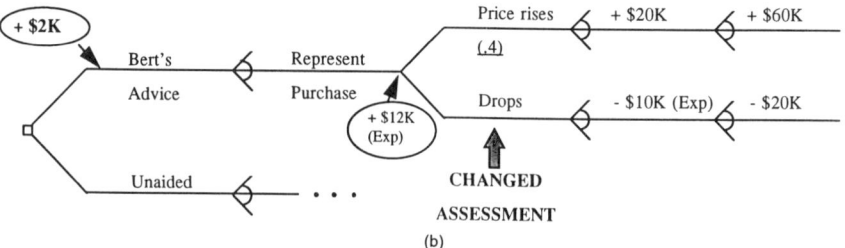

Evaluation of help goes from $10K to $2K

Figure 8.4 Worked example of evaluation of intervention (follow Uncle Bert's advice). (a) Initial aid evaluation; (b) evaluation after one interim event.

To make concepts more concrete, Figure 8.4 shows a simplified investment example slightly different from the above case. E's task is to evaluate Bert's stock advice for D, who has used it once to invest $50K in a stock he will hold for five years. The price has dropped $25K after six months. Although the market has fallen as a whole, E suspects D would be doing better on this occasion if he had relied on his own judgement.

Figure 8.4(a) summarizes E's earlier, conventional pre-choice decision analysis (using familiar decision tree conventions). Based on this, E had judged that, in a situation like this, Bert's advice could be expected to yield $10K more, when D sells in five years' time, than not using it. More specifically, his investments will produce $20K more than default investments, whereas the consequence of D's unaided choices would be only $10K. Embedded in the analysis is an assessment significant for present purposes: a 50% probability that a representative investment will rise over six months, in which case it would be doing an expected $20K better than whatever investment D would have made on his own, and $60K better by the time he sells; a complementary 50% probability of a six-months drop, doing $10K worse and $20K worse respectively. (Analytic simplifications not distinctive of the present problem, such as equivalent substitution, are not discussed here.)

Figure 8.4(b) shows E's updated analysis for this class of decisions, after he notes that the value of the stock has declined after six months. All that changes is the probability of a price rise, which has dropped from 50% to 40% and results in the expected consequence of D using the aid to be an investment worth $12K more than the default investment. This is still $2K better than not using the aid (whose expected consequence has not changed). $2K is now the evaluation of Bert's advice.

The critical step is the changed probability of the early price drop. Since this event is treated as binary, the logic of revision is the same as tossing a suspect coin and having it come down heads. It all depends on prior judgement. A prior 50% probability of heads might derive from different prior assessments of the generating process, with different diagnostic implications. In the 'normal' situation where E was confident the coin was unbiased, one heads would have no detectable impact on his prior judgement of its being a fair coin. If, on the contrary, he had been absolutely convinced it was either a two- headed or a two-tailed coin, observing one heads would move the posterior probability of heads from 50 to 100%.

General argument In the stock investment case and, more generally, the impact of one interim observation—or many repetitions—depends on the nature and firmness of prior judgements. The more repetitions (i.e. the larger the sample of observations) the less critical prior judgement becomes, and the less important to take it into account. A classical (non-personalist) analysis of data becomes more feasible, with the advantage that findings can be reported without controversial subjectivity.

In those cases where the choice is repetitive (as in the application of a standardized career planning aid), the prospect of developing a sample of data points sufficient to swamp prior judgement is much enhanced. Furthermore, the repetitive choice permits control samples to be built up, i.e. of alternative choices.

By extending the logic discussed above, and illustrated in the worked example, the required utility of an action or process can be modelled by focusing on any given stage in the chain. The possible effects at that stage are related probabilistically backward to the action or process option and forward to the ultimate utility. The required decision effectiveness (expected utility) is inferred straightforwardly through conditioned assessment.

8.2.4 Developing empirical model inputs

The modelling approach discussed above has been silent on how model inputs are to be derived. The default case is that they be supplied judgementally by E, on informal observation of unfolding action aftermath. However, the evaluation is enhanced by deliberate generation of empirical evidence on which to base these inputs. This requires consideration of what type of data it is feasible and productive to gather, how data gathering should be planned and how resulting data should be analysed.

Much of this is a matter of understanding the specific domain in question, such as career planning, but there are significant methodological issues, including sampling and experimentation, susceptible to statistical development.

Implementing the evaluation of an OBE (decision skill, process, action, etc.) with the causal model requires the following steps:

1. Selecting the intermediate stage on which to condition utility (e.g. actions resulting from process) and characterizing the stage by one or more measurable variables utility conditioning.
2. Assessing E's distribution on the conditioning variable, given the OBE.
3. Assessing D's utility conditional on each value of the variable (e.g. the action leads to marked success).

Identifying conditioning variables

The stage or stages (action, outcome, etc.) to key a conditioned assessment to are limited, of course, to those stages already reached at the time of evaluation. Among those available, the earlier tend to be easier to link to the decision OBE (step 2 above), but more difficult to link to utility (step 3). (How to formalize optimal choice of stage?) Later stages, not yet traversed, can be predicted from earlier stages in a two-step conditioning. Here are some features of any given stage, which are intended to be measurable, diagnostic of utility and diverse (since independence enhances diagnosticity).

Decider property Measures would include skills and attributes, such as D having substantive knowledge, realistically appraising possible outcomes and avoiding judgemental biases.

Decision procedure The most popular approach to evaluating procedure in the decision research literature is to observe how closely the procedure follows some claimed ideal of good decision-making practice. The preferred procedure need not resemble a decision theory paradigm. For cognitive reasons, a decision-making technique, such as recognition-primed response (Klein 1989), which makes no explicit attempt at maximizing expected utility, may fare best for D (even though E may conclude that through E's expectation of D's utility). [Arguable?]

Action Domain experts can evaluate D's actions directly, but there is the danger they confuse their goals with D's. Decision analysis can also be used based on E's hindsight knowledge at evaluation time (unavailable in the regular decision analysis case).

Outcome Reported *regret* over an outcome presumes comparison with the outcome of an alternative action and may need adjusting for cognitive hindsight bias. *Reversals* are often indications of a bad decision (e.g. career change, divorce). Unlike regrets, reversals can be verified by observation, but reversals may be diagnostic of good outcomes, rather than bad. *Objective indicators* (e.g. monetary return for business decisions; promotions for job choices) are commonly used as surrogates for utility in multi-attribute decision analysis (Keeney and Raiffa 1976). *External judges* may add insight on what happened and why it was good or bad.

Distribution over conditioning variable

Predicting intermediate outcomes (conditioning variables) probabilistically can be viewed as a conventional Bayesian updating task, with one or more instances as sample data—the more the better. These instances are almost inevitably uncontrolled 'natural' happenings (not necessarily to be treated as random). [Any technical issues?] To account for the impact of confounding events (e.g. deciding to drive home via a route which resulted in a collision with a fire truck), one might evaluate a standardized intervention, such as a computerized career decision aid, using a large sample of users and non-users.

Predicting utility from measures

Assessing conditional utilities draws on well-established inference tools, based on systematic or happenstance empirical data or purely on 'expert' judgement. [Special issues?]

8.3 CONCLUSIONS

8.3.1 A call to arms

I hope this chapter may stimulate mathematical statisticians to continue the Lindley tradition of developing theoretical underpinnings for decision methodology. In particular, they can impose needed rigour on the informal musings of non-mathematicians like myself. Gradually mature and useful tools decision evaluation may emerge, to serve the practical needs of society.

Various loose ends on decision effectiveness requiring mathematical/statistical development have been noted in brackets and readers can no doubt find more. General issues include: the effectiveness of a decision process or an intervention in it, as E's expectation, conditional on his knowledge at the time of evaluation of its impact on D's utility; the effectiveness of an action as the evaluator's expectation, conditional on his knowledge at the time of evaluation of its impact of the decider's utility; deriving decision process effectiveness as E's expectation of action effectiveness.

A candidate test-bed for decision evaluation, of social importance in its own right, would be a longitudinal study to evaluate a decision aid in common use (such as the widely used 'SIGI-PLUS' computerized career planning aid).

Beyond hindsight evaluation of decision effectiveness, there are many other issues of statistical technique and theory stimulated by the needs of practical decision aiding. Some would extend the collaboration between analyst and statistician on plural evaluation described in section 8.1.2. More generally there is the need to establish the circumstance under which user-friendly intuition-intensive procedures correspond to well-established model-intensive approaches, and how serious any approximation errors are. Although statistical issues of the type suggested here can be pursued autonomously, my experience is that an interactive build–test–build–test approach, interleaving normative, descriptive and applied efforts works best.

ACKNOWLEDGEMENTS

Work on which this chapter is based was supported by the Decision Risk and Management Science Program of the National Science Foundation and by the Child Development Program of the National Institutes of

Health. Vince Campbell, Kathryn Laskey and Jonathan Baron provided valuable input. In particular, Figure 8.2 of the central causal model of decision effectiveness and several suggested empirical measures are due to Vincent Campbell.

REFERENCES

Adams, M. J. and Feehrer, C. E. (1991) Thinking and decision-making, in J. Baron and R. V. Brown, (eds, *Teaching Decision Making to Adolescents*, Hillsdale, NJ; Erlbaum.

Baron, J. and Brown, R. V. (1991) Toward improved instruction in decision making to adolescents: A conceptual framework, in J. Baron and R. Brown, eds, *Teaching Decision-Making to Adolescents*. New York; Erlbaum.

Beyth-Marom, R., Fischhoff, B., Quadrel, M. J. and Furby, L. (1991) Teaching decision-making to adolescents: a critical review, in J. Baron and R. V. Brown, eds, *Teaching Decision- making to Adolescents*, Erlbaum.

Bromage, R. C., Brown, R. V., Chinnis, Jr, J. O., Cohen, M. S. and Ulvila, J. W. (1983) *Decision aids for Submarine Command and Control, Phase III, Concept implementation*, Technical Report 83-2, Falls Church, VA, Decision Science Consortium, Inc.

Brown, R. V. (1987) Decision analytic tools in government, in Karen B. Levitan, ed. *Government Infrastructures*, Westport, CT, Greenwood Press.

Brown, R. V. (1989) Toward a prescriptive science and technology of decision aiding, *Annals of Operations Research, Volume on Choice Under Uncertainty* **19**, 467–83.

Brown, R. V. (1991) Commentary on Politser 'Do the biggest gains come from the smallest decision trees?', *Journal of Behavioral Decision Making*, **4**, 139–41.

Brown, R. V. (1993) Impersonal probability as an ideal assessment based on accessible evidence: a viable construct? *Journal of Risk and Uncertainty*, **7**, 215–235

Brown, R. V. and Campbell, V. N. (1993) *Evaluating Career Decisions in the Light of Outcomes: By Their Fruits Ye Shall Know Them*. Working paper, June.

Brown, R. V. and Lindley, D. V. (1978) *Reconciling Incoherent Judgments (RIJ)—Toward Principles of Personal Rationality* (Technical Report TR 78-8-72), McLean, VA, Decisions and designs, July (NTIS No. AD A059639).

Brown, R. V. and Lindley, D. V. (1986) Plural analysis: multiple approaches to quantitative research, *Theory and Decision*, **20**, 133–54.

Brown, R. V., Lindley, D. V., Ulvila J. W. and von Winterfeldt, D. (1984) *Plural Analysis: Developing Multiple Approaches to Quantitative Research (Phase I)* (Technical Report 84-3), Falls Church, VA, Decision Science Consortium, August (NTIS No.)B87–116331/A05).

Brown, R. V. and Vari, A. (1992) Towards an agenda for prescriptive decision research: The normative tempered by the descriptive. *Acta Psychologica*, **80**, 33–47.

Keeney, R. L. and Raiffa, H. (1976) *Decisions with Multiple Objectives*, New York, John Wiley & sons.

Klayman, J. (1988) On the how and why (not) of learning from outcomes, in *Human Judgment: The SJT View*, B. Berndt, Brehmer C. R. B. Joyce, New York, John eds. Wiley & Sons.

Klein, G. A. (1989) Recognition-primed decisions, *Advances in Man–Machine Systems Research*, **5**, 47–92, JAI Press.

Laskey, K. B. (1990) Design of plural analysis: Allocation of resources between two or more studies, Working Paper. Reston, Decision Science Consortium, Inc.

Laskey, K. B. and Campbell, V. N. (1991) Evaluation of an intermediate level decision analysis course, in *Teaching Decision-making to Adolescents*, J. Baron and R. V. Brown, eds, Hillsdale, NJ, (Erlbaum).

Lindley, D. V. (1982) The improvement of probability judgments, *Journal of the Royal Statistical Society*, A **145**, 117–26.

Lindley, D. V. (1983) Reconciliation of probability distributions, *Operations Research 31*, 866–80.

Lindley, D. V. (1985) Reconciliation of discrete probability distributions, in *Bayesian Statistics, 2*, J. M. Bernardo et al. eds, Amsterdam, North-Holland, pp. 375–90.

Lindley, D. V. (1987a) The reconciliation of decision analyses, *Operations Research*, **34**, 2.

Lindley, D. V. (1987b) Using expert advice on a skew judgmental distribution, *Operations Research* 35(5).

Lindley, D. V., Tversky, A. and Brown, R. V. (1979) On the reconciliation of probability assessments. *Journal of the Royal Statistical Society*, A142(2), 146–80.

Nakdimon, Shlomo, (1989) *First Strike: The Decision to Bomb an Iraqi Reactor*, Jerusalem, Yediot Ahoronot.

Politser, Peter (1991) *Journal of Behavioral Decision Making*, **4**, 121–38.

Raiffa, H. (1968) *Decision Analysis*, Reading, MA, Addison-Wesley.

Ulvila, J. W., and Brown, R. V. (1982) Decision analysis comes of age, *Harvard Business Review*, Sept.–Oct. pp. 130–41. (Reprinted in Dickson, D. N., ed. (1983), *Using Logical Techniques for Making Better Decisions*, New York, John Wiley Sons, and in *Advanced Management Report*, 4 (4, 5, and 6), 1982–83.

Von Winterfeldt, D. and Edwards, W. (1986) *Decision Analysis and Behavioral Research*, New York, Cambridge University Press.

Watson, S. R., and Brown, R. V. (1975) *Case Studies in the Value of Decision Analysis* (Technical Report 75–10), McLean, VA, Decisions and Designs, Oct. (NTIS No. AD A019263).

Watson, S. R. and Brown, R. V. (1978) The valuation of decision analysis, *J. Royal Statistical Society, Series A* **141**, 69–78.
Winkler, R. L. and Makridakis, S. (1983) The combination of forecasts, *Journal of Royal Statistical Society*, **146**, 150–7.

2018 Lakebreeze Way
Reston
VA 22091–4021
USA

CHAPTER 9

The Signature as a Covariate in Reliability and Biometry

Sylvia Campodónico* and Nozer D. Singpurwalla[†]
The George Washington University, Washington, D.C.

9.1 INTRODUCTION AND OVERVIEW

Defects in certain physical and biological items often manifest themselves as oscillatory motions such as vibrations, or as electrical signals such as leakage currents and voltages. The measurement of such motions and signals is referred to as *signature data*, and the interpretation of such data is known as *signature analysis*. For reasons that will become clear, it is common to interpret signature data, which is a time series, in the frequency domain through its power spectrum, which is then referred to as *the signature*.

The purpose of a signature analysis is to identify particular types of defects, if any, and to assess their relative impact on the item's survival. As currently practised, the identification procedure appears to be well developed, both in the engineering and the biomedical sciences, in the sense that an examination of the spectrum is used to compare two items, or to classify an item as being defective or not. However, not much, if anything, appears to have been done with respect to the assessment of relative impact of defects revealed by the spectrum. For example, a physician can examine an electrocardiogram and make an overall judgement about a patient's heart condition, but cannot use it to quantify the impact of the observed

---
* Supported by the Association of American Railroads.
[†] Support by Contract N00014-85-K0202, Office of Naval Research, Grant DAAH04-93-G-0020, Army Research Office and Grant AFOSR-F49620-92-J-0030 Air Force Office of Scientific Research.

Aspects of Uncertainty edited by P. R. Freeman and A. F. M. Smith. © 1994 John Wiley & Sons Ltd.

electrocardiogram on the patient's life length. The aim of this chapter is to propose an approach which attempts to quantify the assessment process by treating the power spectrum as a covariate in survival analysis. In the sequel, Bayesian approaches for estimating the power spectrum using a regression model and its state-space representation, via a hierarchical model, are described.

The organization of this chapter is as follows. In section 9.2 we introduce some notation and terminology, and for the benefit of readers not familiar with frequency domain methods, overview a commonly used approach for estimating the power spectrum of a time series. The estimated power spectrum enables one to identify hidden periodicities in the data-generating process; as previously stated, such periodicities are indicators of defects. In section 9.3 we motivate our work by describing two scenarios that involve the identification of defects using the signature. These scenarios have been chosen from the engineering and the biomedical contexts. In section 9.4 we describe two Bayesian approaches for estimating the power spectrum, and in section 9.5 we propose a procedure for using the spectrum as a covariate in survival analysis. Sections 9.4 and 9.5 constitute the main contributions of this chapter. Section 9.6 describes an application of our ideas for assessing the service life of locomotive traction motors using a signature of its vibration data.

9.2 THE SPECTRUM AND ITS LEAST-SQUARES ESTIMATION

Let $Y(t)$ denote the observed value of a variable of interest, such as the displacement due to vibration or the voltage of a heart muscle, at time t, where $t = 1, 2, \ldots, T$. Suppose that T is odd, and that $Y(t)$ is described via a regression model

$$Y(t) = f(t) + \varepsilon(t), \qquad (2.1)$$

where $f(t)$ is an *unknown periodic* function with a *known* and fixed period T. For example, T could be the time taken by a motor to complete a revolution, or the time taken by the heart muscle to complete a pumping cycle. The former is determined by the speed of the motor and the latter via empirical observations or biomedical knowledge. The observation error $\varepsilon(t)$ is assumed to be normally distributed with mean 0 and variance σ^2, where σ^2 is unknown. Also, the sequence $\{\varepsilon(t)\}$ is assumed to be uncorrelated. The purpose of a spectral analysis of the observations is to identify *hidden periodicities* in $f(t)$; that is, periodicities in addition to the one that is known to be T. To do this we approximate $f(t)$ via a *Fourier series model* of the form

$$a(0) + \sum_{j=1}^{q} \left(a(m_j) \cos\left(\frac{2\pi}{T} m_j t\right) + b(m_j) \sin\left(\frac{2\pi}{T} m_j t\right) \right), \qquad (2.2)$$

where

$$1 \leq q \leq \frac{T-1}{2} \quad \text{and} \quad M = \{m_1, m_2, \ldots, m_q\} \subseteq \mathscr{T} = \left\{1, 2, \ldots, \frac{T-1}{2}\right\}.$$

The coefficients $a(m_j)$ and $b(m_j)$ are the *amplitudes* of the respective cosine and sine curves with *period* T/m_j and *frequency* m_j/T; these are unknown, and have to be estimated from the $Y(t)$'s. The summand q, and M—a subset of $\mathscr{T}$—are chosen arbitrarily; they reflect the analyst's beliefs about the number and the order of the hidden periodicities. Thus for example, the choice $q = 1$ and $M = \{1\}$ reflects the belief that there are no hidden periodicities. In this case,

$$f(t) = a(0) + a(1)\cos\left(\frac{2\pi}{T}t\right) + b(1)\sin\left(\frac{2\pi}{T}t\right)$$

is known as the *simple one-harmonic model*. When $q = (T-1)/2$ and $M = \mathscr{T}$, (2.2) is an exact representation of any $f(t)$, and is referred to as the *full seasonality model*; this reflects the belief that there are a total of $(T-1)/2$ periodicities, of orders

$$T, T/2, T/3, \ldots, \quad \text{and} \quad T\Big/\left(\frac{T-1}{2}\right) > 2.$$

The case of T even follows analogously, but with suitable modifications. If $M = \mathscr{T}$, and if the $a(m_j)$'s and the $b(m_j)$'s are assumed known, then a plot of $\rho^2(m_j) = a^2(m_j) + b^2(m_j)$, versus the frequency m_j/T, for $j = 1, 2, \ldots, q$, is called the *spectrum* of $f(t)$. When the $a(m_j)$'s and the $b(m_j)$'s are replaced by their estimates, then the corresponding plot is known as the *estimated spectrum* of $f(t)$.

9.2.1 Least-squares estimation of the spectrum

A commonly used approach for estimating the spectrum is by the method of least squares (Anderson 1971, p. 105). It is easy to verify that the least-squares estimators of the unknown coefficients $a(0)$, $a(m_j)$ and $b(m_j)$, for $j = 1$ to q and $m_j \in M$, are

$$\hat{a}(0) = \frac{1}{T}\sum_{t=1}^{T} Y(t)$$

$$\hat{a}(m_j) = \frac{2}{T}\sum_{t=1}^{T} Y(t)\cos\left(\frac{2\pi m_j}{T}t\right) \tag{2.3}$$

$$\hat{b}(m_j) = \frac{2}{T} \sum_{t=1}^{T} Y(t) \sin\left(\frac{2\pi m_j}{T} t\right).$$

To obtain the least-squares estimator of σ^2 we must have $q < (T-1)/2$, when such is the case the estimator of σ^2 is

$$\hat{\sigma}^2 = \frac{\sum_{t=1}^{T} Y^2(t) - T\bar{Y}^2 - \frac{1}{2}T \sum_{j=1}^{q} [\hat{a}^2(m_j) + \hat{b}^2(m_j)]}{T - (2q+1)},$$

where

$$\bar{Y} = \frac{1}{T} \sum_{t=1}^{T} Y(t).$$

A plot of $\hat{\rho}^2(m_j) = \hat{a}^2(m_j) + \hat{b}^2(m_j)$ versus the frequency m_j/T, for $j = 1, 2, \ldots, q$, gives us the estimated spectrum of $f(t)$. Here $\hat{\rho}^2(m_j)$ is a measure of how closely the suitably chosen trigonometric function with period T/m_j fits the observed data. Thus a large value of $\hat{\rho}^2(m_j)$, for $m_j > 1$, suggests a hidden periodicity of orders T/m_j. Therefore, in a signature analysis, one is on the look-out for those periods T/m_j at which $\hat{\rho}^2(m_j)$ is large, because such periods give the analyst a clue about the nature and the magnitude of defects.

Observe that in order to obtain the least-squares estimate of the spectrum, we must have all the T observations. Consequently, a sequential estimation of the spectrum is possible only when *batches* of T observations are available. The Bayesian formulation of section 9.4, which results in a sequential estimation of the spectrum, is not based on this requirement of batched data.

9.2.2 Properties of the least squares estimators

Because of the orthogonality of the sine and cosine terms, all of the estimated quantities given above are unbiased and uncorrelated. The latter feature implies that a large value of $\hat{\rho}(m_j)$ will not effect its neighbouring values, giving the analyst an uncontaminated picture of the magnitudes of the hidden periodicities. There is a disadvantage to the method of least squares. Specifically, when choosing $q < (T-1)/2$, the analyst may fail to include elements of M that may happen to be significant, vis-à-vis the hidden periodicities. Also, $T(\hat{\rho}^2(m_j))/2\sigma^2$ has a non-central chi-square distribution with the degrees of freedom that is fixed at 2, no matter what the value of T. In the Bayesian approach suggested by us since q could equal $(T-1)/2$ there is no danger in overlooking any elements of the set M that may be indicative of the hidden periods.

9.3 THE SIGNATURES OF VIBRATIONS AND CARDIOGRAMS

9.3.1 Vibrations of rotating machinery

Engineers dealing with rotating machinery are concerned with vibrations, because vibrations are responsible for an excessive wear of bearings, damage to electrical insulations, and general discomfort to humans (Wowk 1991, p. 3). Figure 9.1, shows the *event tree* of an electrical motor tracing the effect of high vibration on motor failure.

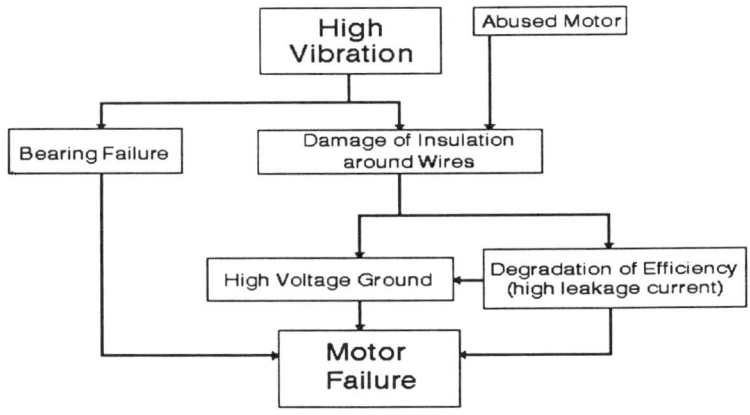

Figure 9.1 Event tree showing the effects of vibration on motor failure.

Vibrations in electrical motors are caused by sources such as rotor imbalance, shaft misalignment, gearmesh and bearing defects, to name a few. These sources can occur either individually or collectively. Whereas it is literally impossible to avoid vibrations, it is possible to minimize them by suitable controllers. Thus it is common practice to test motors for high vibration signals before commissioning them for use. The purpose of such tests are twofold: the first is to classify a motor as being defective or not, and if defective to identify and eliminate the causes of the defects; the second is to assess the relative impact of the causes of the (low) vibrations of acceptable motors on their service life. Whereas the former exercise appears to be well understood, the latter appears to be purely judgemental and in need of a formal development.

Vibration signals are generally recorded in the time domain and their signature estimated using the technique of section 9.2, or its non-statistical versions such as Cooley and Tukey's *Fast Fourier Transform* (1965). The reason for examining a signature of the vibration data is that the signature is able to provide a snapshot of several defects, which otherwise may be difficult to directly observe. With rotating machinery, the running speed of the motor, expressed in revolutions per minute (rpm) or as cycles per second (known as *hertz*), is the smallest frequency of the spectrum. The smallest frequency, known as the *fundamental frequency*, and its multiples, known as *harmonics of the fundamental frequency*, are plotted on the horizontal axis. The amplitudes of the spectrum, that is the values $\hat{\rho}^2(\cdot)$, corresponding to the above frequencies are plotted on the vertical axis. The frequency at which the amplitude takes a large value suggests the nature of a defect, and the magnitude of the amplitude determines the severity of the defect. For example, with rotating machinery, a large amplitude at one times the motor running speed, which we denote by $1(\times \text{rpm})$ suggests an imbalance, whereas a large amplitude at $2(\times \text{rpm})$ suggests a misalignment of the motor. For more details along the above lines see *Effective Machinery Measurements using Dynamic Signal Analyzers* (1990); see also section 9.6.

Whereas the state of the art of vibration analysis provides us, via the spectrum, a vehicle to identify the nature and the magnitude of the defects, it does not provide us with a mechanism to assess the relative impacts of the identified defects on the motor's service life. For example, is a small amount of imbalance more serious to service life than a large amount of misalignment? Do the effects of imbalance cancel out those of misalignment, or do they complement each other? An approach to addressing the above and related issues, is by treating the spectrum as a covariate in reliability studies, and performing the type of investigations that are common to survival analysis; see section 9.5.

9.3.2 The electrocardiogram

The analysis of signature data is also performed in the biomedical context. For example, the electrocardiogram (ECG), which measures the changes in electric potential produced by contractions of the heart, is used by physicians to assess the condition of the heart muscle. The ECG is a time domain plot of the voltage; its shape and pattern reveal deficiencies. A frequency domain representation of the ECG has only recently been used by physicians to compare patients with a normal heart muscle with those having a heart disease. As an illustration, Figure 9.2 shows the estimated spectra of two ECG's for a patient, each with three leads on the head, denoted by X, Y and Z, one prior to heart surgery and the other after. The latter reflects a dampening of the amplitudes at all frequencies, suggesting the positive impact of the surgical intervention.

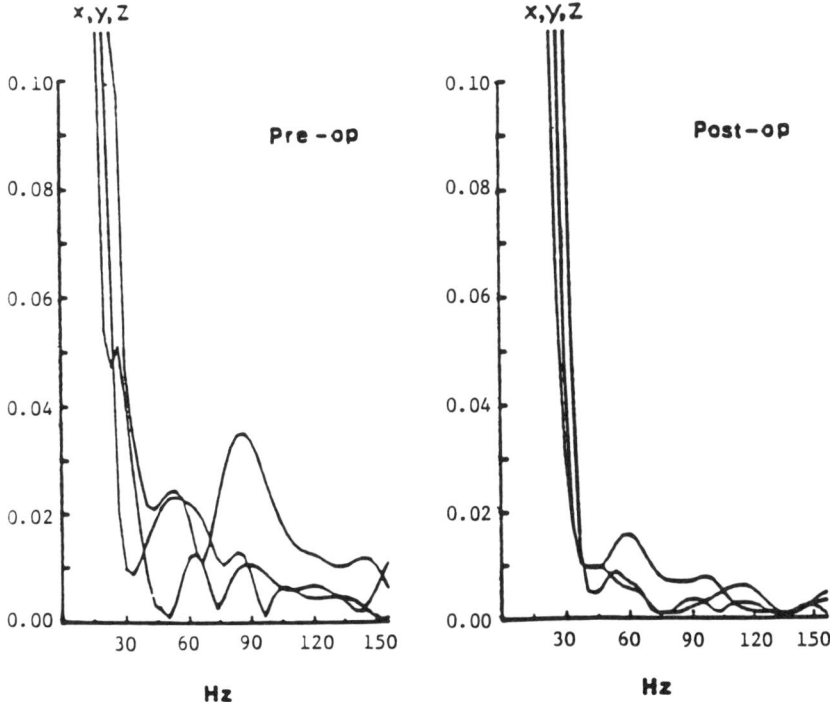

Figure 9.2 The estimated spectra of pre- and post-surgery ECGs (from Pierce *et al.* (1989).

A few criteria have been proposed for interpreting the power spectrum of an ECG. One criterion pertains to examining the ratio of the area of the estimated spectrum at a selected range of frequencies to the total area under the spectrum. The selected range of frequencies depends on the disease being investigated. In the case of tachycardia the frequency range of 60–120 hertz has been recommended (Pierce *et al.* 1989), whereas in the case of arrhythmia the range is 20–50 hertz (Cain *et al.* 1990). In general, large amplitudes at high frequencies are indicators of potential medical problems.

9.4 BAYESIAN ESTIMATION OF THE SPECTRUM

In this section, the Bayesian estimation of the spectrum of an observed time series, with a known and fixed period T, is developed through two approaches, one based on the regression model of section 9.2, and the other based on its state-space formulation. The case of *evolutionary spectra* (Priestley 1965), which can arise, for example, when the period changes with time, will be considered in a subsequent paper. The advantages of a Bayesian approach, besides coherence, are an ability to incorporate prior knowledge about the number and the order of the hidden periodicities at all the harmonics of the fundamental frequency, and the ability to sequentially estimate the spectrum based on the receipt of individual pieces of data, not just batched data. Furthermore, the issue of inconsistency, which calls for a smoothing of the estimated spectrum via arbitrarily chosen smoothing windows, is no more germane. Bretthorst (1988) also considers a Bayesian approach for estimating the spectrum; however, he focuses on the case of an unknown period T.

9.4.1 Bayesian estimation via a regression model

For the regression model of section 9.2, if the vectors F'_t, Θ', and the scalar ϕ are defined as follows:

$$F'_t = \left[1 \cos\left(\frac{2\pi}{T} m_1 t\right) \sin\left(\frac{2\pi}{T} m_1 t\right) \cos\left(\frac{2\pi}{T} m_2 t\right) \sin\left(\frac{2\pi}{T} m_2 t\right) \cdots \right.$$
$$\left. \cos\left(\frac{2\pi}{T} m_q t\right) \sin\left(\frac{2\pi}{T} m_q t\right) \right],$$

$$\Theta' = [a(0) \ a(m_1) \ b(m_1) \ a(m_2) \ b(m_2) \ldots a(m_q) \ b(m_q)],$$

and the scalar $\phi = (\sigma^2)^{-1}$, then (2.1) and (2.2) can be written as

$$(Y(t) \mid \Theta, \phi) \sim \mathcal{N}(F'_t \Theta, \phi^{-1}),$$

where Θ' denotes the transpose of Θ and $X \sim \mathcal{N}(\mu, \sigma^2)$ denotes the fact that X is normal with mean μ and variance σ^2. The Bayesian paradigm requires the specification of a joint prior distribution for the unknown parameters Θ and ϕ. For this, we assume that the vector Θ has a multivariate normal distribution with mean vector Λ_0 and covariance $\phi^{-1} R_0$, written

$$(\Phi \mid \phi) \sim \mathcal{N}(\Lambda_0, \phi^{-1} R_0) \quad \text{and that} \quad \phi \sim \text{Gamma}\left(\frac{n_0}{2}, \frac{d_0}{2}\right),$$

where the notation $X \sim \text{Gamma}(\alpha, \beta)$ denotes the fact that X has a gamma distribution with a shape parameter α and a scale parameter β. The prior parameters Λ_0 and R_0, n_0 and d_0 have to be specified by an analyst; some general guidelines which dictate these choices will be given in the section on choice of prior parameters (p. 128).

Once the above has been done, standard Bayesian updating machinery (West and Harrison 1989, p. 666) can be used to obtain the posterior distribution of Θ, from which the hidden periodicities can be inferred. Specifically, upon observing $Y_t = [Y(1) \ Y(2) \ldots Y(t)]'$, the posterior of Θ, given ϕ, is of the form

$$\left(\Theta \mid \phi, \mathbf{Y}_t\right) \sim \mathcal{N}\left(\Lambda_t, \phi^{-1} R_t\right), \tag{4.1}$$

where

$$\Lambda_t = (\mathbf{F}_t \mathbf{F}'_t + R_0^{-1})^{-1} (\mathbf{F}_t \mathbf{Y}_t + R_0^{-1} \Lambda_0) \quad \text{and} \quad R_t = (\mathbf{F}_t \mathbf{F}'_t + R_0^{-1})^{-1},$$

with

$$\mathbf{F}'_t = \begin{bmatrix} F'_1 \\ F'_2 \\ \vdots \\ F'_t \end{bmatrix}, \quad \text{a } t \times (2q+1) \text{ matrix.}$$

The posterior of ϕ is

$$(\phi \mid \mathbf{Y}_t) \sim \text{Gamma}\left(\frac{n_t}{2}, \frac{d_t}{2}\right), \tag{4.2}$$

where $n_t = n_0 + t$ and

$$d_t = d_0 + \mathbf{Y}'_t \mathbf{Y}_t + \Lambda'_0 R_0^{-1} \Lambda_0 - (\mathbf{F}_t \mathbf{Y}_t + R_0^{-1} \Lambda_0)'(\mathbf{F}_t \mathbf{F}'_t + R_0^{-1})^{-1}(\mathbf{F}_t \mathbf{Y}_t + R_0^{-1} \Lambda_0).$$

Averaging out ϕ, the posterior of Θ is

$$(\Theta \mid \mathbf{Y}_t) \sim T_{n_t}\left(\Lambda_t, R_t \frac{d_t}{n_t}\right); \tag{4.3}$$

i.e. $(\Theta \mid \mathbf{Y}_t)$ has a multivariate Student's T-distribution in dimension $(2q+1)$ with $n_0 + t$ degrees of freedom, mode Λ_t and a scale matrix $R_t(d_t/n_t)$. In particular the ith element of Θ has a Student's T-distribution with a mode equal to the ith element of Λ_t, and a scale equal to the ith diagonal entry of $[R_t(d_t/n_t)]$. Note that, unlike least-squares estimation, a Bayesian estimation of the spectrum does not require that we have observations in batches of T.

Our ultimate objective of course is to estimate the amplitudes of the power spectrum. For this we need to obtain the posterior distribution (given $\mathbf{Y}_t$) of the vector

$$[(a^2(m_1) + b^2(m_1))(a^2(m_2) + b^2(m_2)) \ldots (a^2(m_q) + b^2(m_q))].$$

For any general specification of the matrix R_0, it is not possible to obtain a closed-form expression for the necessary posterior. However, its numerical evaluation via a Monte Carlo simulation using a technique such as say the 'Gibbs sampler' (Gelfand and Smith, 1990) is feasible. This attractive possibility is currently being explored and codified. However, it can be shown (see Appendix B), that the marginal posterior distribution of

$$\rho^2(m_j) \stackrel{\text{def}}{=} a^2(m_j) + b^2(m_j), \quad j = 1, \ldots, q,$$

is of the form

$$\pi_{\rho^2(m_j)}(z \mid \mathbf{Y}_t) =$$

$$\int_0^z \frac{\tau(\sqrt{z-x},\sqrt{x}) + \tau(-\sqrt{z-x},\sqrt{x}) + \tau(\sqrt{z-x},-\sqrt{x}) + \tau(-\sqrt{z-x},-\sqrt{x})}{4\sqrt{x(z-x)}} dx$$

(4.4)

where $\tau(x, y)$ denotes the density at (x, y) of a bivariate Student's T-distribution, having a mode $[\Lambda_t(a(m_j))\ \Lambda_t(b(m_j))]'$, and a scale matrix $(d_t/n_t)R_t(a(m_j), b(m_j))$. In our case, $\Lambda_t(a(m_j))$ and $\Lambda_t(b(m_j))$ are the elements of the vector Λ_t that correspond to the expected values of $a(m_j)$ and $b(m_j)$ respectively, and $R_t(a(m_j), b(m_j))$ is the 2×2 submatrix of R_t associated with the elements $a(m_j)$ and $b(m_j)$.

A plot of $E[\rho^2(m_j) \mid \mathbf{Y}_t]$, the mean of the distribution given by (4.4), or $M[\rho^2(m_j) \mid \mathbf{Y}_t]$, its mode, versus m_j/T, for $j = 1, 2, \ldots, q$, could be used as a Bayes estimate of the spectrum of the time series. The value of $E(\rho^2(m_j) \mid \mathbf{Y}_t)$ can be evaluated (see Appendix B), for $j = 1, 2, \ldots, q$, as

$$E[\rho^2(m_j) \mid \mathbf{Y}_t] = (\Lambda_t(a(m_j)))^2 + \frac{d_t}{n_t - 2} r_t(a(m_j))$$

$$+ (\Lambda_t(b(m_j)))^2 + \frac{d_t}{n_t - 2} r_t(b(m_j)), \quad (4.5)$$

where, as before, $\Lambda_t(a(m_j))$ and $\Lambda_t(b(m_j))$ are the elements of the vector Λ_t that correspond to the expected values of $a(m_j)$ and $b(m_j)$ respectively, and $r_t(a(m_j))$ and $r_t(b(m_j))$ are the diagonal elements of the matrix R_t associated with $a(m_j)$ and $b(m_j)$ respectively; and $n_t/2$ and $d_t/2$ are the shape and scale parameters respectively, of the posterior gamma density of ϕ. One should bear in mind, however, that more formally, a Bayes estimate of the spectrum is a q-dimensional surface representing the posterior distribution of the vector $[\rho^2(m_1)\ \rho^2(m_2) \ldots \rho^2(m_q)]$.

As an alternative to the above, if all the off-diagonal elements of the matrix R_0 are chosen to be 0, then the $\rho^2(m_j)$'s, given ϕ, $j = 1, 2, \ldots, q$, are a priori independent. Consequently, the marginal posterior distributions of the $\rho^2(m_j)$'s, were ϕ to be known, can be approximated by a non-central chi-square distributions with two degrees of freedom and a non-centrality parameter that depends on Λ_t and R_t. Averaging out with respect to the posterior distribution of ϕ (a gamma density with scale parameter $n_t/2$ and shape parameter $d_t/2$), we can obtain the joint distribution of the vector $[\rho^2(m_1)\ \rho^2(m_2) \ldots \rho^2(m_q)]$ (see Appendix C). Observe that once the above averaging is done the $\rho^2(m_j)$'s, $j = 1, 2, \ldots, q$, are no longer independent.

Choice of prior parameters

The vector Λ_0 reflects prior opinion on the order and the magnitudes of the hidden periodicities, and is specified by the analyst. The elements of the covariance matrix ($\phi^{-1} R_0$) reflect the strength of conviction about the prior beliefs. Thus for example, if the analyst's prior belief is that there are no hidden periodicities, then all the elements of Λ_0 would be zero. This is a neutral choice that one would often make. In the case of rotating machinery, if the analyst suspects an imbalance, then the second and the third elements of Λ_0 which correspond to $a(m_1)$ and $b(m_1)$, would be assigned a non-zero value. With $\Lambda_0 = \theta$, the elements of the covariance matrix would typically be chosen to reflect large variances that decrease with the harmonics of the fundamental frequency, and a zero covariance for the coefficients associated with any two harmonics. The decreasing variances reflect the often observed fact that the observed spectra of many time series tend to reveal no amplitudes at high frequencies. For the paired coefficients associated with any harmonic, the corresponding variances could be taken to be equal, so that

$$\text{Var}[a(m_j)] = \text{Var}[b(m_j)] = V(m_j),, \quad \text{for } j = 1, \ldots, q,$$

where, as stated before, $V(m_j)$ is a decreasing function of m_j; an example could be $V(m_j) = \phi^{-1} \exp\{-km_j\}$, for some positive k. We note that the $V(m_j)$'s are specified up to the unknown factor ϕ. Let $r_0(\cdot)$ be the decreasing function such that $V(\cdot) = \phi^{-1} r_0(\cdot)$. The *a priori* correlation between the paired coefficients associated with any harmonic could be chosen to be a neutral 0.5, so that for any j, $\text{Cov}[a(m_j), b(m_j)] = 0.5 V(m_j)$. Thus, a suggested specification for the covariance matrix $\phi^{-1} R_0$ which attempts to incorporate the above considerations is of the form:

$$\phi^{-1} R_0 = \phi^{-1} \begin{bmatrix} r_0(0) & 0 & 0 & & & & & 0 \\ 0 & r_0(m_1) & 0.5 r_0(m_1) & & & & & \\ 0 & 0.5 r_0(m_1) & r_0(m_1) & & & & & \\ & & & r_0(m_2) & 0.5 r_0(m_2) & & & \\ & & & 0.5 r_0(m_2) & r_0(m_2) & & & \\ & 0 & & & & \ddots & & \\ & & & & & & r_0(m_q) & 0.5 r_0(m_q) \\ & & & & & & 0.5 r_0(m_q) & r_0(m_q) \end{bmatrix}.$$

For the parameters n_0 and d_0 of the gamma prior for ϕ, a plausible choice is to let both of these be equal to 0; this results in the prior being non-informative, but one that does not affect the updating mechanism indicated above.

9.4.2 Bayesian estimation via a state-space model

A state-space version of the regression model considered in sections 9.2 and 9.4.1 is based on a formulation by Roberts and Harrison (1984), whose motivation was to produce forecast functions for periodic time series. With suitable modifications this formulation can also be used to obtain a Bayesian estimate of the spectrum of a periodic time series. To see this, suppose that the observation and the system equations of the state-space representation are in the form of a dynamic linear model (West and Harrison 1989, p. 254) as follows. Observation equation:

$$Y(t) = a_t(0) + a_t(m_1) + a_t(m_2) + \ldots a_t(m_q) + \varepsilon(t);$$

System equations:

$$a_t(0) = a_{t-1}(0) + v_t(0)$$

$$a_t(m_1) = a_{t-1}(m_1)\cos\frac{2\pi}{T}m_1 + b_{t-1}(m_1)\sin\frac{2\pi}{T}m_1 + v_t(m_1)$$

$$b_t(m_1) = -a_{t-1}(m_1)\sin\frac{2\pi}{T}m_1 + b_{t-1}(m_1)\cos\frac{2\pi}{T}m_1 + \omega_t(m_1)$$

$$a_t(m_2) = a_{t-1}(m_2)\cos\frac{2\pi}{T}m_2 + b_{t-1}(m_2)\sin\frac{2\pi}{T}m_2 + v_t(m_2)$$

$$b_t(m_2) = -a_{t-1}(m_2)\sin\frac{2\pi}{T}m_2 + b_{t-1}(m_2)\cos\frac{2\pi}{T}m_2 + \omega_t(m_2)$$

$$\vdots$$

$$a_t(m_q) = a_{t-1}(m_q)\cos\frac{2\pi}{T}m_q + b_{t-1}(m_q)\sin\frac{2\pi}{T}m_q + v_t(m_q)$$

$$b_t(m_q) = -a_{t-1}(m_q)\sin\frac{2\pi}{T}m_q + b_{t-1}(m_q)\cos\frac{2\pi}{T}m_q + \omega_t(m_q).$$

In matrix notation, the observation and the system equations can also be written as

$$Y(t) = F'\Theta(t) + \varepsilon(t)$$
$$\Theta(t) = G\Theta(t-1) + \mathcal{W}(t) \tag{4.6}$$

where F, $\Theta(t)$ and $\mathcal{W}(t)$ are $(2q+1)$ vectors of the form

$$F' = [1\ 1\ 0\ 1\ 0\ \ldots\ 1\ 0],$$

$$\Theta(t)' = [a_t(0)\ a_t(m_1)\ b_t(m_1)\ a_t(m_2)\ b_t(m_2) \ldots a_t(m_q)\ b_t(m_q)]$$

$$\mathcal{W}(t)' = [v_t(0)\ v_t(m_1)\ \omega_t(m_1)\ v_t(m_2)\ \omega_t(m_2) \ldots v_t(m_q)\ \omega_t(m_q)],$$

and G is a $(2q+1)$ by $2q+1$ matrix given by

$$G = \begin{bmatrix} 1 & & & & & & \\ & \cos\frac{2\pi}{T}m_1 & \sin\frac{2\pi}{T}m_1 & & & 0 & \\ & -\sin\frac{2\pi}{T}m_1 & \cos\frac{2\pi}{T}m_1 & & & & \\ & & & \cos\frac{2\pi}{T}m_2 & \sin\frac{2\pi}{T}m_2 & & \\ & & & -\sin\frac{2\pi}{T}m_2 & \cos\frac{2\pi}{T}m_2 & & \\ & 0 & & & & \ddots & \\ & & & & & & \cos\frac{2\pi}{T}m_q & \sin\frac{2\pi}{T}m_q \\ & & & & & & -\sin\frac{2\pi}{T}m_q & \cos\frac{2\pi}{T}m_q \end{bmatrix}.$$

To see that the above formulation results in a Fourier series model of the type given by (2.1) and (2.2), we recursively invoke the system equation of (4.6) t times to write the state variable $\Theta(t)$ as

$$\Theta(t) = G^t \Theta(0) + \Omega(t),$$

where the matrix

$$G^t = \underbrace{G \times G \times \ldots \times G}_{t \text{ times}}$$

is of the form

$$G^t = \begin{bmatrix} 1 & & & & & & \\ & \cos\frac{2\pi m_1 t}{T} & \sin\frac{2\pi m_1 t}{T} & & & 0 & \\ & -\sin\frac{2\pi m_1 t}{T} & \cos\frac{2\pi m_1 t}{T} & & & & \\ & & & \cos\frac{2\pi m_2 t}{T} & \sin\frac{2\pi m_2 t}{T} & & \\ & & & -\sin\frac{2\pi m_2 t}{T} & \cos\frac{2\pi m_2 t}{T} & & \\ & 0 & & & & \ddots & \\ & & & & & & \cos\frac{2\pi m_q t}{T} & \sin\frac{2\pi m_q t}{T} \\ & & & & & & -\sin\frac{2\pi m_q t}{T} & \cos\frac{2\pi m_q t}{T} \end{bmatrix}.$$

and

$$\Omega(t) = G^{t-1}\mathscr{W}(1) + G^{t-2}\mathscr{W}(2) + \ldots + G\mathscr{W}(t-1) + \mathscr{W}(t).$$

With the above in place, the observation equation of (4.6) can also be written in terms of a Fourier series of the form

$$Y(t) = a_0(0) + \sum_{j=1}^{q}(a_0(m_j)\cos\left(\frac{2\pi}{T}m_j t\right) + b_0(m_j)\sin\left(\frac{2\pi}{T}m_j t\right) + \Omega(t) + \varepsilon(t). \quad (4.7)$$

To proceed further, we assume that in (4.6)

$$(\varepsilon(t) \mid \phi) \sim \mathcal{N}(0, \phi^{-1}),$$
$$(\mathcal{W}(t) \mid \phi) \sim \mathcal{N}(0, \phi^{-1} W(t)), \quad (4.8)$$
$$(\Theta(0) \mid \phi) \sim \mathcal{N}(\Delta_0(0), \phi^{-1} R_0(0)),$$
$$(\phi) \sim \text{Gamma}(n_0/2, d_0/2),$$

with $W(t)$, $\Lambda_0(0)$, $R_0(0)$, n_0 and d_0 specified by the analyst. Guidelines analogous to those of the section on choice of prior parameters (p. 128) can be used to specify $\Lambda_0(0)$, $R_0(0)$, n_0 and d_0, with the understanding that the notation used here is a generalized one, so that, for any k and t,

$$\Lambda_t(k) = E(\Theta(k) \mid \phi, \mathbf{Y}_t] = E(\Theta(k) \mid \mathbf{Y}_t] \quad \text{and} \quad \phi^{-1} R_t(k) = \text{Var}[\Theta(k) \mid \phi, \mathbf{Y}_t],$$

where $\mathbf{Y}_t = [Y(1)\ Y(2)\ \ldots\ Y(t)]'$. The specification of the elements of $W(t)$ should be motivated by the fact that for a continuous time series, the discrete Fourier transform is an approximation, whose goodness improves as the time interval between observations decreases (Anderson 1971, pp. 95–102). Thus with widely spaced observations, the elements of $W(t)$ that correspond to the variances should be large and vice versa. A general rule of thumb would be to make the elements of $W(t)$ a fraction of $R_0(0)$ so that $W(t) = k R_0(0)$, for all t, with $k \in (0, 1)$. If the observations are closely spaced, then we can use a relatively small value for k, say $k = 0.1$.

Having observed $\mathbf{Y}_t$, we use standard Bayesian updating techniques to obtain the posterior distribution of $\Theta(t)$, and thence the *smoothed posterior of* $\Phi(0)$ (West and Harrison 1989, pp. 122–4) given as

$$(\Theta(0) \mid \phi, \mathbf{Y}_t) \sim \mathcal{N}(\Lambda_t(0), \phi^{-1} R_t(0)),$$

where, for $k < t$, and

$$\Lambda_t(k) = \Lambda_k(k) + \mathbb{B}_k[\Lambda_t(k+1) - G\Lambda_k(k)]$$
$$R_t(k) = R_k(k) + \mathbb{B}_k[R_{k+1}(k) - R_t(k+1)]\mathbb{B}'_k,$$

with

$$\Lambda_t(t) = G\Lambda_{t-1}(t-1) + \mathbb{A}_t[Y(t) - F'G\Lambda_{t-1}(t-1)],$$
$$R_t(t) = R_{t-1}(t) - \mathbb{A}_t\mathbb{A}'_t[1 + F' R_{t-1}(t) F],$$
$$\mathbb{A}_t = \frac{R_{t-1}(t) F}{1 + F' R_{t-1}(t) F},$$

THE SIGNATURE AS A COVARIATE 133

$\mathbb{B}_t = R_t(t)G'(R_t(t+1))^{-1}$ and $R_{t-1}(t) = GR_{t-1}(t-1)G' + W(t)$.

After averaging out ϕ, the smoothed posterior distribution of $\Theta(0)$ is

$$(\Theta(0) \mid \mathbf{Y}_t) \sim T_{n_t}(\Lambda_t(0), \frac{S_t}{S_0} R_t(0)),$$

where

$$n_t = n_{t-1} + 1,$$

$$d_t = d_{t-1} + S_{t-1} \frac{(Y(t) - F'G\Lambda_{t-1}(t-1))^2}{1 + F'R_{t-1}(t)F} \quad \text{and} \quad S_t = \frac{d_t}{n_t}.$$

The smoothed posterior distribution of $\Theta(0)$ can be used to obtain the distribution of the vector

$$\left[\left(a_0^2(m_1) + b_0^2(m_1) \right) \left(a_0^2(m_2) + b_0^2(m_2) \right) \ldots \left(a_0^2(m_q) + b_0^2(m_q) \right) \right],$$

from which we can obtain a Bayes estimate of the spectrum. Considerations analogous to those at the end of section 9.4.1, in particular equation (4.4), come into play.

9.5 THE SPECTRUM AS A COVARIATE

It was stated in section 9.1, that the main purpose of a signature analysis is to identify defects and to assess the relative impact of the identified defects on the life span of an item. The different types of defects reveal themselves as large amplitudes at select frequencies of the spectrum. Thus the signature, which can be viewed as a snapshot of several defects, can be used as a covariate for assessing life spans. The purpose of this section is to propose some approaches which formalize the above notion. But first we need some additional notation.

Suppose that X denotes the length of life, or the service life, of an item and suppose that an estimate of the signature of some variable which is believed to influence X, such as vibration or the electric potential of a heart muscle, has been obtained. Referring to the development at the end of sections 9.4.1 and 9.4.2, a Bayes estimate of the signature is the posterior distribution of the vector

$$\rho^2 \stackrel{\text{def}}{=} [\rho^2(m_1) \ \rho^2(m_2) \ \ldots \ \rho^2(m_q)]',$$

or as its summarization, the vector

$$\rho^* \stackrel{\text{def}}{=} [\rho^*(m_1) \ \rho^*(m_2) \ \ldots \ \rho^*(m_q)]',$$

where $\rho^*(m_j) = E(\rho^2(m_j) \mid \mathbf{Y}_t)$, given by (4.5), is the mean of the marginal posterior distribution of $\rho^2(m_j)$, (4.4); alternatively, we could also let

$$\rho^*(m_j) = M[\rho^2(m_j) \mid \mathbf{Y}_t],$$

the mode of the distribution (4.4). A plot of $\rho^*(m_j)$ versus m_j/T, for $j = 1, 2, \ldots, q$, can then be viewed as a Bayesian analogue of the plot of $\hat{\rho}^2(m_j)$ versus m_j/T of section 9.2, which was based on the method of least squares. Since, in many applications, each $\rho^2(m_j)$ is indicative of a particular type of defect, our idea is to treat each $\rho^2(m_j)$ or $\rho^*(m_j)$ as a covariate in either an 'accelerated life model', or a 'proportional hazards model' (Crowder et al. 1991, p. 69) of the types indicated in sections 9.5.1 and 9.5.2 below.

A final, practical, issue that remains to be pointed out, pertains to our ability always to observe the failure time X associated with an estimated signature. In many applications, particularly those of engineering, a signature is obtained at the time the item is newly built or is reconditioned to behave like a new one. Unlike medicine, wherein say an electrocardiogram can be ordered on a patient at any time, it is rare to obtain the signature of an item that has experienced use. An exception is the case of conditioning monitoring (Lydersen 1988). Furthermore, most engineering units have long life spans and often serve as bootstraps in larger systems; thus their failure times and service lives are hard to keep track of. The same may also be true in biometry, wherein the issues of censorship and withdrawal pose difficulties of accounting and tractability, though not to the same extent as in engineering. Thus, in practice, it is not always possible to associate an estimated signature with its observed X. However, it is often the case, especially in engineering, that when measurements for a signature are taken, observations on certain *auxiliary variables* which are indicative of future life can also be taken. In the case of traction motors, observations on the 'leakage current', which is a good predictor of service life, are taken at the same time that data on vibrations are obtained. Defects in a motor induce leakage currents; these defects are magnified with use, resulting in larger leakage currents, which eventually render the motor unusable. Similarly, in medicine, when a physician orders an electrocardiogram, it is very likely that other tests, such as cholesterol levels, blood pressure, etc. will also be ordered. The latter variables are indicative of life span, and may serve as the auxiliary variables mentioned above. In what follows, we let Z denote the value taken by an auxiliary variable, and when data on X are either not available, or are heavily censored, our proposal is to regress Z on the $\rho^2(m_j)$'s, or the $\rho^*(m_j)$'s, using an accelerated life model. This idea is explored first, in section 9.5.1 below.

9.5.1 An accelerated life model based on the amplitudes of a spectrum

Let $\psi(\rho^2)$ and $\psi(\rho^*)$ be functions of the vectors ρ^2 and ρ^* respectively, and let $S_Z(z \mid \psi(\cdot))$ denote the survival function of Z given the covariate

THE SIGNATURE AS A COVARIATE

$\psi(\,\cdot\,)$. There are several possible versions of the accelerated life model, one of which is to assume that

$$S_Z(z \mid \psi(\,\cdot\,)) = L(z\psi(\,\cdot\,) \mid 0, \tau^2) \qquad (5.1)$$

where $L(x \mid 0, \tau^2)$ denotes the survival probability to x under a lognormal density with parameters 0 and τ^2. This assumption implies that given $\psi(\,\cdot\,)$, $W = Z\psi(\,\cdot\,)$ has a lognormal distribution with parameters 0 and τ^2, so that $\ell n Z \sim \mathcal{N}(-\ell n \psi(\,\cdot\,), \tau^2)$.

Following a practice that is common in survival analysis, if for some unknown constants $C_j, j = 1, 2, \ldots, q$, we let

$$\psi(\rho^2) = \exp\left\{\sum_{j=1}^{q} C_j \rho^2(m_j)\right\},$$

and take logs, we obtain the log linear model

$$\ell n Z = -\sum_{j=1}^{q} C_j \rho^2(m_j) + \ell n\, W, \qquad (5.2)$$

where $\ell n\, W \sim \mathcal{N}(0, \tau^2)$.

Similarly, should we choose to work with the $\rho^*(m_j)$'s, we would have the log linear model

$$\ell n Z = -\sum_{j=1}^{q} C_j \rho^*(m_j) + \ell n\, W. \qquad (5.3)$$

Let Z_i denote the observed value of the auxiliary variable for the ith unit tested, $i = 1, 2, \ldots, n$, and let ρ_i^2 and ρ_i^* denote the corresponding signature vectors. Then, given the sequence of observations $Z_1, \ldots, Z_n$, and conditional on the vectors ρ_i^2's being known, the coefficients $C_1, \ldots, C_q$, can be estimated via a standard Bayesian analysis of the log linear model. (Box and Tiao 1992, p. 113)

$$\ell n Z_i = -\sum_{j=1}^{q} C_j \rho_i^2(m_j) + \varepsilon_i, \qquad (5.4)$$

or alternatively, by using

$$\ell n Z_i = -\sum_{j=1}^{q} C_j \rho_i^*(m_j) + \varepsilon_i, \qquad (5.5)$$

where $\varepsilon_i \sim \mathcal{N}(0, \tau^2)$. Once the above have been done, we must, in the case of (5.4), uncondition with respect to the distributions of the ρ_i^2's, and for this the Gibbs sampler may again come into play. In the case of (5.5), the unconditioning step is of course not necessary, and this is an advantage of working with the marginal posterior means (or the modes) $\rho_i^*(m_j)$'s. With both (5.4) and (5.5), we need to assign a prior distribution for the vector $(\tau^2, \mathbf{C})$, where $\mathbf{C} = (C_1, \ldots, C_q)$. The natural conjugate prior for $\mathbf{C}$ is a multivariate normal whose parameters reflect the analyst's prior opinion about the presence and the relative impact of each defect on the auxiliary variable. The analyst may use covariate information, if any, to assign this prior. For example, in the medical context, the patient's life-style could be a covariate. A neutral choice would be to assume that the prior means are zero, that the prior variances large and the covariances small. The natural conjugate prior for τ^2 will be an inverted gamma distribution, independent of the distribution of $\mathbf{C}$. Except for the unconditioning with respect to the distribution of $\mathbf{\rho}^2 = (\rho_1^2, \rho_2^2, \ldots, \rho_n^2)$, the analysis proceeds in a standard manner.

Before closing this section, it is helpful to remark that the log linear model proposed above can be expanded to include interaction terms, and that the accelerated life model need not be restricted to an assumption of a lognormal survival function. For example, one may prefer to describe the uncertainty of Z via a Weibull survival function; the consequence is that the analyses would involve further computational burdens. No matter what, it is clear that the incorporation of a signature as a covariate would impose computational demands on the user, and simulation is a possible way out. The application discussed in section 9.6 involves a use of the accelerated life model described above.

9.5.2 A proportional hazards model based on the amplitudes of a spectrum

The proportional hazards model can be used whenever information on life times is available (either complete or censored data). Let X_i denote the life span of the ith item, $i = 1, 2, \ldots, n$, and the vectors ρ_i^2 and ρ_i^* the outcomes of its signature. In what follows, we adopt a 'counting process' formulation of Cox's proportional hazards model [cf. (Gill 1984)]. Accordingly, if δ_i denotes the censoring time of the ith item, we define the indicator variable

$$W_i(t) = I\{X_i \geq t, \delta_i \geq t\} = 1$$

if the ith item is under observation just before time t, and is 0 otherwise, and the indicator process $\{N_i(t); t \geq 0\}$, where

$$N_i(t) = I\{X_i \leq t, T_i \leq \delta_i\}, \quad i = 1, \ldots, n.$$

Then, according to the proportional hazards model, given ρ_i^2, $W_i(t)$ and $N_i(t)$, it can be argued that a likelihood function for estimating the vector $\mathbf{C}$ defined in section 9.1 is of the form

$$\mathscr{L}(\mathbf{C}') = \prod_{t \geq 0} \prod_{i=1}^{n} \left(\frac{W_i(t) \exp\{\mathbf{C}' \rho_i^2\}}{\sum_{j=1}^{n} W_j(t) \exp\{\mathbf{C}' \rho_j^2\}} \right)^{dN_i(t)}, \quad (5.6)$$

THE SIGNATURE AS A COVARIATE 137

where $dN_i(t) = 1$, if the process $N_i(t)$ jumps at time t; $dN_i(t) = 0$, otherwise. The Bayesian estimation of C follows from standard considerations once a prior is assigned to it, and the result averaged out with respect to the posterior distribution of ρ^2. Clearly, this exercise will have to be numerically performed. If in (5.6), were ρ_i^2 be replaced by ρ_i^*, then the averaging step mentioned above will not be necessary. Finally, the likelihood (5.6) is general enough to accommodate the case of time varying signatures which would be appropriate when one needs to entertain evolutionary spectra.

9.6 APPLICATION: SERVICE LIFE OF TRACTION MOTORS

9.6.1 Defects revealed by vibration tests

Newly reconditioned locomotive traction motors have been tested for defects using vibration signals taken at different running speeds. The motors are run uncoupled, that is, no electrical load is imposed on them. Here we shall discuss the lateral vibration of eight motors when the running speed is fixed at 900 rpm. This running speed is chosen because intermediate speeds such as 900 rpm are prone to higher stability than speeds that are too low or too high.

As stated in section 9.3.1, the frequency at which a large amplitude occurs suggests the nature of a defect, and the magnitude of the amplitude determines the severity of the defect. For rotating machinery, a large amplitude at 1($\times$ rpm) suggests an imbalance or a bent shaft, whereas a large amplitude at 2($\times$ rpm) suggests a misalignment of the motor. Also, large amplitudes at 1($\times$ rpm) 2($\times$ times) 3($\times$ rpm) . . . , 10($\times$ rpm) suggest loose components. In particular, a peak at 3($\times$ rpm) is a likely indicator of the looseness of the bearing on the shaft, whereas a peak at 4($\times$ rpm) suggests a looseness of the bearing on the housing. If the gear in the motor has k teeth, then a large amplitude at k($\times$ rpm) suggests gearmesh. Finally, if the number of ball-bearings in the motor is n, then large amplitudes at approximately 0.4n ($\times$ rpm) and 0.6n ($\times$ rpm) suggest a defect on the outer and inner races of the ball-bearing respectively; those frequencies generally occur in the range from 4 to 10($\times$ rpm). For the traction motors tested, the number of teeth in the gear is 18 and the number of balls in the ball-bearing is 13.

9.6.2 Analysis of vibration data

In this application, we have chosen the subset of frequencies ranging from 1 to 10 times the motor running speed, as well as the frequency 18($\times$ rpm) to account for gearmesh. That is, in the notation of section 9.2 and 9.4, we choose $q = 11$ and $M = \{1, 2, \ldots, 10, 18\}$. Using the regression model of section 9.4.1 to estimate the spectrum, we consider the relationship

$$Y(t) = a(0) + \sum_{j=1}^{10} \left(a(j) \cos\left(\frac{2\pi}{T} jt\right) + b(j) \sin\left(\frac{2\pi}{T} jt\right) \right) + \varepsilon(t),$$

and based on the discussion of (p. 136), the prior normal density of the vector

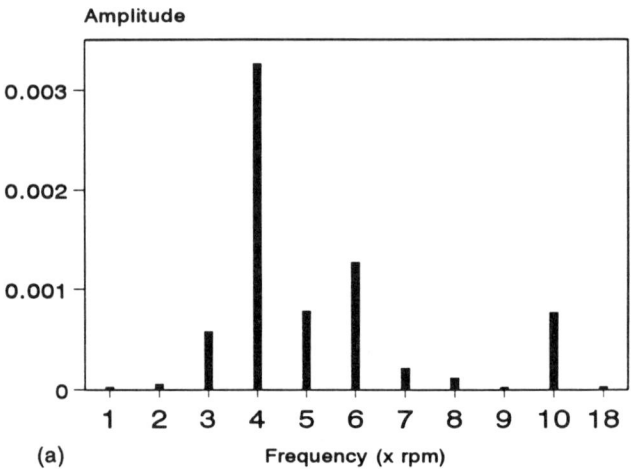

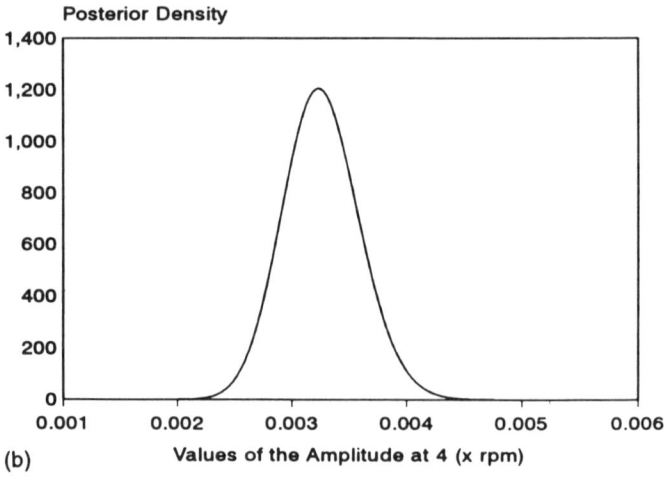

Figure 9.3 (a) Bayesian estimation of the spectrum for motor No. 4. (b) Posterior density of the amplitude at 4 (× rpm) for motor No. 4.

$$\Theta' = [a(0) \quad a(1) \quad b(1) \quad a(2) \quad b(2) \quad \ldots \quad a(10) \quad b(10)]$$

is chosen to have a mean of $\Lambda_0 = 0$, and a covariance matrix $\phi^{-1} R_0$, where R_0 is a diagonal matrix whose diagonal elements are (1, 0.95, 0.95, 0.9,

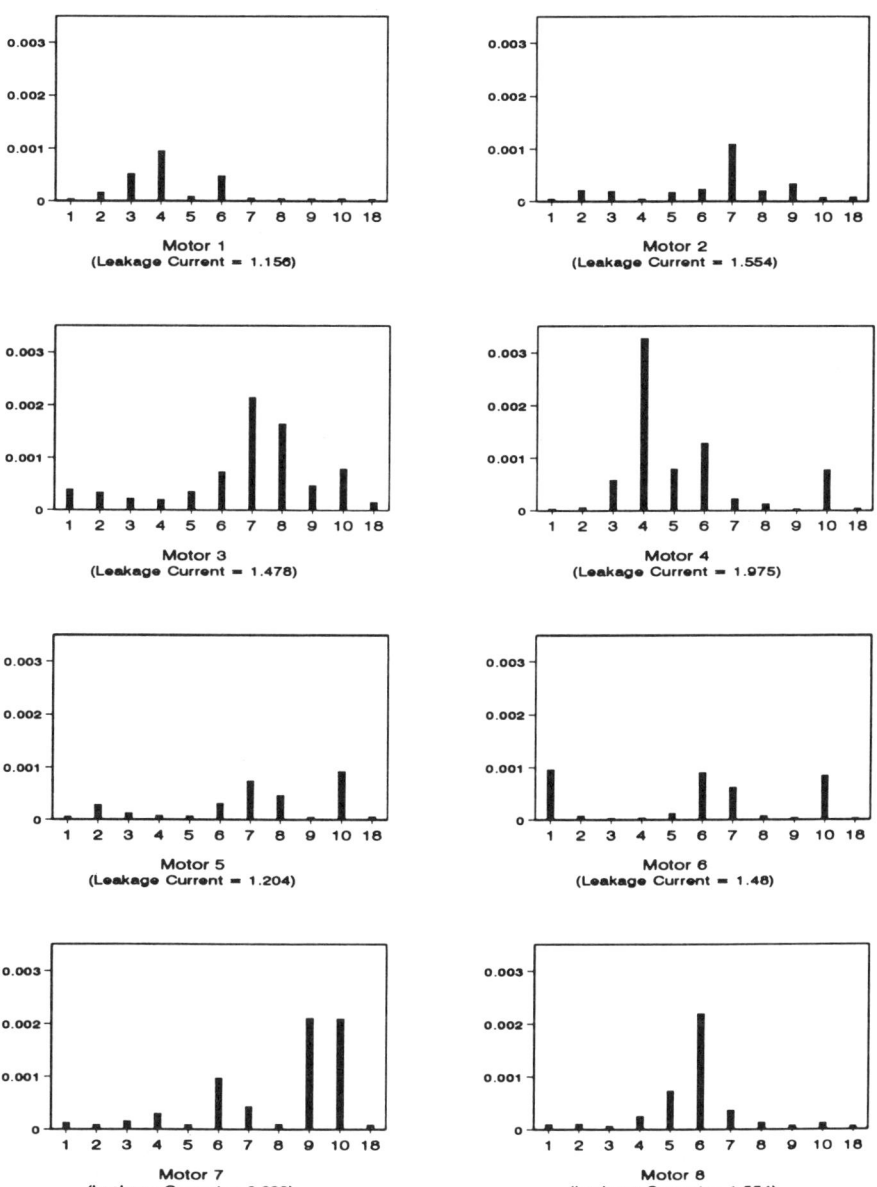

Figure 9.4 Bayesian estimate of the spectra for eight motors.

0.9, 0.85, 0.85, 0.8, 0.8, 0.75, 0.75, 0.7, 0.7, 0.65, 0.65, 0.6, 0.6, 0.55, 0.55). The parameters of the gamma density of ϕ are chosen as $n_0 = d_0 = 0$.

Our Bayes estimate of the power spectrum is based on equation (4.5), which is the marginal posterior mean of the amplitudes $\rho^2(m_j)$'s, where $\rho^2(m_j) = a^2(m_j) + b^2(m_j)$, $j \in M$. In Figure 9.3(a) we show the estimated spectrum for motor No. 4. The marginal posterior distributions of the amplitudes can be obtained via equation (4.4), so that probability intervals can be assigned for any desired amplitude. As an example, in Figure 9.3(b) we show the marginal posterior density of the amplitude at 4($\times$ rpm) for motor No. 4.

The estimated spectra for all the eight motors are shown in Figure 9.4. Also shown are the observed leakage currents for each of the eight motors. Observe that the smallest leakage currents are associated with motor Nos 1 and 5, whereas the largest are associated with motor Nos 4 and 7. An examination of their corresponding estimated spectra indicates a predominance of large values of the amplitudes of motor Nos 4 and 7, supporting the belief that higher levels of vibration result in large leakage currents.

Having estimated the spectra, which as we have said before is a snapshot of the condition of the motors, we want to use it as a covariate for assessing the motor's performance. Our goal is to identify the relative impact of the estimated amplitudes on the motor's performance, so that pertinent actions can be taken in the rebuilding process to eliminate those defects which generate large amplitudes in the spectrum.

9.6.3 Assessing the relative impact of defects on leakage current

Life history information on the traction motors for which vibration data were collected is not yet available, and will not be available in the conceivable future. Consequently, we use the leakage current as an indicator of failure behaviour and, via the accelerated life model of section 9.5.1, we study the relative impact of the various defects on the leakage current. Using the notation of section 9.5, we consider the model (see equation (5.5))

$$\ell n Z_i = - \sum_{j=1}^{10} C_j \rho_i^*(m_j) + \varepsilon_i,$$

where Z_i denotes the observed leakage current for the ith motor tested, the $\rho_i^*(m_j)$ are given by (4.5) for the ith motor tested, and with $\varepsilon_i \sim \mathcal{N}(0, \tau^2)$, with the sequence $\{\varepsilon_i\}$, $i = 1, \ldots, 8$, uncorrelated. We next obtain a Bayes estimate of the C_j's, where each $- C_j$ represents a relative contribution of the amplitude at m_j/T to the leakage current. A priori, the C_j's are assumed independent and normally distributed with means 0 and a variance proportional to τ^2; i.e. $(C_j | \tau^2) \sim \mathcal{N}(0, k\tau^2)$. To reflect prior ignorance about the relative contributions of the C_j's we assign a very large value to k, say $k = 10\,000$; the prior on τ^2 itself is assumed to be noninformative. The Bayes estimates of the $-C_j$'s, rescaled to be between 0 and 1, are shown in Figure 9.5.

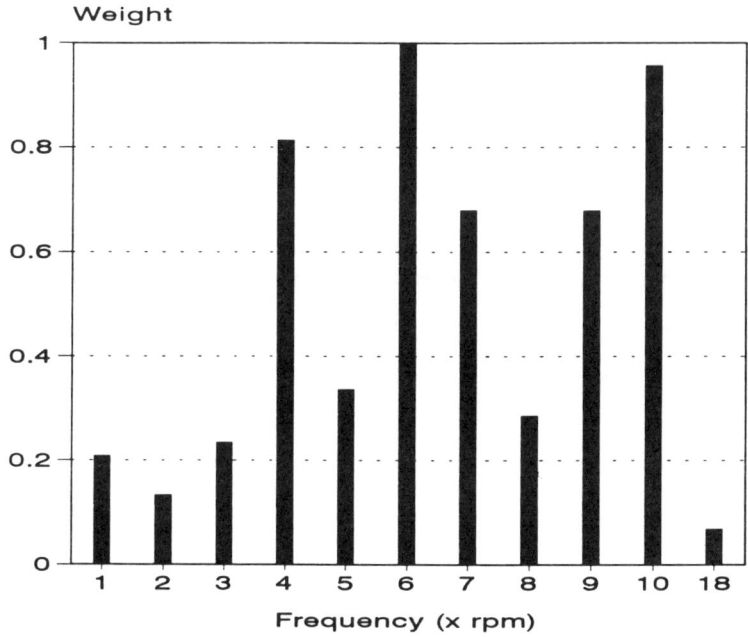

Figure 9.5 Bayes' estimates of the rescaled weights $-C_j$ for the eight motors tested.

Based on this limited analysis, our overall conclusion is that defects which correspond to large frequencies in the spectrum, such as 4, 6, 7, 9 and 10, tend to be more deleterious to the motors than those at lower frequencies, such as 1, 2 and 3. In our application, large amplitudes at the frequencies 5–10 correspond to defects in the inner and the outer race of the bearings, whereas a large amplitude at 4 (× rpm) corresponds to a looseness of the bearing on the housing. Large amplitudes at the frequencies 1–3 suggest an imbalance, or a general looseness in the motor assembly. Thus with traction motors, it appears that the desirability of good bearing races and a tightness of the bearing on the housing supersedes that of a general looseness and imbalance.

APPENDIX A

PRELIMINARIES

In this appendix, we list some probability distributions and their properties referred to in the text.

A1 The random ($p \times 1$) vector X has a p-variate normal distribution with mean vector μ of dimension ($p \times 1$) and covariance matrix Σ of dimension ($p \times p$), if its joint density at x, given μ and Σ, is of the form

$$f_X(x \mid \mu, \Sigma) = ((2\pi)^p \mid \Sigma \mid)^{-1/2} \exp\{-1/2(x - \mu)' \Sigma^{-1} (x - \mu)\},$$

where x is a column vector in $\Re^p$ (the p-dimensional real plane), $|\cdot|$ represents the determinant operator, and x' stands for the transpose of x.

A2 The random variable ϕ (of dimension 1) has a gamma distribution with shape parameter α and scale parameter β, if its density at ϕ is

$$f_\phi(\phi \mid \alpha, \beta) = \frac{\beta^\alpha}{\Gamma(\alpha)} \exp\{-\beta\phi\}\phi^{\alpha-1},$$

where $\phi > 0$, and $\Gamma(\cdot)$ is the gamma function given by

$$\Gamma(x) = \int_0^\infty s^{x-1} \exp(-s)\, ds.$$

Let $E[\cdot]$ and $\text{Var}[\cdot]$ denote the expected value and variance operators respectively. Then,

$$E[\phi \mid \alpha, \beta] = \frac{\alpha}{\beta}, \quad \text{Var}[\phi \mid \alpha, \beta] = \frac{\alpha}{\beta^2} \quad \text{and} \quad E[\phi^{-1} \mid \alpha, \beta] = \frac{\beta}{\alpha - 1} \quad \text{if } \alpha > 1.$$

A3 The random ($p \times 1$) vector X has a Student's T-distribution with p degrees of freedom, with a mode μ (of dimension ($p \times 1$)), and scale matrix R of dimension ($p \times p$), if its joint density at x is given by

$$t_X(x) = \frac{\Gamma((n+p)/2)}{(\pi n)^{p/2} \Gamma(n/2) \mid R \mid^{1/2}} (1 + n^{-1}(x - \mu)' R^{-1}(x - \mu))^{-(n+p)/2},$$

for $x \in \Re^p$.

A4 The one-dimension random variable Y has a chi-square distribution with n degrees, if its density at y is of the form

$$\chi^2(y; n) = \frac{1}{2^{n/2} \Gamma(n/2)} \exp\{-\frac{y}{2}\} y^{(n/2-1)}$$

for $y > 0$. We denote $Y \sim \chi^2[n]$.

A5 The one-dimensional random variable Z follows a non-central chi-square distribution with n degrees of freedom and non-centrality parameter δ, if its density at z is:

$$f_Z(z \mid n, \delta) = \exp\left\{-\frac{\delta}{2}\right\} \sum_{i=0}^{\infty} \left(\frac{\delta}{2}\right)^i \frac{1}{i!} \chi^2(z; 2i + n)$$

for $z > 0$. We denote $Z \sim \chi^2[n, \delta]$.

A6 If $X_1 \sim \mathcal{N}(\mu_1, \sigma^2)$ and $X_2 \sim \mathcal{N}(\mu_2, \sigma^2)$, and X_1 and X_2 are independent $\Rightarrow$

$$\frac{X_1^2 + X_2^2}{\sigma^2} \sim \chi^2\left[2, \frac{\mu_1^2 + \mu_2^2}{\sigma^2}\right]$$

(Johnson and Kotz (1970, pp. 130 and 135).

APPENDIX B

POSTERIOR DENSITY OF THE AMPLITUDES OF THE SPECTRUM

Proof of equations (4.4) and (4.5) of section 9.4:

B1 If

$$\left(\begin{bmatrix} a \\ b \end{bmatrix} \mid \phi\right) \sim \mathcal{N}(\mu, \phi^{-1}\Sigma), \text{ with } \mu = \begin{bmatrix} \mu_a \\ \mu_b \end{bmatrix} \text{ and } \Sigma = \begin{bmatrix} r_a & r_{ab} \\ r_{ab} & r_b \end{bmatrix},$$

and $\phi \sim$ Gamma $(n/2, d/2)$, then the density of $Z = a^2 + b^2$, unconditional of ϕ, is of the form:

$f_Z(z) =$

$$\int_0^z \frac{\tau(\sqrt{z-x}, \sqrt{x}) + \tau(-\sqrt{z-x}, \sqrt{x}) + \tau(\sqrt{z-x}, -\sqrt{x}) + \tau(-\sqrt{z-x}, -\sqrt{x})}{4\sqrt{x(z-x)}} dx,$$

where $z > 0$, and $\tau(\cdot, \cdot)$ stands for the density function of a bivariate Student's T-distribution with n degrees of freedom, a mode μ and scale matrix $R = (n/d)\Sigma$.

Proof Let $A = a^2$ and $B = b^2$, then the joint distribution of $\begin{bmatrix} A \\ B \end{bmatrix}$ given ϕ is

$\mathbb{F}_{\begin{bmatrix} A \\ B \end{bmatrix}}(x_1, x_2 \mid \phi) = Pr(A^2 < x_1, B^2 < x_2 \mid \phi)$

$= Pr(-\sqrt{x_1} < a < \sqrt{x_1}, -\sqrt{x_2} < b < \sqrt{x_2} \mid \phi)$

$= Pr(a < \sqrt{x_1}, b < \sqrt{x_2} \mid \phi) - Pr(a < -\sqrt{x_1}, b < \sqrt{x_2} \mid \phi)$

$$-Pr(a < \sqrt{x_1}, b < -\sqrt{x_2} \mid \phi) + Pr(a < -\sqrt{x_1}, b < -\sqrt{x_2} \mid \phi)$$
$$= F(\sqrt{x_1}, \sqrt{x_2} \mid \phi) - F(-\sqrt{x_1}, \sqrt{x_2} \mid \phi) - F(\sqrt{x_1}, -\sqrt{x_2} \mid \phi) + F(-\sqrt{x_1}, -\sqrt{x_2} \mid \phi),$$

where $F(\cdot, \cdot \mid \phi)$ stands for the normal cumulative distribution of $\begin{bmatrix} a \\ b \end{bmatrix}$ given ϕ. Consequently, the conditional density of $\begin{bmatrix} A \\ B \end{bmatrix}$ given ϕ is:

$$f_{[B]}^{[A]}(x_1, x_2 \mid \phi) = \frac{d^2}{dx_1 dx_2} \mathbb{F}_{[B]}^{[A]}(x_1, x_2 \mid \phi);$$

using the chain rule we obtain

$$f_{[B]}^{[A]}(x_1, x_2 \mid \phi) =$$
$$\frac{f(\sqrt{x_1}, \sqrt{x_2} \mid \phi) + f(-\sqrt{x_1}, \sqrt{x_2} \mid \phi) + f(\sqrt{x_1}, -\sqrt{x_2} \mid \phi) + f(-\sqrt{x_1}, -\sqrt{x_2} \mid \phi)}{4\sqrt{x_1 x_2}},$$

where $f(\cdot, \cdot \mid \phi)$ represents the bivariate normal density with mean μ and covariance $\phi\Sigma$. Finally, averaging out ϕ, we obtain the unconditional density of $\begin{bmatrix} A \\ B \end{bmatrix}$ given by

$$f_{[B]}^{[A]}(x_1, x_2) = \frac{\tau(\sqrt{x_1}, \sqrt{x_2}) + \tau(-\sqrt{x_1}, \sqrt{x_2}) + \tau(\sqrt{x_1}, -\sqrt{x_2}) + \tau(-\sqrt{x_1}, -\sqrt{x_2})}{4\sqrt{x_1 x_2}},$$

where $\tau(\cdot, \cdot)$ is the density function of a bivariate Student's T-distribution with n degrees of freedom, mode μ and scale matrix $R = (n/d)\Sigma$. Finally the desired result, the density of $Z = A + B$, is obtained using the convolution formula, namely

$$f_{A+B}(z) = \int_0^z f_{[B]}^{[A]}(z - s, s) \, ds. \qquad \square$$

B2 The expected value of Z is easily obtained by conditioning on ϕ, that is

$$E[Z] = E_\phi[E[Z \mid \phi]] = E_\phi[E[a^2 + b^2 \mid \phi]]$$
$$= E_\phi[(\mu_a)^2 + \phi^{-1} r_a] + E_\phi[(\mu_b)^2 + \phi^{-1} r_b].$$

Given that $\phi \sim$ Gamma $(n/2, d/2)$, then

$$E[\phi^{-1}] = \frac{d/2}{n/2 - 1} = \frac{d}{n - 2},$$

and therefore

$$E[Z] = (\mu_a)^2 + \frac{d}{n-2}r_a + (\mu_b)^2 + \frac{d}{n-2}r_b.$$

APPENDIX C

APPROXIMATION FOR THE POSTERIOR DENSITY OF THE AMPLITUDES OF THE SPECTRUM

This appendix is relevant to the discussion at the end of section 9.4.1.

C1 If

$$(a \mid \phi) \sim \mathcal{N}(\mu_1, \phi^{-1}r), \ (b \mid \phi) \sim \mathcal{N}(\mu_2, \phi^{-1}r),$$

and $\phi \sim$ Gamma $(n/2, d/2)$ with μ_1, μ_2, r, n, and d specified, and a and b are independent given ϕ, then the density function of $Z = a^2 + b^2$ is of the form

$$f_Z(z) = \sum_{i=0}^{\infty} \frac{1}{2r}\left(\frac{Uz}{4r}\right)^i \frac{1}{(i!)^2} \frac{(d/2)^{n/2}}{\Gamma(n/2)} \frac{\Gamma(2i+n/2+1)}{\left(\frac{U+z/r+d}{2}\right)^{2i+n/2+1}},$$

where $U = (\mu_1^2 + \mu_2^2)/r$.

Proof: Because of (A6),

$$\left(\phi \frac{a^2+b^2}{r} \mid \phi\right) \sim \chi^2\left(2, \phi\left(\frac{\mu_1^2+\mu_2^2}{r}\right)\right).$$

Let

$$U = \frac{\mu_1^2 + \mu_2^2}{r} \quad \text{and} \quad Z = a^2 + b^2,$$

then the density of Z, given ϕ, is

$$f_Z(z \mid \phi) = \chi^2\left(\frac{\phi z}{r}; 2, \phi U\right)\left|\frac{\phi}{r}\right|$$

$$= \exp\left\{-\frac{\phi U}{2}\right\} \sum_{i=0}^{\infty} \left(\frac{\phi U}{2}\right)^i \frac{1}{i!} \left[\frac{(\phi/r)^i}{2^{(i+1)}\Gamma(i+1)} \exp\left\{-\frac{\phi z}{2r}\right\} z^i\right] \frac{\phi}{r}$$

$$= \frac{1}{2}\frac{\phi}{r} \exp\left\{-\frac{\phi(U+z/r)}{2}\right\} \sum_{i=0}^{\infty} \left(\frac{\phi^2 Uz}{4r}\right)^i \left(\frac{1}{i!}\right)^2.$$

Averaging out ϕ, for any $z > 0$, we obtain

$$f_Z(z) = \int_0^\infty \frac{1}{2}\frac{\phi}{r}\exp\left\{-\frac{\phi(U+z/r)}{2}\right\} \sum_{i=0}^\infty \left(\frac{\phi^2 Uz}{4r}\right)^i \left(\frac{1}{i!}\right)^2 \frac{(d/2)^{n/2}}{\Gamma(n/2)}\exp\left\{-\frac{\phi d}{2}\right\}\phi^{n/2-1}d\phi$$

$$= \sum_{i=0}^\infty \frac{1}{2r}\left(\frac{Uz}{4r}\right)^i \frac{1}{(i!)^2} \frac{(d/2)^{n/2}}{\Gamma(n/2)} \int_0^\infty \exp\left\{-\frac{\phi(U+z/r+d)}{2}\right\} \phi^{2i+(n/2)} d\phi$$

$$= \sum_{i=0}^\infty \frac{1}{2r}\left(\frac{Uz}{4r}\right)^i \frac{1}{(i!)^2} \frac{(d/2)^{n/2}}{\Gamma(n/2)} \frac{\Gamma(2i+n/2+1)}{\left(\frac{U+z/r+d}{2}\right)^{2i+n/2+1}}. \qquad \Box$$

REFERENCES

Anderson, T. W. (1971) *The Statistical Analysis of Time Series*, New York, NY, John Wiley & Sons.

Box, G. E. and Tiao, G. C. (1992) *Bayesian Inference in Statistical Analysis*. New York, NY, John Wiley & Sons.

Cain, J. E., Lindsay, B. D., Arthur, R. M., Markham, J. and Ambos, H. D. (1990) 'Noninvasive detection of patients prone to life-threatening ventricular arrhythmias by frequency analysis of electrocardiographic signals', in *Cardiac Electrophysiology: from Cell to Bedside* D. P. Zipes and J. Jalife, eds, W. B. Saunders.

Crowder, M. J., Kimber, A. C., Smith, R. L. and Sweeting, T. J. (1991) *Statistical Analysis of Reliability Data*, New York, NY, Chapman & Hall.

Bretthorst, G. L. (1988) *Bayesian Spectrum Analysis and Parameter Estimation*, New York, NY, Springer-Verlag.

Cooley, J. W. and Tukey J. W. (1965) 'An algorithm for the machine calculation of complex Fourier series', *Math. Comput.* **19** 297.

Effective Machinery Measurements using Dynamic Signal Analyzers (1990) Application Note 243-1, Hewlett-Packard Co.

Gelfand, A. E. and Smith, A. F. (1990) 'Sampling-based approaches to calculating marginal densities', *Journal of the American Statistical Association*, **85** (410), 398–409.

Gill, R. D. (1984) 'Understanding Cox's regression model: A martingale approach', *Journal of the American Statistical Association*, **79** (386), 441–7.

Johnson, N. L. and Kotz, S. (1970) *Distributions in Statistics: Continuous Univariate Distributions-2*, Boston, MA, Houghton Mifflin.

Lydersen, S. (1988) Reliability testing based on deterioration measurements. *Doktor Ingeniøravhandling* **1988**, 32, Trondheim, Institutt for Matematiske fag, NTH.

McCormick, G. (1983) *Nonlinear Programming: Theory, Algorithms, and Applications*, New York, NY, John Wiley & Sons.

Pierce, D. L., Easley, A. R., Windle, J. R. and Engel T. R. (1989) 'Fast Fourier transformation of the entire low amplitude late QRS potential to predict ventricular tachycardia', *Journal of the American College of Cardiology*, **14** (7), 1731–40.
Priestley, M. B. (1965) 'Evolutionary spectra and non-stationary processes', *Journal of the Royal Statistical Society*, **B27**, 204–37.
Roberts, S. A. and Harrison, P. J. (1984) 'Parsimonious modelling and forecasting of seasonal time series', *European Journal of Operations Research*, **16**, 365–77.
West, M. and Harrison, J. (1989) *Bayesian Forecasting and Dynamic Models*, New York, NY, Springer-Verlag.
Wowk, V. (1991) *Machinery Vibration: Measurement and Analysis*, New York, NY, McGraw-Hill.

Department of Operational Research and Statistics
George Washington University
707 22nd Street NW
Washington DC 20052
USA

CHAPTER 10

Residual Analysis and Outliers in Bayesian Hierarchical Models

Kathryn Chaloner
Department of Applied Statistics, University of Minnesota

10.1 THE REALIZED ERRORS

Realized errors in non-hierarchical linear models will first be briefly discussed to provide a framework for the extension to hierarchical models. In a linear model let $\theta = (\theta_1, \ldots, \theta_p)^T$ be the p unknown regression coefficients and let $y_1, \ldots, y_n$ be independent observations such that $y_i = x_i^T \theta + \sigma \varepsilon_i$. The $x_1, \ldots, x_n$ are explanatory variables and σ is the unknown standard deviation. The $\varepsilon_1, \ldots, \varepsilon_n$ are the realized errors, or residuals, and are a sample from a normal distribution with mean zero and standard deviation one. Zellner (1975) first suggested their use for diagnostic purposes. He derived the posterior distribution of the ε_i^*'s where $\varepsilon_i^* = \sigma \varepsilon_i$, and the posterior distribution of their sample skewness and kurtosis to examine the normality assumption. Zellner and Moulton (1985) used the posterior distribution of the ε_i^*'s to construct a residual plot. Chaloner and Brant (1988) defined outliers as observations for which the absolute value of ε_i is surprisingly large and calculated the exact posterior probability of any observation being an outlier. They also calculated the joint probability of any two observations being outliers. These methods were demonstrated to be able to detect outliers, even when masking is present.

Chaloner and Brant (1988) define an outlier as any observation with $|\varepsilon_i| > k$ for a suitable choice of k. If the ε_i are independent observations from a normal population and we have m observations, then k can be defined as

---

Aspects of Uncertainty edited by P. R. Freeman and A. F. M. Smith. © 1994 John Wiley & Sons Ltd.

$$k = \Phi^{-1}\{0.5 + \frac{1}{2}(0.95^{1/m})\}. \qquad (1)$$

The prior probability of no outliers is then 0.95. After observing the data the posterior probability that each observation is an outlier can be calculated. Any observation with a posterior probability, $pr(|\varepsilon_i| > k | y)$, larger than the prior probability of $2\Phi(-k)$ would be a possible outlier.

A slightly different but related approach is used in Chaloner (1991), combining the idea of realized errors with Cox and Snell's (1968) concept of generalized residuals. This gives a method for detecting outliers in a class of models where the observations are independent but not necessarily normally distributed.

In section 10.2 the posterior distribution of the realized errors in a one-way hierarchical model is derived, conditional on the variances being known. An example of Sharples (1990) is used in section 10.3 to illustrate the use of the method when the variances are unknown.

10.2 THE ONE-WAY MODEL

A realized error analysis for a simple hierarchical model with known variances will be discussed in this section. Exact expressions for the posterior means and variances of the realized errors will be derived. Extensions to more complicated models and models with unknown variances are straightforward but computationally more involved.

Conditional on θ_i for $i = 1, \ldots, p$ suppose that observations y_{ij} are independent for each i and $j = 1, \ldots, n_i$ and are normally distributed:

$$y_{ij} | \theta_i \sim N(\theta_i, \sigma_w^2).$$

The prior distribution on the θ_i's is such that, conditional on μ, the θ_i's are independent:

$$\theta_i | \mu \sim N(\mu, \sigma_b^2).$$

The θ_i's are therefore exchangeable group means, σ_b^2 and σ_w^2 are the between- and within-group variances respectively. As σ_w^2 and σ_b^2 are known a prior distribution is specified on μ alone.

Two sets of residuals can be defined and examined: the within-group residuals and the between-group residuals. The within-group residuals measure how far y_{ij} is from its mean:

$$\varepsilon_{ij} = \frac{(y_{ij} - \theta_i)}{\sigma_w}.$$

The between-group residuals measure how far away the group mean θ_i is from its mean:

$$\varepsilon_i = \frac{(\theta_i - \mu)}{\sigma_b}.$$

Both sets of residuals are, prior to conditioning on the data y, independent, and are a sample from a standard normal distribution. Observations y_{ij} can be written as

$$y_{ij} = \mu + \sigma_b \varepsilon_i + \sigma_w \varepsilon_{ij}.$$

To examine the ε_i's and the ε_{ij}'s their posterior distributions must be calculated. We will assume that the prior distribution of μ is the improper distribution, uniform on the real line, although the results are easily adapted to a proper normal prior distribution.

The results of Lindley (1971) and Lindley and Smith (1972) can be applied to calculate the multivariate normal posterior distribution of the θ_i's. As the within-group residuals are linear in the θ_i's they are also normally distributed.

Define $\bar{y}_{i.}$ to be the mean of the ith group and define

$$\alpha_i = \frac{n_i}{\sigma_w^2 + n_i \sigma_b^2}.$$

Further define

$$\bar{y}_{..} = \frac{\Sigma \alpha_i \bar{y}_{i.}}{\Sigma \alpha_i}.$$

The posterior mean of ε_{ij} is

$$\frac{1}{\sigma_w} \left\{ y_{ij} - \frac{\bar{y}_{..} \sigma_w^2 + \bar{y}_{i.} n_i \sigma_b^2}{\sigma_w^2 + n_i \sigma_b^2} \right\} \quad (2)$$

and the posterior variance of ε_{ij} is

$$\frac{\sigma_b^2}{\sigma_w^2 + n_i \sigma_b^2} + \frac{\sigma_w^2}{(\sigma_w^2 + n_i \sigma_b^2)^2 \Sigma \alpha_i}. \quad (3)$$

The correlation between two residuals in the same group, ε_{ij} and $\varepsilon_{ij'}$ where $j \neq j'$, is one because the only uncertainty is in θ_i. Residuals in different groups are also correlated, but not perfectly.

The between-group residuals, the ε_i's, depend on both the θ_i's and μ and so their posterior distribution is not quite as straightforward to calculate. The calculation can be done, however, using first the distribution of the θ_i's conditional on both y and μ and then the posterior distribution of μ alone. Using again the general results in Lindley and Smith (1972) the distribution of the θ_i's given y and μ is multivariate normal with the θ_i's being independent. The posterior distribution of μ is also normal. The posterior mean of ε_i is calculated as

$$E(\varepsilon_i|y) = \int E(\varepsilon_i|y, \mu)p(\mu|y)d\mu$$

and the posterior variance of ε_i is calculated through the relationship

$$\text{var}(\varepsilon_i|y) = \int [\text{var}(\varepsilon_i|\mu, y) + \{E(\varepsilon_i|\mu, y) - E(\varepsilon_i|y)\}^2]p(\mu|y)d\mu.$$

Omitting the details of this calculation, the posterior mean of the between-group residual ε_i becomes

$$\frac{n_i\sigma_b(\bar{y}_{i.} - \bar{y}_{..})}{\sigma_w^2 + n_i\sigma_b^2}. \tag{4}$$

The posterior variance of ε_i is

$$\frac{\sigma_w^2}{\sigma_w^2 + n_i\sigma_b^2} + \frac{\alpha_i^2 \sigma_b^2}{\sum_{j=1}^{j=p} \alpha_j}. \tag{5}$$

Suppose now that there are n observations in each group, that is the n_i's are equal. The posterior variances simplify considerably. The posterior variance of ε_{ij} is

$$\frac{1}{np}\left(\frac{\sigma_w^2 + np\sigma_b^2}{\sigma_w^2 + n\sigma_b^2}\right). \tag{6}$$

The between-group residuals, in the case where the n_i's are equal, have variance

$$\frac{p\sigma_w^2 + n\sigma_b^2}{p(\sigma_w^2 + n\sigma_b^2)}. \tag{7}$$

10.3 UNKNOWN VARIANCES

It is more usual that the variances are unknown. One way of dealing with this is to use estimates of the variances, $\hat{\sigma}_w^2$ and $\hat{\sigma}_b^2$ say, and substitute these into the expressions for the case where the variances are known. A better alternative would be to do exact calculations where the posterior distributions must be averaged over the joint posterior distribution of σ_w^2 and σ_b^2. Numerical integration in at least one dimension will be required. (For a complete discussion of the posterior distribution of σ_w^2 and σ_b^2 see Box and Tiao, 1973, Ch. 5, and Hill, 1965.) Laplace approximations, as described in Tierney and Kadane (1986) will be used in the following example. The approximations will be implemented using the software of Tierney (1990) after integrating over μ exactly.

10.3.1 Example

With the exception of Sharples (1990) outliers and residuals in Bayesian hierarchical models have not been extensively studied. She used both inflated-variance and mean-shift contaminating mixture models for outliers. A realized error analysis will be demonstrated using a data set generated by Sharples.

The data are from a one-way model with contamination. The contamination is such that within-group errors are sampled, with probability 0.1, from a gamma distribution with mean 5.5. In addition, all the groups have a mean of 25 except for the fifth group which has mean 50. Table 10.1 gives the data with the three outliers indicated with asterisks.

Table 10.1 Table of Sharples generated data, with outliers indicated by an asterisk.

		Group		
1	2	3	4	5
24.80	23.96	18.30	51.42*	34.12
26.90	28.92	23.67	27.97	46.87
26.65	28.19	14.47	24.76	58.59*
30.93	26.16	24.45	26.67	38.11
33.77	21.34	24.89	17.58	47.59
63.31*	29.46	28.95	24.29	44.67

Following Box and Tiao (1973, Ch. 5) a prior distribution

$$p(\mu, \sigma_w^2, \sigma_b^2) \propto \sigma_w^{-2}(\sigma_w^2 + n\sigma_b^2)^{-1}$$

is used with $n = 6$. The joint posterior distribution $p(\sigma_w^2, \sigma_b^2 | y)$, as given by Box and Tiao (1973, p. 252), is

$$p(\sigma_w^2, \sigma_b^2 | y) \propto (\sigma_w^2)^{-(\{p(n-1)/2\} + 1)}(\sigma_w^2 + n\sigma_b^2)^{-p/2}$$

$$\exp\left[-\frac{1}{2}\left\{\frac{p(n-1)m_w}{\sigma_w^2} + \frac{(p-1)m_b}{\sigma_w^2 + n\sigma_b^2}\right\}\right]$$

where m_w and m_b are the within and between mean squares respectively. This posterior distribution will be used in the approximations to the posterior distribution of the residuals.

For the within-group residuals the posterior mean is the posterior mean of equation (2) with respect to $p(\sigma_w^2, \sigma_b^2 | y)$. A normal approximation to $p(\varepsilon_{ij} | y)$ will be used with mean equal to the Laplace approximation to the mean of equation (2). The posterior variance is the sum of two components:

$$\text{var}(\varepsilon_{ij} | y) = \int \text{var}(\varepsilon_{ij} | \sigma_w^2, \sigma_b^2, y) p(\sigma_w^2, \sigma_b^2 | y) d\sigma_w^2 d\sigma_b^2$$

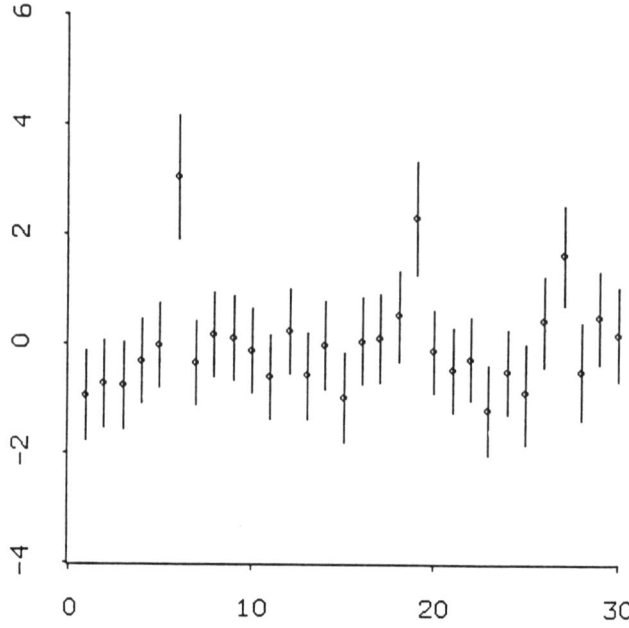

Figure 10.1 Plot of within-group residuals.

$$+ \int \{E(\varepsilon_{ij}|\sigma_w^2, \sigma_b^2, y) - E(\varepsilon_{ij}|y)\}^2 p(\sigma_w^2, \sigma_a^2|y) d\sigma_w^2 d\sigma_b^2. \quad (8)$$

The first term is just the expectation of equation (6) and the second is the variance of equation (2) with respect to $p(\sigma_w^2, \sigma_b^2|y)$. The variance will therefore be approximated by the sum of approximations for each term: the first is the Laplace approximation to the mean of equation (6) which approximates the first term of equation (8), the second is the Laplace approximation to the variance of equation (2) which approximates the second term. A plot of the expected within-group residuals is given as Figure 10.1. Intervals of the mean plus and minus two approximate standard deviations have been added.

Using equation (1) for a total sample size of 30 gives $k = 3.14$. The prior probability of an observation being an outlier is therefore 0.0017. In Figure 10.1 the contaminated observations y_{16}, y_{14} and y_{53} appear to be outlying on the plot. Under normal approximations, these observations have posterior probabilities of 0.4364, 0.0508 and 0.0004 respectively of being outliers, and all other observations have posterior probability of less than 10^{-6}. The posterior probabilities of observations y_{16} and y_{41} are larger than the prior probability of an individual observation being an outlier. Observation y_{16} and y_{41} are definitely suspect as outliers. If y_{16} is decided to be definitely outlying and is deleted and the analysis repeated with 29 obser-

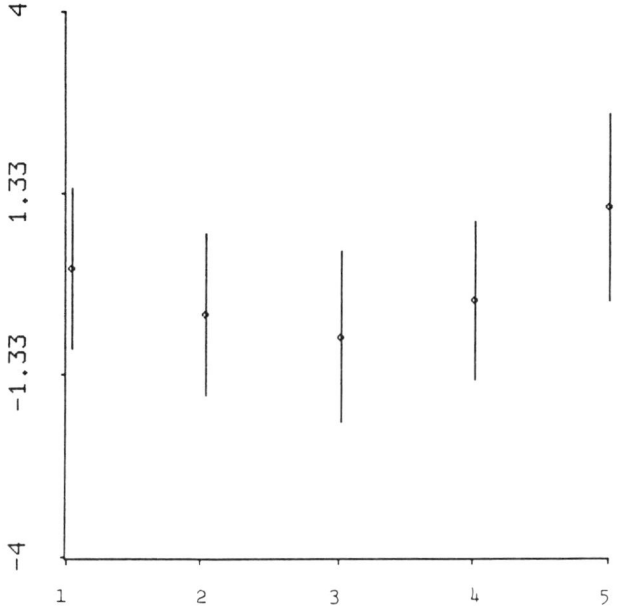

Figure 10.2 Plot of between-group residuals based on all the observations.

vations and the same prior distribution the posterior probability of y_{41} and y_{53} being outliers increases to 0.4278 and 0.0283 respectively, with the next largest posterior probability being less than 2×10^{-4}. The details of this calculation have been omitted, but the likelihood can be found, for example, in Hill (1965).

Similar approximations were used for the between-group residuals using all the data including the three outliers. Figure 10.2 is a plot of the posterior means, using Laplace approximations, augmented with intervals of plus and minus twice the approximate standard errors. As there are only five groups we use equation (1) for $m = 5$ to give $k = 2.57$ and the prior probability of any one group being outlying is 0.0102. The posterior probabilities of each between-group residual being an outlier was calculated and the one for group number 5 has the largest posterior probability of 0.0191 of being an outlier. All the others have posterior probability much smaller. If the analysis is repeated after deleting the three outlying observations, y_{16}, y_{41} and y_{53}, the outlying nature of group 5 is even more apparent as the approximate posterior probability that group 5 is outlying is 0.0429 and the next largest probability, for group 3, is about 10^{-7}.

This analysis agrees with that of Sharples (1990) in identifying the deliberately contaminated observations and group. The calculations are, arguably, simpler than hers.

10.4 DISCUSSION

These ideas easily extend to a general hierarchical linear model. All that is required is to express the observations as functions of parameters and errors, where the errors are independent and have a standard normal distribution prior to conditioning on the data. The posterior distribution of these errors can be used for residual analysis and outlier detection.

The more popular approach to outlier detection is to define outliers as observations which have been generated from a model other than that generating most of the data (Freeman 1981; Pettit and Smith 1985). The usual implementation of this approach is to model outliers as coming from a contaminating model: either a mean-shift or an inflated-variance model.

The alternative idea used here comes from the simple and powerful idea expressed in Zellner (1975). This concept can be widely applied to many models and provides a method for successfully defining and detecting outliers. Defining an outlier as an observation with a large realized error is a precise definition that requires that outliers do actually outlie. This contrasts with the more popular approach of a contaminating mixture where, with quite high probability, outliers do not necessarily outlie. Pettit and Smith (1985) address this problem, in the mixture model, by requiring that outliers are not only generated from a contaminating distribution but also outlie. It is much simpler to take the direct approach and define outliers to be outlying observations. Realized error analysis is also computationally much simpler than mixture models as the computations are based on the usual posterior distribution rather than an average of the posterior distributions omitting each case, and set of cases, which leads to a computational explosion.

A realized error analysis could also be done with a model where the errors are heavy tailed: t-distributions for example, or even contaminated normal models. The analysis would identify outlying observations rather than attempt to identify which observations, irrespective of whether they outlie, come from the contaminating distribution.

ACKNOWLEDGEMENTS

Most of this chapter was written when the author was visiting Carnegie Mellon University with support from a single quarter leave from the University of Minnesota. I thank J. B. Kadane and L. Tierney for helpful comments.

REFERENCES

Box, G. E. P. and Tiao, G. C. (1973) *Bayesian Inference in Statistical Analysis*, Reading, MA, Addison-Wesley.

Chaloner, K. (1991) Bayesian residual analysis in the presence of censoring, *Biometrika*, **78**, 637–44.
Chaloner, K. and Brant, R. (1988) A Bayesian approach to outlier detection and residual analysis, *Biometrika*, **75**, 651–9.
Cox, D. R. and Snell, E. J. (1968) A general definition of residuals (with discussion), *J. Roy. Statist. Soc.*, **B30**, 248–75.
Freeman, P. R. (1981) On the number of outliers in data from a linear model, in *Bayesian Statistics*, J. M. Bernardo, *et al.* eds, Valencia, Valencia University Press, pp. 349–65.
Hill, B. M. (1965) Inference about variance components in the one-way model, *J. Amer. Statist. Soc.* **60**, 806–25.
Lindley, D. V. and Smith, A. F. M. (1972) Bayes estimates for the linear model (with discussion), *J. Roy. Statist. Soc.* **B34**, 1–41.
Lindley, D. V. (1971). The estimation of many parameters, in *Foundations of Statistical Inference*, V. P. Godambe and D. A. Sprott eds, Toronto, Holt, Reinhart and Winston, pp. 435–47.
Pettit, L. I. and Smith, A. F. M. (1985) Outliers and influential observations in linear models, in *Bayesian Statistics 2*, J. M. Bernardo *et al.* eds, Amsterdam, North-Holland, pp. 473–94.
Sharples, L. D. (1990) Identification and accommodation of outliers in general hierarchical models, *Biometrika*, **77**, 445–53.
Tierney, L. and Kadane, J. B. (1986) Accurate approximations for posterior moments and marginal densities. *J. Am. Statist. Assoc.*, **81**, 82–6.
Tierney, L. (1990) *LISP-STAT, an Object-oriented Environment for Statistical Computing and Dynamic graphics*, New York, John Wiley & Sons.
Zellner, A. (1975) Bayesian analysis of regression error terms, *J. Am. Statist. Assoc.*, **70**, 138–44.
Zellner, A. and Moulton, B. R. (1985) Bayesian regression diagnostics with applications to international consumption and income data, *J. Econometrics*, **29**, 187–211.

Department of Applied Statistics
University of Minnesota
352 Classroom Office Building
1994 Buford Avenue
St Paul
MN 55108
USA

CHAPTER 11

The Island Problem: Coherent Use of Identification Evidence

A. P. Dawid
University College London

11.1 INTRODUCTION

Many court cases hinge on identification evidence which links the accused with the scene of the crime through some shared characteristic: for example, a match between fingerprints found on the murder weapon and those of the accused; a correspondence between DNA found in blood-stains at the scene, and that of the accused; rifling marks on a bullet which could have been produced by the accused's gun; characteristics of an incriminating typed letter which match those of letters typed on a certain typewriter; or eye-witness evidence of age, sex, height, etc. which describe the accused. In recent years there has been much discussion of the use of probabilistic and statistical arguments to quantify the strength and import of such evidence.

Suppose, for simplicity, that the evidence implies that the culprit must possess a certain characteristic C, which otherwise occurs in a general member of the population with (typically small) probability P. Let C_0 be the event that the accused possesses C. A naïve argument, which is nevertheless implicit in the way such evidence is often interpreted in court, identifies $pr(\text{innocent} | C_0)$ with $pr(C_0 | \text{innocent})$ thus yielding $pr(\text{guilty} | C_0) = 1 - P$. If the characteristic C occurs in only one person in 10 000 (as might occur, say, for fingerprint or DNA profiling evidence), possession of C would thus be considered to lead to a probability of 99.99% that the suspect is guilty. This misapplication of conditional probability has been termed 'the prosecutor's fallacy'.

Aspects of Uncertainty edited by P. R. Freeman and A. F. M. Smith. © 1994 John Wiley & Sons Ltd.

The coherent Bayesian argument is as follows. Let π be the probability of guilt taking all other evidence into account, and assume that this evidence and the identification evidence are conditionally independent, given either innocence or guilt. Since $pr(C_0|\text{guilty}) = 1$ and $pr(C_0|\text{innocent}) = P$, the likelihood ratio in favour of guilt is $1/P$, and Bayes's theorem readily yields

$$pr(\text{guilty}|\text{all evidence}) = \pi/(\pi + P - \pi P). \qquad (1.1)$$

In the special case $\pi = 1/2$, we obtain $1/(1 + P)$, which, for small P, well approximates $1 - P$, showing that the 'prosecutor's fallacy' is not in fact fallacious if (and only if), before the identification evidence is considered, the suspect is considered equally likely to be innocent or guilty. This might be reasonable if the other evidence is sufficiently convincing, but would not be appropriate in the absence of other evidence.

11.2 THE ISLAND PROBLEM

Eggleston (1983, App. 3) considers the case of an island, cut off from the outside world, on which a murder has been committed. There remain $N + 1$ inhabitants, of whom one must be the murderer, and the other N innocent. Evidence at the scene of the crime may be summarized in the assertion C_S that the murderer possesses a certain characteristic C, which normally occurs in a general member of the population with probability P. A suspect is arrested and brought to trial: the sole evidence against him, C_0, is that he has C. What is the probability that he is guilty?

11.2.1 The Bayes approach

Let G_0 denote the guilt of the accused. Since there is no prior reason to distinguish the suspect from any of the other N islanders, we take the prior probability of guilt to be $\pi = pr(G_0) = 1/(N + 1)$. Substitution into Bayes' formula (1.1) then yields

$$pr(G_0 \mid C_0) = 1/(1 + NP). \qquad (2.1)$$

This result can be readily explained in court as follows. In the population there is one murderer, who possesses C, and also NP innocent people possessing C (of course this is really an expectation, but we pass rapidly over this subtlety). The suspect is thus placed, by the evidence, in a group of $1 + NP$ otherwise indistinguishable individuals, of whom 1 is the murderer. Hence, the probability that he is the murderer is $1/(1 + NP)$.

Eggleston takes $P = 0.004$, $N = 100$, for which this Bayesian approach yields

$$pr(G_0 \mid C_0) = 0.714. \qquad (2.2)$$

11.2.2 The Supreme Court approach

In the infamous case of *People* v. *Collins* (1968), the original conviction was based on the 'prosecutor's fallacy', using the value $P = 1/(12 \text{ million})$. On appeal, the Supreme Court of California overturned this verdict. The argument, transferred to the island, was as follows.

Let M be the (unknown) number of islanders possessing characteristic C. Before any evidence is obtained, we take M to have the binomial distribution $\text{Bin}(N + 1; P)$. We know the accused to have C, so that $M \geq 1$. The Supreme Court evaluated $pr(M > 1 \mid M \geq 1)$, the conditional probability that there is at least one other islander possessing C. This is

$$pr(M > 1 \mid M \geq 1) = \frac{1 - (1 - P)^{N+1} - (N + 1)P(1 - P)^N}{1 - (1 - P)^{N+1}}, \quad (2.3)$$

for which a Poisson approximation gives $1 - \lambda e^{-\lambda}/(1 - e^{-\lambda})$ with $\lambda = (N + 1)P$. In the Collins case, the Supreme Court considered that a reasonable population size might be $N + 1 = 12$ million, thus yielding

$$\lambda = 1 \quad \text{and} \quad pr(M > 1 \mid M \geq 1) = 0.42.$$

In the light of this non-negligible probability that the defendants were not the only ones who possessed the incriminating evidence C, the Supreme Court found that the case had not been established beyond a reasonable doubt.

For Eggleston's figures $N = 100$, $P = 0.004$, we obtain $pr(M > 1 \mid M \geq 1) = 0.19$. A 20% chance of there being another islander possessing C is enough to raise a reasonable doubt in the guilt of the accused.

An extension

The above line of reasoning can be developed further as follows. With possession of C the only incriminating evidence,

$$pr(G_0 \mid C_0, M = m) = 1/m.$$

We have already argued that, given C_0, $M \sim \text{Bin}(N + 1; P)$ conditioned on $M \geq 1$. Hence

$$pr(G_0 \mid C_0) = E(1/M \mid M \geq 1) \quad (2.4)$$

with $M \sim \text{Bin}(N + 1; P)$. Evaluating this for $N = 100$, $P = 0.004$ yields

$$pr(G_0 \mid C_0) = 0.902. \quad (2.5)$$

11.2.3 Eggleston's approach

Following Yellin (1979), Eggleston argues as follows. We know the accused has C. The remaining N islanders each possess C with probability P.

Assuming independence, the number of these possessing C has the binomial distribution $\text{Bin}(N; P)$. We thus obtain

$$pr(G_0 \mid C_0) = E(1/M)$$

with $M \sim 1 + \text{Bin}(N; P)$, yielding

$$pr(G_0 \mid C_0) = \{1 - (1 - P)^{N+1}\}/(N + 1)P. \tag{2.6}$$

Evaluating for $N = 100$, $P = 0.004$ gives

$$pr(G_0 \mid C_0) = 0.824. \tag{2.7}$$

11.3 WHICH ANSWER?

The three approaches described above all yield different answers (2.1), (2.4), (2.6) for $pr(G_0 \mid C_0)$. Which—if any—is correct? In order to investigate further, we must be a little more careful in specifying what is known, what is assumed and what is required.

The total evidence can be split into three components:

1. H: the murder has been committed;
2. C_S: the murderer possesses C;
3. C_0: the accused possesses C.

We need to calculate $pr(G_0 \mid H, C_S, C_0)$.

As argued above, $pr(G_0 \mid H) = 1/(N + 1)$. The 'background' evidence H does not affect any other probabilities to be calculated, and will henceforth be omitted from the notation, being taken as given throughout all further discussion.

11.3.1 The Bayes approach

Consider in detail the assumptions implicit in the Bayesian argument of section 11.2.1. We want $pr(G_0 \mid C_S, C_0)$, the posterior probability of guilt. This can be found by Bayes's formula applied conditional on C_S, thus using 'prior probability' $pr(G_0 \mid C_S)$, and 'likelihoods' $pr(C_0 \mid G_0, C_S)$ and $pr(C_0 \mid \bar{G}_0, C_S)$. The previous argument thus applies if

(A1) $pr(G_0 \mid C_S) = 1/(N + 1)$;
(A2) $pr(C_0 \mid G_0, C_S) = 1$;
(A3) $pr(C_0 \mid \bar{G}_0, C_S) = P$.

Assumption (A1) requires that, before knowing that the accused possesses C, the evidence from the scene of the crime that the murderer must do so is not in itself incriminating to the accused, and so the accused remains exchangeable with the other N islanders. This seems entirely reasonable.

Assumption (A2) is likewise valid, by definition: if the murderer has C, and the accused is the murderer, then the accused must have C.

Finally, consider (A3). If this is regarded as defining P, there can of course be no discussion. However, this might differ from $P_0 = pr(C_0)$, for two reasons. First, there may be dependence between C_0 and C_S: for example, it may have been known in advance that exactly k islanders possess C, in which case $P_0 = k/(N + 1)$, and (assuming that the impact of $\bar{G}_0$ is merely to specify that the murderer and the accused are distinct) $P = (k - 1)/N$. Second, even if we treat distinct individuals as exhibiting C independently of one another, C_0 may be incriminating of itself. For example, in Eggleston's narrative C combines motive, opportunity and access to a weapon, so that presumably $P = pr(C_0 \mid \bar{G}_0) < pr(C_0) = P_0$. However, if C = 'possesses red hair', say, then $P = P_0$ could be reasonable.

In summary, we have vindicated the Bayesian formula (2.1) when (A3) is regarded as defining P. In particular it holds when

(K1) the characteristic C is not in itself incriminating;
(K2) distinct individuals possess C each with probability P, independently.

Note that, in this case, the pair of variables ('has C?', 'is guilty?') is exchangeable across all individuals. Balding and Donnelly (1994) base a more general analysis on this weaker requirement.

11.3.2 Alternative approach

Assume the conditions (K1) and (K2) at the end of section 11.3.1, and let M denote the number of islanders possessing C. Then, by exchangeability, $pr(G_0 \mid C_S, C_0, M) = 1/M$, and so

$$pr(G_0 \mid C_S, C_0) = E(1/M \mid C_S, C_0). \tag{3.1}$$

By Bayes' theorem,

$$pr(M = m \mid C_S, C_0) \propto pr(C_0 \mid M = m, C_S) \times pr(M = m \mid C_S). \tag{3.2}$$

We consider further the terms on the right-hand side of (3.2).

We have

$$pr(M = m \mid C_S) \propto pr(M = m) \times pr(C_S \mid M = m), \tag{3.3}$$

and the first term in (3.3) is given by the binomial formula

$$\binom{N+1}{m} P^m (1 - P)^{N+1-m}.$$

The second is $m/(N + 1)$. We thus obtain

$$pr(M = m \mid C_S) = \binom{N}{m-1} P^{m-1}(1 - P)^{N-(m-1)} \quad (m = 1, 2, \ldots, N+1), \quad (3.4)$$

so that

$$M \mid C_S \sim 1 + \text{Bin}(N; P). \quad (3.5)$$

Also by exchangeability, which persists even conditional on C_S, the first term in (3.2) is also $m/(N + 1)$. Combining this with (3.4) gives, from (3.2),

$$pr(M = m \mid C_S, C_0) \propto m \binom{N}{m-1} P^{m-1}(1 - P)^{N-m+1} \quad (m = 1, 2, \ldots, N+1), \quad (3.6)$$

so that, using (3.1),

$$pr(G_0 \mid C_S, C_0) = \frac{\sum_{m=1}^{N+1} \binom{N}{m-1} P^{m-1}(1 - P)^{N-m+1}}{1 + \sum_{m=1}^{N+1} (m-1) \binom{N}{m-1} P^{m-1}(1 - P)^{N-m+1}}$$

$$= 1/(1 + NP), \quad (3.7)$$

agreeing with the Bayes approach.

11.3.3 Discussion

The above development shows where the arguments of sections 11.2.2 and 11.2.3, leading to formulae (2.4) and (2.6), are in error. Whereas $M \sim 1 + \text{Bin}(N; P)$ given either C_S or C_0, we actually have to condition on C_S and C_0. Eggleston conditions M on C_0 alone to obtain (2.6).

The 'Supreme Court' argument of section 11.2.2 treats the evidence as equivalent to $M \geq 1$. This is not valid even if the evidence is confined to C_0, because $pr(C_0 \mid m)$ is increasing with m, so that evidence C_0 favours larger values of M, as opposed to merely eliminating $M = 0$. Essentially the same error is made in Eggleston's approach, in ignoring the additional impact of C_S. In arguing for the need to condition properly on C_0, thus introducing one weighting factor of $m/(N + 1)$ into the distribution of M, he neglects the need for a further such factor to account for C_S.

When all factors are properly accounted for, we thus see that the above approach reproduces the solution more easily found by the direct application of Bayes's theorem, and reconciles the seemingly conflicting answers found in section 11.2 above.

11.3.4 Lindley's example

Discussing a very similar example, Lindley (1987) considers the effect of additional evidence that might be available. Specifically, he considers C_1:

THE ISLAND PROBLEM 165

individual 1 (not the suspect) has C; and C^*: at least one other individual has C. Application of Bayes's theorem yields

$$pr(G_0 \mid C_S, C_0, C_1) = 1/\{2 + (N-1)P\}; \qquad (3.8)$$

$$pr(G_0 \mid C_S, C_0, C^*) = \{1 - (1-P)^N\}/\{1 + NP - (1-P)^N\}. \qquad (3.9)$$

Lindley remarks that (3.9) exceeds (3.8), and leaves it to the reader to explain why 'the evidence "Smith is left-handed" and "There are left-handers, one of whom is called Smith" have different evidential value'. With $N = 100$ and $P = 0.004$, (3.8) gives 0.417 and (3.9) gives 0.452.

Once again, insight into the impact of such evidence is afforded by an analysis similar to that of section 11.3.2. We have

$$pr(G_0 \mid C_S, C_0, C_1, M) = pr(G_0 \mid C_S, C_0, C^*, M) = 1/M,$$

as before, and have again to consider the impact of the relevant evidence on the distribution of M. We have already accounted for the evidence (C_S, C_0) in section 11.3.2, yielding the conditional distribution (3.6) for M. Now we have to include a further term, respectively either

$$pr(C_1 \mid C_S, C_0, M = m) = (m-1)/N \qquad (3.10)$$

or

$$pr(C^* \mid C_S, C_0, M = m) = \begin{cases} 1 & \text{if } m \geq 2 \\ 0 & \text{if } m = 1. \end{cases} \qquad (3.11)$$

The distinction between (3.10) and (3.11) parallels very closely that between the respective weighting terms implicit in the Eggleston and Supreme Court analyses, as discussed in section 11.3.3. In this case, each of (3.10) and (3.11) is the correct additional weighting factor to account for the appropriate evidence, C_1 or C^*. Clearly, (3.10) again weights more heavily the larger values of m than does (3.11), and hence, on taking the expectation of $1/M$, leads to a smaller probability of guilt. It may be verified that the appropriate calculations using (3.6) and (3.10) or (3.11) do yield (3.8) or (3.9).

Lindley's 'paradox' is thus resolved by realizing that, on investigating Smith and discovering him to be left-handed, this evidence, being more probable if M is large, thus favours larger M and so entails a smaller probability of guilt; whereas investigating the whole population so as to learn whether or not there are any further left-handers, and discovering that there are, merely eliminates $M = 1$, but does not otherwise affect uncertainty about M.

11.4 SEARCHING FOR SUSPECTS

We have so far been a little careless as to how the suspect was identified. Our analysis would be valid if a random islander were arrested, and then found to possess C. However, this is unrealistic. More reasonable is some sort of search strategy. At its simplest, in the knowledge that the murderer must have C, the police might search through the population of the island one by one, in random order, and arrest the first individual possessing C. Does such a search strategy affect the probability that the accused is guilty?

My own intuitive first response to this question, backed up by vague invocations of 'the irrelevance of optional stopping' and the likelihood principle, was 'No': Bayesian inference should not be affected by the detailed structure of the experiment, so long as the data are the same. Closer analysis, however, suggests that, in the current context, this is not so, and that account should indeed be taken of the search strategy.

11.4.1 Analysis

We again assume conditions (K1) and (K2). Suppose a search is conducted entirely at random, or in any way, deterministic or stochastic, which is independent of the values of C and the identity of the murderer: by exchangeability, all such strategies will be equivalent. We may as well suppose the population numbered from 1 to $N + 1$, and searched in order. Let C_i denote that individual i possesses C, and G_i that he is guilty. Let K be the first individual to exhibit C, who is then accused, and $G_0 \equiv G_K$ the event that he is guilty. Note that $C_0 \equiv C_K$ is certain under this strategy, but we shall sometimes condition on it, for clarity.

Given C_S, if we know $K = k$ the first $k - 1$ individuals are cleared of suspicion, and it is as if individual k was chosen at random from the remaining $N - k + 2$, containing 1 murderer and $N - k + 1$ innocents, and found to exhibit C. The reasoning of section 11.3.1 then yields

$$pr(G_0 \mid C_S, C_0, K = k) = 1/\{1 + (N + 1 - k)P\}. \tag{4.1}$$

Clearly, for $k > 1$ this exceeds (2.1).

Suppose now that K is unknown, but that the above strategy is known to have operated to select the suspect. We want $pr(G_0 \mid C_S, C_0)$. This could be found by taking the expectation of (4.1) with respect to the distribution of K given C_S, again yielding a value exceeding (2.1). However, it is easier to argue directly as follows:

$$pr(G_0 \text{ and } K = k \mid C_S, C_0) = pr(G_k \text{ and } K = k \mid C_S)$$
$$= pr(G_k \mid C_S) \times pr(K = k \mid C_S, G_k).$$

The first term is $1/(N + 1)$, by symmetry, and the second is $(1 - P)^{k-1}$. Hence

$$pr(G_0 \text{ and } K = k \mid C_S, C_0) = (1 - P)^{k-1}/(N + 1). \tag{4.2}$$

Summing over k from 1 to $N + 1$ we obtain

THE ISLAND PROBLEM 167

$$pr(G_0 \mid C_S, C_0) = \{1 - (1 - P)^{N+1}\}/(N + 1)P. \quad (4.3)$$

Note that (4.3) now agrees with Yellin's formula (2.6) (yielding 0.824 for $N = 100$, $P = 0.004$), rather than the Bayes formula (2.1) (yielding 0.714). To appreciate why this should be, we can again reason as in section 11.3.2. As before, we have $M \mid C_S \sim 1 + \text{Bin}(N; P)$. Previously, the further evidence C_0 changed this distribution of M, introducing the factor $m/(N + 1)$ to reflect the fact that, for a given suspect, C_0 is more likely the larger is M. But in the current context we know that, whatever the value of M, the selected suspect will exhibit C. Hence the evidence C_0 is uninformative about M, which thus still has the distribution $1 + \text{Bin}(N; P)$ used in deriving Yellin's formula.

An interesting variant of the above analysis arises if we assume that, instead of distinct innocent islanders initially exhibiting C independently, the actual number m of those exhibiting C is known (although not their identities). Then the search is equally likely to pick up any of the $(m + 1)$ islanders with C, so that the probability that the suspect is the murderer is $1/(m + 1)$, whatever the extent of search required to identify him. Since $P = m/N$ is the probability that an innocent exhibits C, Bayes' formula (2.1) does apply in this case. The presence or absence of the 'size-bias' terms $m/(N + 1)$, which affected previous answers, can have no effect when M is not random.

11.4.2 Lindley's example

Similarly, Lindley's example shows how the impact of evidence is affected by the manner of its disclosure: the finding C_1 that a random individual other than the suspect possesses C is not the same as the finding C^* that $M \geq 2$—which information might, for example, be obtained by searching through the population until one discovers another individual possessing C. This search interpretation of C^* offers another explanation as to why (3.9) exceeds (3.8), since (unless one is lucky enough to find this other individual immediately) the process of search will eliminate some of the population from suspicion, equivalent to reducing N in (3.8), thus increasing its value—compare the argument following (4.1).

A full search scenario for Lindley's example would involve searching until an individual exhibiting C is found, who is then accused; and then continuing the search until a further such individual is found. We have

$$pr(G_0 \text{ and } K = k \text{ and } C^* \mid C_S, C_0)$$
$$= pr(G_0 \text{ and } K = k \mid C_S, C_0) pr(C^* \mid G_0, K = k, C_S, C_0).$$

The first term is given by (4.2), and the second is $1 - (1 - P)^{N+1-k}$, so that

$$pr(G_0, C^*, K = k \mid C_S, C_0) = \{(1 - P)^{k-1} - (1 - P)^N\}/(N + 1), \quad (4.4)$$

whence

$$pr(G_0, C^* \mid C_S, C_0) = \{1 - (1 - P)^{N+1} - (N + 1)P(1 - P)^N\}/(N + 1)P. \quad (4.5)$$

Also,

$$pr(\bar{G}_0, C^* \mid C_S, C_0) = pr(\bar{G}_0 \mid C_S, C_0) = [(N + 1)P - \{1 - (1 - P)^{N+1}\}]/(N + 1)P, \quad (4.6)$$

from (4.3), so that, when the suspect is identified by a search,

$$pr(C^* \mid C_S, C_0) = \{1 - (1 - P)^N\}, \quad (4.7)$$

which also follows as $pr(M \geq 2 \mid C_S, C_0)$ on noting that, by the argument below (4.3), the distribution of M given (C_S, C_0) is $1 + \text{Bin}(N; P)$.

From (4.5) and (4.7) we obtain

$$pr(G_0 \mid C_S, C_0, C^*) = \frac{1 - (1 - P)^{N+1} - (N + 1)P(1 - P)^N}{(N + 1)P\{1 - (1 - P)^N\}} \quad (4.8)$$

as the appropriate probability in this extended search scenario. For $N = 100$, $P = 0.004$ we get 0.467. Of course, a full search of all the population would identify the total number m of individuals exhibiting C, and yield $1/m$ as the relevant probability that the suspect is guilty.

11.5 THE PRESUMPTION OF INNOCENCE

Suppose that a random individual is selected, found to have C, and put on trial. We argued in section 11.2 that it is appropriate to take as prior probability $\pi = pr(G_0) = 1/(N + 1)$. But from the point of view of the judge and jury, this is surely incorrect. Prosecutions are not brought lightly, and the very fact that a suspect is on trial suggests, rationally, that the police must have some evidence against him, so that $\pi > 1/(N + 1)$, perhaps by a large factor. Applying Bayes' formula (1.1) would lead to a correspondingly higher posterior probability of guilt.

The legal doctrine of 'the presumption of innocence' seems to run counter to this, and suggests that judge and jury should, at the start of the trial, continue to treat the suspect just the same as anyone else, thus employing $\pi = 1/(N + 1)$, as already treated at length. Is this in conflict with a fully rational (coherent) analysis?

Suppose that all parties are aware that a prosecution will be brought against an individual only if he is found to possess characteristic C. The very fact that the prosecution is brought is thus equivalent to the evidence C_0, so that the 'prior' probability of guilt, as assessed by judge and jury, is $1/(1 + NP)$, as given by (2.1). But having thus accounted for C_0 in establishing the prior, it would clearly constitute inadmissible double counting to use this evidence once again when it is eventually revealed in court. Indeed, under the known selection process, $pr(C_0 \mid \bar{G}_0, C_S) = 1$, rather than P as supposed in (A3), while we have seen that $pr(G_0 \mid C_S) = 1/(1 + NP)$ rather than $1/(N + 1)$ in (A1). The likelihood ratio induced by the evidence C_0 is thus unity, and the posterior probability of guilt remains $1/(1 + NP)$—exactly as if the selection process had not been taken into account at all. This is a special case of a more general phenomenon: when all the evidence affecting the decision to prosecute is to be presented in court, the selection process can be entirely ignored in calculating the posterior probability of guilt: it affects both prior and likelihood in ways which totally cancel out. A similar point was made by Dawid (1976) in the context of medical diagnosis. In this sense the presumption of innocence is fully in accord with Bayesian reasoning: before any evidence is adduced, we should treat the accused as exchangeable with all other members of the population, and take the evidence into account on the same basis.

Of course, in real cases some of the evidence governing a decision to prosecute will not be admissible in court, or will not, for one reason or another, be brought before the court. One could argue that it is not then fully rational to ignore the fact of prosecution. However, other criteria than rationality, such as fairness and the avoidance of prejudice, govern the conduct of cases at law. If evidence is not brought explicitly, it should not be taken into account in any way. Once again, a Bayesian analysis of the evidence actually presented, ignoring entirely the selection of cases for prosecution, corresponds to the desired process of law.

In particular, the above argues for the use of the Bayes formula (2.1) even in cases where, for example, the suspect has been identified by a search of the population—which evidence would generally be inadmissible, although it affects the probability of guilt. As we have seen, a fully rational analysis of such a process would increase the posterior probability of guilt above (2.1), so that Bayes' theorem errs on the side of caution.

Similarly, the impact of evidence such as 'Smith is left-handed' has been seen to depend on how that evidence is established. We have three formulae, (3.8), (3.9) and (4.8). We know (3.8) < (3.9), and it appears that (3.9) < (4.8), although this is less easy to establish. So, to err on the side of caution, it would seem appropriate to use formula (3.8). In general, choosing any individual and discovering him to exhibit C will favour larger values of M, and thus produce a smaller probability of guilt, than searching for such an individual. Hence, to err on the side of caution, it is advisable to ignore any search process, and assume any individual examined (be he the suspect or anyone else) has been selected at random.

11.6 CONCLUSION

Coherent analysis of identification evidence can sometimes proceed by the use of Bayes's formula. This is appropriate when the evidence is obtained on randomly selected individuals, but not when it is found as the result of a search procedure. However, use of Bayes' formula errs on the side of caution, and is recommended in general as consonant with the presumption of innocence.

REFERENCES

Balding, D. J. and Donnelly, P. J. (1994) Inference in forensic identification, *J. Roy. Statist. Soc*, **A** (to appear).
Dawid, A. P. (1976) Properties of diagnostic data distributions, *Biometrics*, **32**, 647–58.
Eggleston, R. (1983) *Evidence, Proof and Probability*, 2nd edn, London, Weidenfeld and Nicolson.
Lindley, D. V. (1987) The probability approach to the treatment of uncertainty in artificial intelligence and expert systems, *Statistical Science*, **2**, 17–24.
People v. *Collins* (1968) *California Reporter*, **66**, 242–53.
Yellin, J. (1979) Book review of *Evidence, Proof and Probability*, 1st edn, *J. Economic Literature* (Pittsburgh), 583.

Department of Statistical Science
University College London
Gower Street
LONDON
WC1E 6BT
UK

CHAPTER 12

Utility: Probability's Younger Twin?

Simon French
University of Leeds

12.1 INTRODUCTION

The Bayesian paradigm[†] for decision analysis is based upon the twin concepts of subjective probability and subjective utility, which are combined through subjective[‡] expected utility (SEU) to support decision-making. Twin concepts: but are probability and utility twins of equal stature? Specifically, is probability, historically the senior twin in years and the first to emerge in most axiomatizations of SEU, a more fundamental concept?

Bayesian statisticians view probability as more fundamental for two reasons. The first, as easy to dismiss as it is to understand, is psychological. Bayesian statistics is primarily concerned with modelling and reporting uncertainty. Probability dominates in their modelling to the extent that utility often enters subliminally, if at all. It is hard to perceive one concept as being as fundamental as another if the first is hardly used as a tool while the other permeates all analyses. The second reason does give one pause for thought, however. The measurement of utility, and the encoding of that measurement within the axiomatizations of SEU, requires probability (see e.g. Savage 1972; DeGroot 1970; Lindley 1985). Thus, goes the argument, probability must be the more fundamental.

Dennis Lindley and I have discussed this on several occasions (e.g. Lindley 1990, 1992). This chapter is not written to refute his views, but rather to record the thoughts that his remarks have catalysed. It matters

---

[†] I should emphasize, I guess, that I mean the Bayesian normative paradigm used to underpin prescriptive analyses (Bell *et al.* 1988; French 1992).
[‡] Henceforward, the word 'subjective' will be dropped.

Aspects of Uncertainty edited by P. R. Freeman and A. F. M. Smith. © 1994 John Wiley & Sons Ltd.

not a penny whether probability is more fundamental than utility. What matters is that we understand both concepts so that we use them as correctly and as fully as possible. As usual, his thought-provoking words have helped me to a better understanding of the foundations of the Bayesian paradigm: whether it is a correct or useful understanding I leave for him and others to judge.

12.2 TWO FAMILIES OF AXIOMATIZATIONS

Two distinct families may be discerned among the many axiomatizations of probability, utility and SEU. The defining property of the first is that the axiomatizations seek to define and encode an external view of rational choice behaviour of an individual. On the other hand, the second family seek to help an individual organize his or her judgements of belief and preference in such a way that rational choice behaviour can be constructed.

Savage's axiomatization (Savage 1972) of SEU is typical of those that begin from an external view of choice behaviour. Influences stemming from models of *rational economic man* within economics are apparent. Savage begins by assuming that a decision-maker faces a choice between strategies which are modelled as functions from a state space to a consequence space:

strategy: state → consequence.

The state space comprises the range of uncertain futures beyond her (certain) control and the consequence space comprises the possible outcomes that may result from the interaction between a strategy and a possible future. Savage assumes that the decision-maker has (or would like

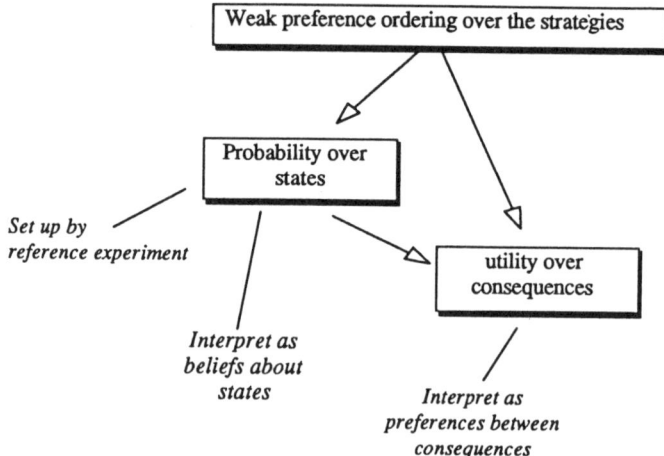

Figure 12.1 SEU as a model of rational choice behaviour.

to have?) a weak preference ordering over the strategies, between which she must choose. Moreover, he axiomatizes consistency properties which she would wish her choice between strategies to obey in order that they be considered rational choices. From these axioms and the assumption that the strategy space is sufficiently rich, namely that it has a reference experiment embedded in it, he deduces that her weak preference ordering of the strategies can be modelled by expected utilities. In constructing the SEU model, he first deduces a probability distribution over the state space and then a utility function over the consequence space. In Savage's approach, the probability distribution is *interpreted* as modelling the decision-maker's beliefs about the relative likelihood of the possible states and the utility function is *interpreted* as modelling her preferences between consequences (see Figure 12.1).

Thus Savage's justification of the Bayesian approach to decision analysis is roughly as follows:

1. An answer to the question: what does rational choice behaviour look like?
2. A demonstration that such rational behaviour can be represented by SEU models built from component probabilities and utilities.
3. The implication that if a decision-maker wants to *appear* to behave rationally, then combining probabilities and utilities through SEU will achieve this.

In contrast to Savage's and similar rational choice modelling approaches to axiomatizations of the SEU models, there are constructivist approaches as exemplified by, e.g. DeGroot (1970) or French (1986). Whereas rational behaviour approaches observe and model the choices from a viewpoint *outside* the decision-maker, constructivist views build the SEU from the decision-maker's viewpoint, i.e. from *inside* the decision-maker. They assume that the decision-maker can introspect her beliefs about states and her preferences between consequences. Constructivist axiomatizations encode consistency properties that she would like these beliefs and preferences to exhibit and, furthermore, suggest how her preferences between strategies should be constructed. From the axioms and the assumption of the existence of a reference experiment, they demonstrate that her beliefs about states can be modelled by a probability distribution, that her preferences between consequences can be modelled by an utility function and that these can be combined through SEU to provide a rational construction of her preferences between strategies.

Thus a constructivist justification of the Bayesian approach to decision analysis is roughly as follows:

1. Answers to the questions: what do rational beliefs about states and what do rational preferences between consequences look like?

2. A deduction of the representation of these beliefs and preferences by probabilities and utilities respectively.
3. The implication that, if a decision-maker wants to *construct* her preferences between strategies rationally, then combining probabilities and utilities through SEU will achieve this.

I have a strong preference for constructivist approaches because, both as a decision-maker or as an analyst helping others, I need methods that help sort out a person's *thinking* not their external behaviour. Analysts should help their clients' gain understanding so that through that understanding they can make better substantiated inferences and take more informed decisions. Analysts need to work with their client's beliefs and preferences not their choices. Most of the following, therefore, will be cast in a constructivist framework, although much applies to both.

12.3 MODELLING DECISION SITUATIONS

Within the Bayesian framework, one should realize that neither probability nor utility are the most fundamental concepts. Nor, for that matter, are beliefs, preferences or choices. None of our approaches or analyses could be applied if we did not model decisions by assuming that the decision-maker must choose between strategies, the consequence of which will be

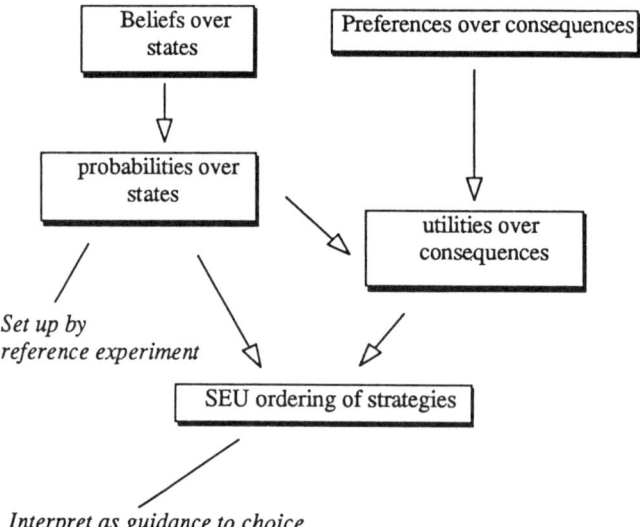

Figure 12.2 Constructivist view of SEU.

determined through the interaction of the chosen strategy and future events beyond the decision-maker's control. Savage modelled this by

strategy: state → consequence.

Others have used decision tables, decision trees and influence diagrams. The model has been extended to allow that the state space be conditional on the strategies. But underpinning all representations used in Bayesian analyses is the separation of the situation into three component parts: strategies, states and consequences. (Of course, there is a plethora of terminology which sometimes hides this common form of representation.)

If the model of the situation is separated into these three components, one is inevitably led to a model of decision-making in which there is some representation of uncertainty (probability, belief function or whatever), some representation of preference between outcomes (utility function or whatever) and a mechanism for bringing them together to guide choice (expected utility or whatever). One can argue whether the representations of belief and preference can be completely independent (Kadane and Winkler 1988; Kahneman and Tversky 1979; Tversky and Kahneman 1988), but some sort of separation is implicit in Anglo-American approaches to statistics and decision. Indeed, this separation within the modelling of the underlying situation is so fundamental to our approaches that it is seldom if ever discussed. The collection of readings edited by Bell *et al.* (1988), for instance, is remarkably silent on the subject. Yet in recent years many other approaches to modelling decision situations have emerged: e.g. some of the soft systems methodologies (Rosenhead 1989). It is tempting to dismiss these as muddled; but perhaps we should be alarmed that users do not.

However, to return to the point of discussion and to draw a dubious metaphor: probability and utility are 'scales on the axes' within the model, the small world which we construct to represent a decision situation. Other representations of decision situations, in some sense, use rotations and transformations of those axes to provide a different 'coordinate system'. Our claim that the Bayesian paradigm is best is equivalent to claiming that we are using 'principal axes'. A claim that probability is more fundamental than utility is a claim that the former is the 'first principal axis'.

12.4 THE REFERENCE EXPERIMENT

So far in this chapter I have conspicuously avoided the question that it seeks to answer: is probability more fundamental than utility? The kernel of the answer to this lies, I believe, in the reference experiment.

In essence, in both the rational choice behaviour and the constructivist developments of the Bayesian paradigm the utility of a consequence, x, is measured or constructed by assuming that the decision-maker can choose between options of the form:

Option A: x for certain;
Option B: x_{best} if some event E in a reference experiment occurs;
    x_{worst} if its complement E^c occurs.

Here x_{best} and x_{worst} are, respectively, the best and worst possible consequence in the decision problem and the reference experiment refers, for example, to the drawing from a urn containing two different colours of ball in some proportion or the spin of a probability wheel (Lindley 1985; French 1986). Savage (1972) and DeGroot (1970) express the reference experiment in terms of uniform partitions of increasing fineness: but to all who see their axiomatizations it is clear what they and others are doing through the reference experiment. They are introducing into the analysis uncertainties, the *probabilities* of which the decision-maker feels she knows or is happy to assign. Thus utility can only be defined and assessed once probability has been. Ergo, probability is the more fundamental concept. But perhaps that is too naïve a conclusion.

The reference experiment is not part of the underlying decision situation. Its introduction provides an *Archimedean axiom* (Krantz et al. 1971), a scale against which to assess the decision-maker's preferences between the consequences. Moreover, the same scale is used to assess her beliefs about states in the decision problem. The assessment of her probabilities requires her to choose between options of the form:

Option C: x_{best} if some event A in the state space occurs;
    x_{worst} if its complement A^c occurs;
Option D: x_{best} if some event E in a reference experiment occurs;
    x_{worst} if its complement E^c occurs.

In other words, the decision-maker's probability distribution over the state space and her utility function over the space of consequences have precisely the same status. Her probability distribution can be constructed before her utility function as in DeGroot (1970); or *vice versa* as in Ferguson (1967) or French (1986).

The events in the reference experiment correspond to outcomes of a mind experiment, whereas events in the state space correspond to possible future happenings in the real world. It was partly for this reason that in French (1982) I provided an axiomatization in which the decision-maker might construct a σ-field over the reference experiment, but a much less rich structure over the state space. Recently Sarin and Wakker (1992) have gone further, classing events in the reference experiment as *unambiguous*, whereas events in the state space might be *ambiguous* or *unambiguous*. By making this distinction, they seek to emphasize that some uncertain events such as the outcome of a throw of a die or whether it rains tomorrow are essentially simple, whereas others such as the successful conclusion of a pharmaceutical R & D programme are much more complex. The judgements required of the decision-makers for these complex, ambiguous events are assumed to obey a restricted of axioms compared with those of unambiguous events. In consequence, their development provides a full SEU model for strategies, the consequences of which depend only on unambiguous events. When ambiguous events are involved, a weaker representation is obtained. I cite Sarin and Wakker's work not because it convinces me: it does not—I'll stick with the standard Bayesian paradigm. But their work does emphasize the distinction between probabilities over the reference experiment and probabilities over the state space.

Thus Figure 12.2 should be revised to show that *both* subjective probabilities over the state space and utilities over the consequences are based upon the reference experiment. See Figure 12.3.

In fact, it is possible to derive the SEU model without recourse to a reference experiment. Wakker (1989) has derived the SEU representation basing his development on additive value models. In this the Archimedean axiom does not involve an explicit reference experiment, and probabilities emerge after utilities as normalized weights. But it should be noted that the judgements his theory requires of the decision-makers are complex and do not separate uncertainties from preferences. His deviation is closer to modelling a rational economic man than a constructivist approach in the sense used here.

It is also interesting that assessing and developing utility functions, Archimedean axioms other than that inherent in introducing the reference experiment can be important (although strictly unnecessary). Value difference or multi-attribute value functions can be developed without any

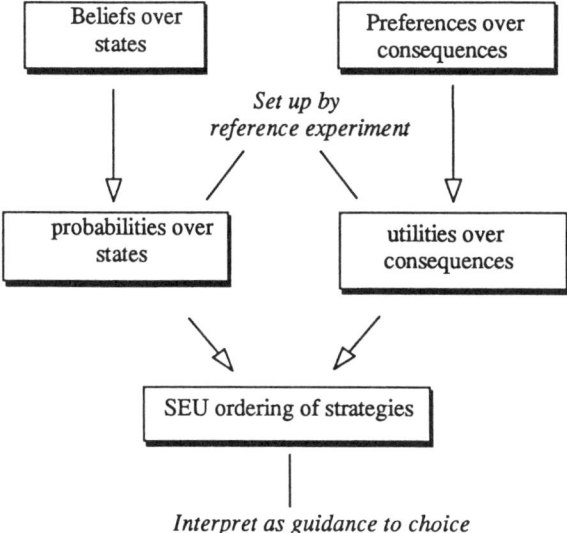

Figure 12.3 Constructivist view of the SEU (revised).

need for choices involving uncertainty. Only when risk attitude needs to be modelled in preparation for the construction of expected utilities, is it necessary to introduce the reference experiment (French 1986).

12.5 THE IMPORTANCE OF BEING BOTH BELIEF AND PREFERENCE ANALYSTS

So it seems there are no grounds for assuming probability (over the state space) to be a more fundamental concept than utility. But that, of course, does not mean that it is not the more important in a particular analysis. On particular occasions, being a belief analyst may be a more important and necessary skill than being a preference analyst. Many Bayesian statisticians use utility in their analyses only very occasionally. Their work concerns inference without the focus of a decision. They report posterior distributions and avoid offering estimates or conducting hypothesis tests for which they would need utility (or loss) functions: see, for example, Box and Taio (1973). Such approaches to inference can be justified simply by following a constructivist approach to the modelling of belief without any need to mention, much less model preference. One can point to the work of Jeffreys (1961) and, possibly, De Finetti (1974, 1975) here. Within their axiomatic theories probability is fundamental: indeed, utility plays no formal role.

However, their avoidance of the use of utility may be illusory. Presumably they must avoid all aspects of experimental design (Verdinelli 1992). More importantly, they offer too narrow a view of Bayesian statistics in that they cannot offer useful support to many of the major problems facing society (Lindley 1992).

Generally I believe that we must be both belief and preference analysts: see, for example, Keeney (1992) and Lindley (1992). The main reason I find our paradigm so valuable is that it allows the analyst to help decision-makers to think their way through problems by considering their beliefs and preferences separately. Attention can switch between focusing on complex uncertainties and on complex value judgements, and back again, without either confusing the other. The Bayesian paradigm does model decision situations with two 'principal axes'. If we ignore either of those axes, we will end up with a one-dimensional paradigm that another can and will displace.

12.6 CONCLUDING REMARKS

With the exception of the remarks in the previous section, this chapter does perhaps beg the question: so what? Does any of the above argue for a change in the way we practise Bayesian analyses? Probably not: but I do believe it has implications for the way we market our approach. There are many decision-makers and scientists who want and need the help and insight that well-structured analyses can bring. Only if we are clear on the foundations of our approach can we assure them that Bayesian analyses do bring that insight by explaining how and why they do. Moreover, one only has to read some of the vast psychological and behavioural science literature to realize that many of those who need help find the majority of our rigorous axiomatic justifications obscure and unconvincing. In my experience, many are more convinced by a justification that explicitly separates and exhibits a reference experiment, articulated perhaps as a probability wheel, and uses this as a 'ruler' or 'scale' to measure off their uncertainties and preferences. Furthermore, they instinctively feel that their concerns to understand and come to terms with their preferences for the risks and conflicts between sub-objectives before them is every bit as important as their need to deal with uncertainties. According utility and probability equal status matches their instinct and helps persuade them of the good sense of our paradigm.

REFERENCES

Bell, D. E., Raiffa, H. and Keeney, R. L. (eds) (1988) *Decision Making: Descriptive, Normative and Prescriptive Interactions*, Cambridge University Press.
Box, G. E. P. and Taio, G. C. (1973) *Bayesian Inference in Statistical Analysis*. Reading, Mass., Addison Wesley.
De Finetti, B. (1974) *Theory of Probability*, volume 1, Chichester, John Wiley & Sons.
De Finetti, B. (1975) *Theory of Probability*, volume 2, Chichester, John Wiley & Sons.
DeGroot, M. H. (1970) *Optimal Statistical Decisions*, New York, McGraw-Hill.
Ferguson, T. S. (1967) *Mathematical Statistics: A Decision Theoretic Approach*. New York, Academic Press.
French, S. (1982) 'On the axiomatisation of subjective probability', *Theory and Decision*, **14**, 19–33.
French, S. (1986) *Decision Theory: An Introduction to the Mathematics of Rationality*. Chichester, Ellis Horwood.
French, S. (1992) 'Uncertainty and imprecision: modelling and analysis' (paper presented at the 4th International Meeting of Statistics in the Basque Country). School of Computer Studies, University of Leeds.
Jeffreys, H. (1961) *Theory of Probability*, 3rd edn, Oxford University Press.
Kadane, J. B. and Winkler, R. L. (1988) 'Separating probability elicitation from utilities', *J. Amer. Statist. Assoc.*, **83**, 357–63.
Kahneman, D. and Tversky, A. (1979) 'Prospect theory: an analysis of decision under risk', *Econometrica*, **47**, 263–91.

Keeney, R. L. (1992) *Value Focused Thinking: A Path to Creative Decision Making*, Cambridge, Mass., Harvard University Press.

Krantz, D. H., Luce, R. D., Suppes, P. and Tversky, A. (1971) *Foundations of Measurement*, volume 1, New York, Academic Press.

Lindley, D. V. (1985) *Making Decisions*, 2nd edn, Chichester, John Wiley & Sons (1st edn, 1971).

Lindley, D. V. (1990) '1988 Wald Memorial Lectures: the present position in Bayesian statistics (with discussion)', *Statist. Sci.*, **5**, 44–89.

Lindley, D. V. (1992) 'Is our view of Bayesian statistics too narrow? (with discussion)' in *Bayesian Statistics 4*, J. M. Bernardo, J. O. Berger, A. P. Dawid and A. F. M. Smith, eds, Oxford University Press, pp. 1–15.

Rosenhead, J. (ed.) (1989) *Rational Analysis for a Problematic World*, Chichester, John Wiley & Sons.

Sarin, R. K. and Wakker, P. (1992) 'A simple axiomatisation of non-additive expected utility', *Econometrica*, **60**, 1255–72.

Savage, L. J. (1972) *Foundations of Statistics*, 2nd edn, New York, Dover (1st edn, New York, John Wiley & Sons, 1954).

Tversky, A. and Kahneman, D. (1988) 'Rational choice and the framing of decisions' in Bell *et al.* (1988), pp. 167–92.

Verdinelli, I. (1992) 'Advances in Bayesian experimental design', in *Bayesian Statistics 4*, J. M. Bernardo, J. O. Berger, A. P. Dawid and A. F. M. Smith, eds, Oxford University Press, pp. 467–81.

Wakker, P. (1989) *Additive Representations of Preferences: a New Foundation of Decision Analysis*, Dordrecht, Kluwer.

School of Computer Studies
The University
LEEDS
LS2 9JT
UK

CHAPTER 13

Fully Bayesian Hierarchical Analysis for Exponential Families via Monte Carlo Computation

E. I. George,[†] U. E. Makov,[‡] and A. F. M. Smith[§]
[†]*University of Texas at Austin*
[‡]*University of Haifa*
[§]*Imperial College London*

13.1 MOTIVATION

This chapter discusses the implementation of Bayesian hierarchical analyses for exponential family models using Monte Carlo methods to perform the necessary calculations. These kinds of analyses facilitate the simultaneous estimation of several parameters of the same type. For example, in actuarial risk assesment, the parameters being the mean losses of several policy holders, in agricultural trials involving many varieties, the parameters being the varietal means, etc. The statistical analysis is aimed at combining the information from the various sources of data (policy holders, varieties, etc.) thus exploiting the relationship between the parameters.

The idea of 'pulling in' records from related sources goes back to the beginning of this century. (For details, see Kempthorne 1971.) From a methodological perspective, we note in particular Stein's estimation rule (Stein 1955), and the parametric empirical Bayes solutions (Efron and Morris 1973, 1975; Morris 1983; Kass and Steffey 1989) which followed the

---

Aspects of Uncertainty edited by P. R. Freeman and A. F. M. Smith. © 1994 John Wiley & Sons Ltd.

non-parametric empirical Bayes of Robbins (1955, 1964). Early expositions of the hierarchical Bayesian approach were given by Lindley (1971), who considered a one-way random effects model with Gaussian disturbances, and Lindley and Smith (1972), who considered a three-stage hierarchical analysis for the normal-linear model. (For an historical background to hierarchical Bayesian models, see Good 1980.) Deely and Lindley (1981) gave a critical assessment of parametric empirical Bayes methods from a Bayesian point of view.

We shall consider the following situation. The members of a sequence of random variables $\{(x_i, \theta_i): i = 1, \ldots, p\}$ are identically distributed. The density of x_i given θ_i is $f(x_i | \theta_i)$. Each θ_i has a density $\pi(\theta_i | \lambda)$ indexed by a hyperparameter λ whose density is given by $\pi(\lambda)$. The hierarchical model assumed here consists of three stages:

Stage 1. Conditionally on $\{\theta_i\}$ and λ, the data $\{x_i\}$ are independent; x_i having density $f(x_i | \theta_i)$ independent of λ and all other θ's other than θ_i.

Stage 2. Conditionally on λ, the parameters $\{\theta_i\}$ are iid with density $\pi(\theta | \lambda)$.

Stage 3. The hyperparameter λ has a density $\pi(\lambda)$.

Having observed $x^{(p)} = (x_1, \ldots, x_p)$, inference about θ_i is based on the posterior distribution

$$\pi(\theta_i | x^{(p)}) = \int \pi(\theta_i | x_i, \lambda) \pi(\lambda | x^{(p)}) d\lambda. \tag{1}$$

The first term in the integrand is the posterior density of θ_i given λ and the data. Here only x_i, which is directly associated with θ_i, is used. (This is proved by Deely and Lindley 1981.) The second term is the posterior density of the hyperparameter given the entire data, which can be written

$$\pi(\lambda | x^{(p)}) = \frac{\left(\prod_{i=1}^{k} f(x_i | \lambda)\right) h(\lambda)}{\int \left(\prod_{i=1}^{k} f(x_i | \lambda)\right) h(\lambda) d\lambda} \tag{2}$$

where $f(x_i | \lambda)$ is the mixed distribution

$$f(x_i | \lambda) = \int f(x_i | \theta_i) \pi(\theta_i | \lambda) d\theta_i. \tag{3}$$

Our focus in this chapter is on the case where stage 1 is confined to the exponential family

$$f(x | \theta) = a(x) \exp[x\theta - \psi(\theta)] \quad \text{with} \quad \psi(\theta) = \ln \int a(x) \exp(x\theta) dx \tag{4}$$

and when, in the second stage, the prior is conjugate, given by

$$\pi(\theta \mid \lambda) = \exp[\lambda_1 \theta - \lambda_2 \psi(\theta)] k(\lambda_1, \lambda_2) \qquad (5)$$

with

$$k(\lambda_1, \lambda_2)^{-1} = \int \exp[\lambda_1 \theta - \lambda_2 \psi(\theta)] d\theta \quad \text{and} \quad \lambda = (\lambda_1, \lambda_2)$$

(see Diaconis and Ylvisaker 1979, for the choice of hyperparameters which renders (5) a proper density).

Under these assumptions (see Deely and Lindley 1981, for details), the posterior distribution of θ_i is given by

$$\pi(\theta_i \mid x^{(p)}) = \exp[x_i \theta_i - \psi(\theta_i)] N/D \qquad (6)$$

where

$$N = \iint \exp[\lambda_1 \theta_i - \lambda_2 \psi(\theta_i)] k(\lambda_1 + x_i, \lambda_2 + 1) g(\lambda_1, \lambda_2, x^{(p)}) d\lambda_1 d\lambda_2$$

$$D = \iint g(\lambda_1, \lambda_2, x^{(p)}) d\lambda_1 d\lambda_2$$

$$g(\lambda_1, \lambda_2, x^{(p)}) = \prod_{i=1}^{p} \frac{k(\lambda_1, \lambda_2)}{k(\lambda_1 + x_i, \lambda_2 + 1)} \pi(\lambda_1, \lambda_2),$$

and is typically difficult to evaluate. However, as will be shown below, the evaluation of (6) can be straightforward using Monte Carlo methods.

The difficulties in implementing the Bayesian model have given rise to various approximation strategies. We note the use of analytic methods (Lindley 1980; Kass, et al. 1988; Tierney and Kadane 1986) and the use of numerical methods (Smith et al. 1985; Smith, et al. 1987) all of which are computationally prohibitive when the number of parameters and their dimensions become very large and, in any case, seem to require too much sophisticated insight to be suitable for routine use by practitioners. Parametric empirical Bayes techniques, see Morris (1983, 1986), offered a convenient methodology for the applied practitioner, but not without a cost. At the root of the parametric empirical Bayes methodology lies the approximation

$$\hat{\pi}(\theta_i \mid x^{(p)}) = \pi(\theta_i \mid x^{(p)}, \hat{\lambda}(x^{(p)})), \qquad (7)$$

where $\hat{\lambda}(x^{(p)})$ is the maximum likelihood estimate of λ based on $x^{(p)}$, or something similar. It is 'as if' the data are used to estimate λ and this estimate replaces the true value in the Bayes calculations. Using this approach and assuming a squared error loss function, $E(\theta_i \mid x^{(p)})$ is approximated by $E(\theta_i \mid x_i, \hat{\lambda}(x^{(p)}))$ as an estimator of θ_i. While this approximation is asymptotically acceptable, for moderate sample sizes it fails to take into account the uncertainty about $\hat{\lambda}$, which enters into (1) through the posterior density $\pi(\lambda \mid x^{(p)})$. Consequently the approximation allows at best the assessment of the moments of θ_i (although we note that

$V(\theta_i | x_i, \hat{\lambda}(x^{(p)}))$ underestimates the posterior variance) and precludes the full range of inferential procedures possible in Bayesian analysis. Kass and Steffey (1989) obtained a first-order approximation to the posterior variance and suggested a second-order approximation for assessing the accuracy of the first-order approximation. As the approximation in (7) is only justified asymptotically, the attraction of parametric empirical Bayes methods is diminished for small p.

We shall now introduce three examples in which the difficulty in implementing the Bayesian methodology is demonstrated, and discuss some of the approximations suggested in the literature. These examples will be referred to in the next section where we discuss alternative implementations based on Monte Carlo methods.

Example 1 Given θ_i ($i = 1, \ldots, p$), the data x_i ($i = 1, \ldots, p$) are independent and normally distributed $N(\theta_i, 1)$ so that

$$f(x_i | \theta_i) = (2\pi)^{-1/2} \exp\{-\tfrac{1}{2}(x_i - \theta_i)^2\}, \quad -\infty < x_i < \infty. \tag{8}$$

Conditional on $\lambda = (\mu, \sigma)$, the θ_i's are iid normal $N(\mu, \sigma^2)$, namely

$$f(\theta_i | \mu, \sigma) = (2\pi)^{-1/2} \exp\{-\tfrac{1}{2}(\theta_i - \mu)^2/\sigma^2\}, \quad -\infty < \theta_i < \infty. \tag{9}$$

As is well known, $\theta_i | x_i, \mu, \sigma$ is normally distributed

$$N\left(\frac{\mu + \sigma^2 x_i}{1 + \sigma^2}, \frac{\sigma^2}{1 + \sigma^2}\right).$$

The marginal distribution (3), here $f(x_i | \mu, \sigma)$, is also normal $N(\mu, 1 + \sigma^2)$.

Many authors have considered this one-way random effects model from a Bayesian perspective. A seminal contribution from a methodological point of view was Lindley's discussion of Stein (1962), with its introduction of what has come to be called the Lindley–Stein estimator. Another important contribution, but focusing more on inference for variance components than for mean parameters, is that of Hill (1965). Detailed contributions can also be found in Box and Tiao (1968). Perhaps the most influential paper was Lindley (1971), foreshadowed in Lindley (1969) and subsequently extending into Lindley and Smith (1972).

However, Lindley (1971) frankly acknowledges the computational difficulties of full Bayesian implementation and opts, instead, for posterior summaries in the form of modes. Recognition that the conditional distributional forms arising from the modelling hierarchy lend themselves to an iterative modal estimation strategy is made more explicit in Lindley and Smith (1972), although the algorithm at that time was not referred to as 'iterated conditional modes'.

Estimation of a more explicitly empirical Bayes flavour appears in Efron and Morris (1973, 1975), further refined in Morris (1983, 1986).

FULLY BAYESIAN HIERARCHICAL ANALYSIS 185

Example 2 Given θ_i ($i = 1, \ldots, p$), the data x_i ($i = 1, \ldots, p$) are independent and Poisson distributed Po($\theta_i t_i$) so that

$$f(x_i \mid \theta_i) = \frac{(\theta_i t_i)^{x_i} \exp(-\theta_i t_i)}{x_i!}, \qquad x_i = 0, 1, \ldots. \tag{10}$$

Conditional on α and β, the θ_i's are iid with a gamma density Gam(α, β), namely

$$f(\theta_i \mid \alpha, \beta) = \frac{\beta^\alpha}{\Gamma(\alpha)} \theta_i^{\alpha - 1} \exp(-\beta \theta_i), \qquad \alpha > 0, \beta > 0, \theta_i > 0. \tag{11}$$

Clearly, $\theta_i \mid x_i, \alpha, \beta$ is distributed Gam($\alpha + x_i, \beta + t_i$). Here, the marginal distribution (3) follows the gamma-Poisson (Raiffa and Schlaifer 1961), presenting formidable difficulties for the evaluation of (1).

Leonard (1976), (for the case $t_i \equiv 1$), assumed that $\ln \beta$ is uniformly distributed and suggested that the α in the Bayes estimate

$$\hat{\theta}_i = \frac{\bar{x}}{\alpha + \bar{x}} x_i + \frac{\alpha}{\alpha + \bar{x}} \bar{x} \tag{12}$$

should be integrated out when α is unknown. This would require the use of numerical integration. Clevenson and Zidek (1975) suggested a minimax, generalized Bayes estimate which shrinks x_i towards zero. Maritz and Lwin (1989) suggested an empirical Bayes point estimator which is based on the estimation of the parameters of a gamma prior from the mixed negative binomial distribution. Gaver and O'Muircheartaigh (1987) obtained an empirical Bayes estimate under the assumption that the prior was either a gamma distribution or a log-Student distribution. Wang and Van Ryzin (1979) applied discrete density smoothing to the empirical Bayes estimation of θ_i.

Example 3 Given θ_i ($i = 1, \ldots, p$), the data x_i ($i = 1, \ldots, p$) are independent and binomially distributed Bin(n_i, θ_i), so that

$$f(x_i \mid \theta_i) = \binom{n_i}{x_i} \theta_i^{x_i} (1 - \theta_i)^{n_i - x_i}, \qquad x_i = 0, \ldots, n_i. \tag{13}$$

Conditional on α and β, the θ_i's are iid with a beta density Be(α, β), namely

$$f(\theta_i \mid \alpha, \beta) = \frac{\Gamma(\alpha + \beta)}{\Gamma(\alpha)\Gamma(\beta)} \theta_i^{\alpha - 1} (1 - \theta_i)^{\beta - 1}, \qquad \alpha > 0, \beta > 0, 0 \leq \theta_i \leq 1. \tag{14}$$

Clearly, $\theta_i \mid x_i, \alpha, \beta$ is distributed Be($\alpha + x_i, \beta + n_i - x_i$). The marginal distribution (3) follows the beta-binomial model (Raiffa and Schlaifer 1972)

$$f(x_i \mid \alpha, \beta) = \binom{n_i}{x_i} \frac{\Gamma(\alpha + \beta)}{\Gamma(\alpha)\Gamma(\beta)} \frac{\Gamma(\alpha + x_i)\Gamma(\beta + n_i - x_i)}{\Gamma(\alpha + \beta + n_i)}. \tag{15}$$

Even if uniform priors are assumed for α and β, substituting (15) into (2) will result in an intractable expression. This led Leonard (1972) to reparametrize the first stage in terms of the log odds γ_i, which is given by $\gamma_i = \ln[(\theta_i)/(1 - \theta_i)]$, and to proceed by assuming that the γ_i's are normally distributed with mean μ and variance σ^2. In the second stage, the mean was assumed to be uniformly distributed over the real line, and $\eta\tau/\sigma^2$ was assumed independent of μ having a chi-squared distribution on η degrees of freedom, where η and τ were specified. This was justified as follows: 'At the first stage, a normal distribution is chosen for γ_i for reasons of technical convenience. Use of the conjugate beta distribution for θ_i would present difficulties in obtaining tractable results not depending on the first-stage parameters. This intractability is caused by the factorials in the proportionality constant.'

Maritz and Lwin (1989) suggested an empirical Bayes point estimator in which the unknown hyperparameters α and β, are replaced by their estimates, derived from (15) either by the method of moments or maximum likelihood.

In section 13.2, we show how the marginal posterior (1) may be calculated by Monte Carlo methods. In section 13.3, these calculations are illustrated on a variety of examples which have been analysed in the literature by partially Bayes approaches.

13.2 MONTE CARLO EVALUATION OF THE POSTERIOR

As mentioned earlier, direct analytic evaluation or even accurate numerical approximation of the posterior

$$\pi(\theta_i \mid x^{(p)}) = \int \pi(\theta_i \mid x_i, \lambda) h(\lambda \mid x^{(p)}) d\lambda$$

in (1) can be prohibitively difficult. Sometimes fruitful alternatives for calculating this posterior are obtained using Monte Carlo methods.

13.2.1 Using a simulation of $h(\lambda \mid x^{(p)})$

A simple Monte Carlo approximation of (1) can sometimes be obtained by simulating a sample

$$\lambda^1, \ldots, \lambda^M \quad \text{where} \quad \lambda^m \sim h(\lambda \mid x^{(p)}), \quad m = 1, \ldots, M \quad (16)$$

from the marginal distribution of the hyperparameters and using

$$\hat{\pi}(\theta_i \mid x^{(p)}) = \frac{1}{M} \sum_{m=1}^{M} \pi(\theta_i \mid x_i, \lambda^m). \quad (17)$$

Ideally, $\lambda^1, \ldots, \lambda^M$ will be an iid sample, although any ergodic sequence which converges in distribution to $h(\lambda \mid x^{(p)})$ will still ensure that

$$\hat{\pi}(\theta_i \mid x^{(\mathrm{p})}) \to \pi(\theta_i \mid x^{(\mathrm{p})}) \qquad (18)$$

as $M \to \infty$. In situations where it is difficult to simulate directly from $h(\lambda \mid x^{(\mathrm{p})})$, the Gibbs sampler (see, for example, Gelfand and Smith 1990, Casella and George 1992 or Smith and Roberts 1993) is a useful alternative which generates $\lambda^1, \ldots, \lambda^M$ as a convergent Markov chain. The practical advantage of the Gibbs sampler is that a multivariate distribution can be simulated as a sequence of univariate distribution simulations.

To calculate characteristics of $\pi(\theta_i \mid x^{(\mathrm{p})})$ such as the posterior mean $E(\theta_i \mid x^{(\mathrm{p})})$, one could use $\hat{\pi}(\theta_i \mid x^{(\mathrm{p})})$ directly, or one could again use a simulation approach. For example, one might simulate a sample from $\pi(\theta_i \mid x^{(\mathrm{p})})$ based on the sample in (16):

$$\theta^1, \ldots, \theta^M \quad \text{where} \quad \theta^m \sim \pi(\theta_i \mid x_i, \lambda^m). \qquad (19)$$

The posterior mean could then be estimated by

$$\hat{E}(\theta_i \mid x^{(\mathrm{p})}) = \frac{1}{M} \sum_{m=1}^{M} \theta_i^m. \qquad (20)$$

Again, only the ergodicity of $\lambda^1, \ldots, \lambda^M$ is required for the consistency of this estimator.

A crucial aspect of this approach is the difficulty or even feasibility of simulating a sample from $\pi(\lambda \mid x^{(\mathrm{p})})$. We now proceed to show how this might be done in the normal–normal set-up of Example 1, and in the gamma–Poisson set-up of Example 2.

Example 1 (continued) For the normal–normal set-up where $\lambda = (\mu, \sigma)$, a commonly used prior is the normal–gamma conjugate prior where

$$\pi(\mu) = N(\xi, \tau^2),$$
$$\pi(\sigma^2) = IG(\nu/2, \nu\lambda/2) \text{(i.e. } \nu\lambda/\sigma^2 \sim \chi_\nu^2),$$
$$\pi(\mu, \sigma^2) = \pi(\mu)\pi(\sigma^2).$$

For this prior, the Gibbs sampler provides a convenient method for sampling from the marginal posterior $h(\lambda \mid x^{(\mathrm{p})}) = \pi(\mu, \sigma \mid x^{(\mathrm{p})})$. The sequence (16), which is here of the form

$$\mu^1, \sigma^1, \ldots, \mu^M, \sigma^M, \qquad (21)$$

is obtained with the Gibbs sampler by alternately simulating from

$$\mu^m \sim \pi(\mu \mid \sigma^{m-1}, x^{(\mathrm{p})})$$

and

$$\sigma^m \sim \pi(\sigma \mid \mu^m, x^{(\mathrm{p})}),$$

after a suitable initialization. For the normal–gamma prior above, these conditional distributions are

$$\pi(\mu \mid \sigma, x^{(p)}) = N\left(\frac{(1+\sigma^2)\xi + p\tau^2\bar{x}}{1+\sigma^2+p\tau^2}, \frac{(1+\sigma^2)p\tau^2}{1+\sigma^2+p\tau^2}\right) \quad (24)$$

where $\bar{x} = (1/p)\sum x_i$, and

$$\sigma^2 \sim \pi(\sigma^2 \mid \mu, x^{(p)}) = IG\left(\frac{p+\nu}{2}, \frac{\sum(x_i-\mu)^2 + \nu\lambda}{2}\right). \quad (25)$$

These can be easily simulated by standard methods.

Example 2 (continued) For the gamma–Poisson set-up where $\lambda = (\alpha, \beta)$, the sequence (16) is of the form

$$\alpha^1, \beta^1, \ldots, \alpha^M, \beta^M. \quad (26)$$

In many cases, this sequence may be obtained with the Gibbs sampler by alternately simulating from

$$\alpha^m \sim \pi(\alpha \mid \beta^{m-1}, x^{(p)}) \quad (27)$$

and

$$\beta^m \sim \pi(\beta \mid \alpha^m, x^{(p)}), \quad (28)$$

after a suitable initialization. Note that if β or α is treated as known (i.e. a point prior is used for β or α), then an iid univariate simulation of either (27) or (28) will suffice. We now consider when (27) and (28) can be simulated fast and efficiently.

To begin with, it is straightforward to show that the posterior of α in (26) is of the form

$$\pi(\alpha \mid \beta, x^{(p)}) \propto L^G(\alpha \mid \beta, x^{(p)})\pi(\alpha \mid \beta), \quad (29)$$

where

$$L^G(\alpha \mid \beta, x^{(p)}) \propto \left(\frac{\beta}{\beta+1}\right)^{p\alpha} \prod_{i=1}^{p} \frac{\Gamma(\alpha+x_i)}{\Gamma(\alpha)}, \quad (30)$$

and the posterior of β in (27) is of the form

$$\pi(\beta \mid \alpha, x^{(p)}) \propto L^G(\beta \mid \alpha, x^{(p)})\pi(\beta \mid \alpha), \quad (31)$$

where

$$L^G(\beta \mid \alpha, x^{(p)}) \propto \left(\frac{\beta}{\beta+1}\right)^{p\alpha} \prod_{i=1}^{p} (\beta+1)^{-x_i}. \quad (32)$$

Here $L^G(\alpha \mid \beta, x^{(p)})$ is log-concave since

$$\frac{\partial^2 \ln(L(\alpha \mid \beta, x^{(p)}))}{\partial \alpha^2} = \sum_{i=1}^{p} \{[\psi'(\alpha+x_i) - \psi'(\alpha)]\} < 0, \quad (33)$$

a consequence of the fact that the trigamma function ψ' is decreasing, see Abramowitz and Stegun (1964). Thus $\pi(\alpha \mid \beta, x^{(p)})$ will be log-concave whenever $\pi(\alpha \mid \beta)$ is log-concave. In such cases, $\pi(\alpha \mid \beta, x^{(p)})$ can be simulated easily using fast adaptive rejection methods such as the ARS algorithm of Gilks and Wild (1992).

To simulate from $\pi(\beta \mid \alpha, x^{(p)})$, it may be more familiar to first simulate $\gamma = \beta/(\beta + 1)$ and then transform back. As a function of γ, (32) becomes

$$L^G(\gamma \mid \alpha, x^{(p)}) \propto \gamma^{p\alpha}(1-\gamma)^{\Sigma_{i=1}^p x_i}, \tag{34}$$

a familiar beta form. Thus, there are many priors $\pi(\gamma \mid \alpha)$ under which $\pi(\gamma \mid \alpha, x^{(p)})$ can be simulated by standard methods. For example, this will be the case when $\pi(\gamma \mid \alpha)$ and hence $\pi(\gamma \mid \alpha, x^{(p)})$ are beta distributions.

13.2.2 Simulating the full posterior via the Gibbs sampler

An alternative to using a direct simulation of $\pi(\lambda \mid x^{(p)})$ as described in section 13.2.1, is to use the Gibbs sampler to simulate the full posterior distribution $\pi(\Theta, \lambda \mid x^{(p)})$ where $\Theta = (\theta_1, \ldots, \theta_p)$ and $\lambda = (\lambda_1, \ldots, \lambda_q)$. This approach entails simulating a sequence

$$\Theta^1, \lambda^1, \ldots, \Theta^M, \lambda^M, \tag{35}$$

where after a suitable 'burn-in sequence', (35) is obtained by alternately simulating $\Theta^m = (\theta_1^m, \ldots, \theta_p^m)$ from

$$\theta_i^m \sim \pi(\theta_i \mid \theta_{-i}^m, \lambda^{m-1}, x^{(p)}), \quad i = 1, \ldots, p, \tag{36}$$

where

$$\theta_{-i}^m = (\theta_1^m, \ldots, \theta_{i-1}^m, \theta_{i+1}^{m-1}, \ldots, \theta_p^{m-1}),$$

and simulating $\lambda^m = (\lambda_1^m, \ldots, \lambda_q^m)$ from

$$\lambda_i^m \sim \pi(\lambda_i \mid \theta^m, \lambda_{-i}^m, x^{(p)}), \quad i = 1, \ldots, q, \tag{37}$$

where

$$\lambda_{-i}^m = (\lambda_1^m, \ldots, \lambda_{i-1}^m, \lambda_{i+1}^{m-1}, \ldots, \lambda_q^{m-1}).$$

Under reasonably weak conditions (see, for example, Roberts and Smith 1993) the sequence (35) is a homogeneous Markov chain which converges in distribution to $\pi(\Theta, \lambda \mid x^{(p)})$.

Several features make the Gibbs sampler particularly attractive here. First, our general hierarchical setting allows for immediate simplification of both

$$\pi(\theta_i \mid \theta_{-i}^m, \lambda^{m-1}, x^{(p)}) \quad \text{and} \quad \pi(\lambda_i \mid \theta^m, \lambda_{-i}^m, x^{(p)}).$$

Conditionally on λ and $x^{(p)}$, the components of Θ are independent so that (36) becomes

$$\pi(\theta_i \mid \theta_{-i}^m, \lambda^{m-1}, x^{(p)}) = \pi(\theta_i \mid \lambda^{m-1}, x_i). \tag{38}$$

Furthermore, conditionally on θ, λ is independent of $x^{(p)}$, so that (37) becomes

$$\pi(\lambda_i \mid \theta^m, \lambda_{-i}^m, x^{(p)}) = \pi(\lambda_i \mid \theta^m, \lambda_{-i}^m). \tag{39}$$

The form of (38) and (39) is often particularly convenient for simulation when $f(x \mid \theta)$ belongs to the exponential family and $\pi(\theta \mid \lambda)$ is conjugate as in (4) and (5). In this case, the posterior of each θ_i in (38) belongs to the same conjugate family, and is of the form

$$\pi(\theta \mid \lambda, x_i) = \exp\{(\lambda_1 + x_i)\theta - (\lambda_2 + 1)\psi(\theta)\} k(\lambda_1 + x_i, \lambda_2 \div 1). \tag{40}$$

Conjugate distributions for exponential families are often standard distributions which can be simulated by fast and effcient methods. In any case, such distributions are always log-concave, and so can be efficiently simulated by the ARS algorithm of Gilks and Wild (1992).

The posterior of λ in (39) will be of the form

$$\pi(\lambda \mid \Theta) \propto L(\lambda \mid \Theta)\pi(\lambda) \tag{41}$$

where

$$L(\lambda \mid \Theta) = \exp\left\{\lambda_1 \left(\sum \theta_i\right) - \lambda_2 \left(\sum \psi(\theta_i)\right)\right\} k(\lambda_1, \lambda_2). \tag{42}$$

The function $L(\lambda \mid \Theta)$ is called a conjugate likelihood distribution by George et al. (1993) who show that $L(\lambda \mid \Theta)$ is log-concave. Thus, when $\pi(\lambda)$ is log-concave, $\pi(\lambda \mid \Theta)$ will also be log-concave, and hence can be simulated by ARS. George et al. (1993) also show that when $p \geq 2$, $L(\lambda \mid \Theta)$ is integrable, i.e. $\int L(\lambda \mid \Theta) d\lambda < \infty$. Thus, it may also be possible to use Gibbs sampling to evaluate the formal Bayes posterior $\pi(\lambda \mid \Theta)$ under the improper prior $\pi(\lambda) \equiv 1$. However, we must warn the reader that when working with improper priors, having proper conditional distributions

$$\pi(\Theta \mid \lambda, x^{(p)}) \quad \text{and} \quad \pi(\lambda \mid \Theta, x^{(p)})$$

does not guarantee that $\pi(\Theta, \lambda \mid x^{(p)})$ will be proper. In such a case, the sequence (35) will not converge.

We now detail this complete Gibbs sampling approach for Examples 2 and 3.

Example 2 (continued) The gamma–Poisson set-up entails $f(x_i \mid \theta_i) = \text{Po}(\theta_i t_i)$ at the first stage of the hierarchy, and $\pi(\theta_i \mid \alpha, \beta) = \text{Gam}(\alpha, \beta)$ at the second stage. Here the posterior (40) is also a gamma distribution,

$$\pi(\theta_i \mid x_i, \alpha, \beta) = \text{Gam}(\alpha + x_i, \beta + t_i), \tag{43}$$

which can be simulated by standard methods. Because the hyperparameters (α, β) are linear in the natural parameter λ of the gamma distribution $\text{Gam}(\alpha, \beta)$, the conjugate likelihood

$$L^G(\alpha, \beta \mid \Theta) = \left(\frac{\beta^\alpha}{\Gamma(\alpha)}\right)^p \left(\prod_{i=1}^k \theta_i\right)^\alpha \exp\left(-\beta \sum_{i=1}^p \theta_i\right), \qquad (44)$$

is log-concave in (α, β), and is integrable for $p \geq 2$. It follows that both of the conditional conjugate likelihoods

$$L^G(\alpha \mid \Theta, \beta) \propto \left(\frac{\beta^\alpha}{\Gamma(\alpha)}\right)^p \left(\prod_{i=1}^k \theta_i\right)^\alpha \qquad (45)$$

and

$$L^G(\beta \mid \Theta, \alpha) \propto \beta^{p\alpha} \exp\left(-\beta \sum_{i=1}^p \theta_i\right) \qquad (46)$$

are log-concave and integrable for $p \geq 2$. Note that $L^G(\beta \mid \Theta, \alpha)$ is a gamma distribution, $\text{Gam}(p\alpha + 1, \Sigma_{i=1}^p \theta_i)$. Thus, the conditional posteriors

$$\pi(\alpha \mid \Theta, \beta) \propto L^G(\alpha \mid \Theta, \beta)\pi(\alpha, \beta) \qquad (47)$$

and

$$\pi(\beta \mid \Theta, \alpha) \propto L^G(\beta \mid \Theta, \alpha)\pi(\alpha, \beta) \qquad (48)$$

will be log-concave when $\pi(\alpha, \beta)$ is log-concave, and will be integrable even if $\pi(\alpha, \beta) \equiv 1$ (when $p \geq 2$).

Example 3 (continued) The beta–binomial setup entails $f(x_i \mid \theta_i) = \text{Bin}(n_i, \theta_i)$ at the first stage of the hierarchy, and $\pi(\theta_i \mid \alpha, \beta) = \text{Be}(\alpha, \beta)$ at the second stage. As is well known, the conditional posterior (40) will also be a beta distribution, namely

$$\pi(\theta_i \mid x_i, \alpha, \beta) \sim \text{Be}(\alpha + x_i, \beta + n_i - x_i), \qquad (49)$$

which can be simulated by standard methods. Because the hyperparameters (α, β) are linear in the natural parameter λ of the beta distribution $\text{Be}(\alpha, \beta)$, the conjugate likelihood

$$L^B(\alpha, \beta \mid \Theta) = \left(\frac{\Gamma(\alpha + \beta)}{\Gamma(\alpha)\Gamma(\beta)}\right)^p \left(\prod_{i=1}^p \theta_i\right)^\alpha \left(\prod_{i=1}^p (1 - \theta_i)\right)^\beta, \qquad (50)$$

is log-concave in (α, β), and is integrable for $p \geq 2$. It follows that both of the conditional conjugate likelihoods

$$L^B(\alpha \mid \Theta, \beta) \propto \left(\frac{\Gamma(\alpha + \beta)}{\Gamma(\alpha)}\right)^p \left(\prod_{i=1}^p \theta_i\right)^\alpha \qquad (51)$$

and

$$L^B(\beta \mid \Theta, \alpha) \propto \left(\frac{\Gamma(\alpha + \beta)}{\Gamma(\beta)}\right)^p \left(\prod_{i=1}^p (1 - \theta_i)\right)^\beta \qquad (52)$$

are log-concave and integrable. Thus, the conditional posteriors

$$\pi(\alpha \mid \Theta, \beta) \propto L^B(\alpha \mid \Theta, \beta)\pi(\alpha, \beta) \qquad (53)$$

and

$$\pi(\beta \mid \Theta, \alpha) \propto L^B(\beta \mid \Theta, \alpha)\pi(\alpha, \beta) \qquad (54)$$

will be log-concave when $\pi(\alpha, \beta)$ is log-concave, and will be integrable even if $\pi(\alpha, \beta) \equiv 1$ (when $p \geq 2$).

A reparametrization of the hyperparameters which we found useful in this example, was the rotation

$$\gamma_1 = \alpha + \beta \quad \text{and} \quad \gamma_2 = \alpha - \beta. \qquad (55)$$

Because (γ_1, γ_2) are also linear in λ, the conditional conjugate likelihoods

$$L^B(\gamma_1 \mid \Theta, \gamma_2) \propto \left(\frac{\Gamma(\gamma_1)}{\Gamma((\gamma_1 + \gamma_2)/2)\Gamma((\gamma_1 - \gamma_2)/2)}\right)^p \left(\prod_{i=1}^p \theta_i\right)^{\gamma_1/2} \left(\prod_{i=1}^p (1 - \theta_i)\right)^{\gamma_1/2} \qquad (56)$$

and

$$L^B(\gamma_2 \mid \Theta, \gamma_1) \propto \left(\frac{\Gamma(\gamma_1)}{\Gamma((\gamma_1 + \gamma_2)/2)\Gamma((\gamma_1 - \gamma_2)/2)}\right)^p \left(\prod_{i=1}^p \theta_i\right)^{\gamma_2/2} \left(\prod_{i=1}^p (1 - \theta_i)\right)^{\gamma_2/2} \qquad (57)$$

are also log-concave and integrable for $p \geq 2$. Here, too, the conditional posteriors

$$\pi(\gamma_1 \mid \Theta, \gamma_2) \propto L^B(\gamma_1 \mid \Theta, \gamma_2)\pi(\alpha, \beta) \qquad (58)$$

and

$$\pi(\gamma_2 \mid \Theta, \gamma_1) \propto L^B(\gamma_2 \mid \Theta, \gamma_1)\pi(\alpha, \beta) \qquad (59)$$

will be log-concave when $\pi(\alpha, \beta)$ is log-concave, and will be integrable even if $\pi(\alpha, \beta) \equiv 1$ (when $p \geq 2$).

This parametrization yielded faster convergence of the Gibbs sampler in our examples in section 13.3 because there was less posterior correlation in (γ_1, γ_2) than in (α, β). Note that when the posterior variances of α and β are equal, (γ_1, γ_2) will be uncorrelated.

13.3 APPLICATIONS OF FULLY BAYESIAN HIERARCHICAL ANALYSIS

In this section, we use the Monte Carlo methods outlined in section 13.2 to evaluate the posterior distribution in hierarchical analyses. For each of the three hierarchical set-ups, the normal–normal, the gamma–Poisson model and the beta–binomial, we analyse a data set which has been previously analysed in the literature by only partially Bayesian methods. For each data set we apply one of the implementations set out in the previous section to obtain fully Bayesian posterior mean estimates. Our results are compared with previous analyses.

13.3.1 A normal–normal application

For the normal–normal model, we revisited the well-known baseball data analysed by Efron and Morris (1975). These data, displayed in Table 13.1 consist of $p = 18$ early season batting averages $p_1, \ldots, p_{18}$ of baseball players in 1970. Efron and Morris apply the variance-stabilising transformation $x_i = \sqrt{45} \arcsin(2p_i - 1)$ to yield approximately normal data with unit variance $x_i \mid \theta_i \sim N(\theta_i, 1)$, where θ_i is the transformed value of the 'true' average. They model the θ_i as $\theta_1, \ldots, \theta_{18}$ conditionally iid $N(\mu, \sigma^2)$. Based on an empirical Bayes argument, the θ_i are then estimated by the Lindley–Stein estimator which shrinks towards the grand mean. These estimates are used to forecast end-of-season averages (a proxy for the 'true' batting averages).

Table 13.1 Baseball data.

Player	Early average	Final average	Bayes estimates	$E\text{–}M$ estimates
1	0.400	0.346	0.366	0.290
2	0.378	0.298	0.349	0.286
3	0.356	0.276	0.333	0.281
4	0.333	0.222	0.316	0.277
5	0.311	0.273	0.299	0.273
6	0.311	0.270	0.299	0.273
7	0.289	0.263	0.282	0.268
8	0.267	0.210	0.266	0.264
9	0.244	0.269	0.248	0.259
10	0.244	0.230	0.248	0.259
11	0.222	0.264	0.231	0.254
12	0.222	0.256	0.231	0.254
13	0.222	0.303	0.231	0.254
14	0.222	0.264	0.231	0.254
15	0.222	0.226	0.231	0.254
16	0.200	0.285	0.214	0.249
17	0.178	0.316	0.197	0.244
18	0.156	0.200	0.179	0.239

Following Example 1, we applied the normal-gamma prior

$$\pi(\mu) = N(\xi, \tau^2), \quad \pi(\sigma^2) = (IG(\nu/2, \nu\lambda/2) \quad \text{and} \quad \pi(\mu, \sigma^2) = \pi(\mu)\pi(\sigma^2)$$

with $\xi = -3.51$ (the transformed value of 0.250), $\tau^2 = 100$, $\nu = 0$ and $\lambda = 1$. (These choices represented the prior opinion of one of us.) Using the method described in section 13.2.1, the posterior means of the 18 players were estimated as follows. A sequence of $M = 20\,000$ hyperparameter values μ^m, σ^m was generated by Gibbs sampling based on (24) and (25). As in (19) and (20), each mean $E(\theta_i | x^{(p)})$ was then estimated by $(1/M)\sum_1^M \theta_i^m$ where $\theta^1, \ldots, \theta^M$ were simulated from $\theta^m \sim \pi(\theta_i | x_i, \mu^m, \sigma^m)$. The retransformed values of these estimates are given in Table 13.1 along with the Efron-Morris estimates. Note how our fully Bayes estimates are shrunk much less than the Lindley-Stein estimator.

13.3.2 A gamma-Poisson application

For the gamma-Poisson set-up we revisited the oil discovery data of Clevenson and Zidek (1975). These data, displayed in Table 13.2, consist of $p = 36$ monthly observations of the number of oil discoveries in Alberta obtained from wildcat exploration. Clevenson and Zidek model each observation x_i as Poisson with mean θ_i, the 'expected number of discoveries that month'. They then consider a variety of shrinkage estimators, some of them partially Bayes for a gamma model on the means. Leonard (1976) also analysed this data, applying an approximation of (12).

Table 13.2 Oilwell discovery data.

Data values	0	1	2	3	5
No. months value observed	19	10	4	2	1
Bayes estimates	0.48	0.89	1.29	1.70	2.52
Clevenson-Zidek estimates	0	0.45	0.89	1.34	2.23
Leonard estimates	0.50	0.88	1.26	1.64	2.41

Using the gamma-Poisson set-up of Example 2 setting $t_i \equiv 1$ in (10), we modelled the θ_i's as an iid sample from (11), namely

$$f(\theta | \alpha, \beta) = \frac{\beta^\alpha}{\Gamma(\alpha)} \theta^{\alpha-1} \exp(-\beta\theta).$$

We applied the (improper) prior $\pi(\alpha, \beta) = \exp\{-\alpha\}/\beta$, and used the method of section 13.2.1 to estimate the posterior mean of θ_i for each of the observed data values. A sequence of $M = 20\,000$ hyperparameter values α^m, β^m was generated by Gibbs sampling based on (29) and (31). The ARS algorithm of Gilks and Wild (1992) was used to simulate α from (29). To simulate β from (31), we simulated $\gamma = \beta/(\beta + 1)$ from the beta distribution obtained by coupling (34) with the induced prior $\pi(\gamma) \propto (\gamma(1-\gamma))^{-1}$.

As in (19) and (20), each mean $E(\theta_i | x^{(p)})$ was then estimated by $(1/M)\sum_1^M \theta_i^m$ where $\theta^1, \ldots, \theta^M$ were simulated from $\theta^m \sim \pi(\theta_i | x_i, \alpha^m, \beta^m)$. These fully Bayes estimates are displayed in Table 13.2 along with those of Clevenson and Zidek, and Leonard. Clearly, our fully Bayes estimates are much more in agreement with those of Leonard. Particularly striking is the disparity of estimates when $x_i = 0$.

We note that the prior used above is log-concave so that the complete Gibbs approach of section 13.2.2 could have also been used. We preferred the approach of section 13.2.1 because it is faster and more efficient. For an illustration of the complete Gibbs approach for a gamma–Poisson model see George et al. (1993).

13.3.3 A beta–binomial application

For the beta–binomial set-up we revisited the toxoplasmosis data analysed by Kass and Steffey (1989), a subset of the toxoplasmosis data analysed by Efron and Morris (1975). These data, displayed in Table 13.3, consist of observations of n_i = number people tested and x_i = number people tested positive for toxoplasmosis in each of $p = 10$ cities in El Salvador. Kass and Steffey applied the beta–binomial set-up of Example 3 to this data with the 'non-informative' (improper) prior $\pi(\alpha, \beta) \propto (\alpha + \beta)^{-2}$ on $\alpha, \beta \geq 0$. This prior is uniform on the parametrization $(\alpha/(\alpha + \beta), \log(\alpha + \beta))$. They obtained posterior mean estimates using a Laplace approximation

Table 13.3 Toxoplasmosis data.

	Data									
Obs no.	1	2	3	4	5	6	7	8	9	10
n_i	51	16	82	13	43	75	13	10	6	37
x_i	24	7	46	9	23	53	8	3	1	23

	Θ estimates									
Prior	θ_1	θ_2	θ_3	θ_4	θ_5	θ_6	θ_7	θ_8	θ_9	θ_{10}
$\exp(-\alpha - \beta)$	0.475	0.456	0.559	0.647	0.534	0.696	0.589	0.370	0.319	0.610
$(\alpha + \beta)^{-2}$	0.533	0.539	0.562	0.581	0.554	0.617	0.568	0.526	0.524	0.579
MLE	0.471	0.438	0.561	0.692	0.535	0.707	0.615	0.300	0.167	0.622
Kass–Steffey	0.531	0.535	0.562	0.584	0.553	0.625	0.570	0.520	0.518	0.580

	α, β estimates	
Prior	α	β
$\exp(-\alpha - \beta)$	2.6	2.3
$(\alpha + \beta)^{-2}$	2016.1	1566.3

Using the same beta–binomial model, we considered the proper prior $\pi(\alpha, \beta) \propto \exp\{-(\alpha + \beta)\}$ and $\pi(\alpha, \beta) \propto (\alpha + \beta)^{-2}$ on $\alpha, \beta \geq 1$ for comparison with Kass and Steffey. (We imposed the minor restriction $\alpha, \beta \geq 1$ to obtain log-concavity to facilitate the simulation below.) Note that under the prior

$$\pi(\alpha, \beta) \propto \exp\{-(\alpha + \beta)\}, \quad E(\alpha) = E(\beta) = 1$$

whereas under $\pi(\alpha, \beta) \propto (\alpha + \beta)^{-2}$ on $\alpha, \beta \geq 1$, $E(\alpha)$ and $E(\beta)$ do not exist.

For each prior, the posterior was evaluated using the complete Gibbs sampling approach described in section 13.2.2. The beta posterior for θ_i in (49) was simulated by standard methods. The posterior distributions of α and β under all the above priors are log-concave, and so were simulated by ARS. We used the rotated parametrization $\gamma_1 = \alpha + \beta$ and $\gamma_2 = \alpha - \beta$ in (55) because the Gibbs sequence output appeared more stable, suggesting improved convergence. In particular, the autocorrelations of the α and β output sequences were much smaller when the rotation was used.

Each mean $E(\theta_i | x^{(p)})$ was then estimated by $(1/M)\sum_1^M \theta_i^m$, where $\theta^1, \ldots, \theta^M$ was the output sequence obtained when simulating $\theta^m \sim \pi(\theta_i | x_i, \alpha^m, \beta^m)$ in (49). These fully Bayes estimates for each of our priors are given in Table 13.3 along with the MLE's and the Kass–Steffey estimates. The posterior means are shrunk much less under $\exp\{-(\alpha + \beta)\}$ than under $\pi(\alpha, \beta) \propto (\alpha + \beta)^{-2}$. This is not surprising since small values of α and β correspond to weaker prior information. That smaller values of α and β obtain from the first prior is illustrated by the posterior means of α and β which we have also displayed in Table 13.3. Note also that the shrinkage obtained by these priors is heterogeneous in that far more shrinkage is obtained from observations with small n_i. Finally, note the close agreement of our estimates under $\pi(\alpha, \beta) \propto (\alpha + \beta)^{-2}$ with those of Kass and Steffey.

REFERENCES

Abramowitz, M. and Stegun, I. A. (ed.) (1964) *Handbook of Mathematical Functions*, National Bureau of Standards, *Applied Mathematics Series No. 55*, Washington, DC, US Government Printing Office.

Box, G. E. P. and Tiao, G. C. (1968) Bayesian estimation of means for the random-effects model, *J. Amer. Statist. Soc.*, **63**, 174–81.

Casella, G. and George, E. I. (1992) Explaining the Gibbs sampler, *The American Statistician*, **46**, 167–74.

Clevenson, M. L. and Zidek, J. V. (1975) Simultaneous estimation of the mean of independent Poisson laws, *J. Amer. Statist. Soc.*, **70**, 698–705.

Deely, J. J. and Lindley, D. V. (1981) Bayes empirical Bayes, *J. Amer. Statist. Soc.*, **76**, 833–41.

Diaconis, P. and Ylvisaker, D. (1979) Conjugate priors for exponential families, *Annals of Statistics*, **7**, 269–81.

Efron, B. and Morris, C. (1973) Stein's estimator and its competitors—an empirical Bayes approach, *J. Amer. Statist. Soc.*, **68**, 117–30.

Efron, B. and Morris, C. (1975) Data analysis using Stein's estimator and its generalizations, *J. Amer. Statist. Soc.*, **70**, 311–19.

Gaver, D. P. and O'Muircheartaigh, I. G. (1987) Robust empirical Bayes analysis of event rates, *Technometrics*, **29**, 1–15.

Gelfand, A. E. and Smith, A. F. M. (1990) Sampling based approaches to calculating marginal densities, *J. Amer. Statist. Assoc.*, **85**, 398–409.

George, E. I., Makov, U. E. and Smith, A. F. M. (1993) Conjugate likelihood distributions, *Scand. J. Statist.*, **20**, 147–56.

Gilks, W. R. and Wild, P. (1992) Adaptive rejection sampling for Gibbs sampling, *J. R. Statist. Soc.*, **C41**, 337–48.

Good, I. J. (1980) Some history of the hierarchical Bayesian methodology, in *Bayesian Statistics*, J. M. Bernardo, M. H. DeGroot, D. V. Lindley and A. F. M. Smith, eds, Valencia, Spain, University Press, pp. 489–519.

Hill, B. M. (1965) Inference about variance components in the one-way model, *J. Amer. Statist. Soc.*, **60**, 806–25.

Kass, R. E. and Steffey, D. (1989). Approximate Bayesian inference in conditionally independent hierarchical models (parametric empirical Bayes), *J. Amer. Statist. Soc.*, **84**, 717–26.

Kass, R. E., Tierney, L. and Kadane, J. B. (1988) Asymptotics in Bayesian computation, in *Bayesian Statistics 3*, J. M. Bernardo, M. H. DeGroot, D. V. Lindley and A. F. M. Smith, eds, Oxford, UK, Oxford University Press, pp. 261–78.

Kempthorne, O. (1971) Comments on D. V. Lindley (1971). The estimation of many parameters (with discussion), in *Foundations of Statistical Inference*, V. P. Godambe and D. A. Sprott, eds, Toronto, Holt, Rinehart and Winston, pp 435–55.

Leonard, T. (1972) Bayesian methods for binomial data, *Biometrika*, **59**, 581–9.

Leonard, T. (1976) Some alternative approaches to multiparameter estimation, *Biometrika*, **63**, 69–75.

Lindley, D. V. (1969) Bayesian least squares, *Bull. Int. Statist. Inst.*, **43**(2), 152.

Lindley, D. V. (1971). The estimation of many parameters (with discussion), in *Foundations of Statistical Inference*, V. P. Godambe and D. A. Sprott, eds, Toronto, Holt, Rinehart and Winston, pp. 435–55.

Lindley, D. V. (1980) Approximate Bayesian methods, in *Bayesian Statistics*, J. M. Bernardo, M. H. DeGroot, D. V. Lindley and A. F. M. Smith, eds, Valencia, Spain, University Press, pp. 489–519.

Lindley, D. V. and Smith, A. F. M. (1972) Bayes estimates for the linear model (with discussion), *J. R. Statist. Soc.*, **B34**, 1–41.

Maritz, J. S. and Lwin, T. (1989) *Empirical Bayes Methods*, 2nd edn, London, New York, Chapman and Hall.

Morris, C. (1983) Parametric empirical Bayes inference: theory and applications (with discussion), *J. Am. Statist. Assoc.*, **78**, 47–65.

Morris, C. (1986) Empirical Bayes: a frequency/Bayes compromise, In: *Adaptive Statistical Procedures and Related Topics*, ed. J. Van Ryzin, IMS Lecture Notes, pp. 195–203.

Raiffa, H. and Schlaifer, R. (1972) *Applied Statistical Decision Theory*, 2nd edn, Massachusetts and London, the MIT Press.

Robbins, H. (1955) An empirical Bayes approach to statistics, *Proc. Third Berkeley Symposium on Math. Statist. and Prob.*, University of California Press, **1**, 157–64.

Robbins, H. (1964) An empirical Bayes approach to statistical problems, *Ann. Math. Statist.*, **35**(1), 157–64.

Roberts, G. O. and Smith, A. F. M. (1993) Simple conditions for the convergence of the Gibbs sampler and Hastings–Metropolis algorithms, *Stoch. Proc. and their Applic.* (to appear).

Smith, A. F. M., Skene, A. M., Shaw, J. E. H., Naylor, J. C. and Dransfield, M. (1985) The implementation of the Bayesian paradigm, *Communication in Statistics, Theory and Methods*, **14**, 1079–1102.

Smith, A. F. M., Skene, A. M., Shaw, J. E. H. and Naylor, J. C. (1987) Progress with numerical and graphical methods for Bayesian statistics, *The Statistician*, **36**, 75–82.

Smith, A. F. M. and Roberts, G. O. (1993) Bayesian computations via the Gibbs sampler and related Markov chain Monte Carlo methods (with discussion), *J. R. Statist. Soc.*, **B55**, 3–24.

Stein, C. M. (1955) Inadmissibility of the usual estimator for the mean of a multivariate normal distribution, *Proc. Third Berkeley Symposium on Math Statist and Prob.*, vol 1, University of California Press, pp. 197–207.

Stein, C. M. (1962) Confidence sets for the mean of a multivariate normal distribution, *J. R. Statist. Soc.*, **B24**, 265–96.

Tierney, L. and Kadane, J. (1986) Accurate approximations for posterior moments and marginal densities, *J. Amer. Statist. Soc.*, **81**, 82–6.

Wang, M. C. and Van Ryzin, J. (1979) Discrete density smoothing applied to the empirical Bayes estimation of a Poisson mean, *J. of Statist. Computation and Simulation*, **8**, 207–26.

Department of Management Science and Information Systems
CBA 5.202
University of Texas at Austin
Austin
TX 78712-1175
USA

Department of Statistics
University of Haifa
Mount Carmel
31999 Haifa
ISRAEL

Department of Mathematics
Imperial College of Science, Technology and Medicine
Huxley Building
180 Queen's Gate
LONDON
SW7 2BZ
UK

CHAPTER 14

Revising Exchangeable Beliefs: Subjectivist Foundations for the Inductive Argument

Michael Goldstein
University of Durham

14.1 INTRODUCTION

In de Finetti (1937) reprinted in Kyburg and Smokler (1964), de Finetti gives a subjectivist account of inductive inference. He writes:

... when the subjectivist point of view is adopted, the problem of induction receives an answer which is naturally subjective but in itself perfectly logical, while on the other hand, when one pretends to *eliminate* the subjective factors one succeeds only in *hiding* them (that is, at least, in my opinion), more or less skillfully, but never in avoiding a gap in the logic.

Consider, for example, the familiar problem of tossing coins. What do we mean by the probability that a coin will land heads? De Finetti's analysis replaces the traditional notion of a hypothetical 'frequency limit' of numbers of heads to number of tosses by the following argument.

If our probabilistic beliefs for the sequence of future tosses are exchangeable (meaning that our beliefs would be unaffected by any permutation of the order of the tosses), then our beliefs may be analysed as though the tosses could be considered to form an independent Bernoulli sequence, with common probability P of heads, over which we specify a prior distribution. When we observe some tosses of the coin, we update the prior

---

Aspects of Uncertainty edited by P. R. Freeman and A. F. M. Smith. © 1994 John Wiley & Sons Ltd.

distribution for P using Bayes's theorem, and make probability statements for future tosses under this revised representation.

Variants of this argument are widely held to offer a logically sound foundation for much of current Bayesian practice. However, there are two fundamental difficulties with this account, each arising from a subjective factor which has been *hidden*, but not *eliminated*.

Firstly, in order to create the exchangeability representation from meaningful beliefs about observable quantities, we must make a prior specification which is far too detailed to be within our genuine subjective capabilities, even for the simplest problem such as tossing coins. Thus we hardly ever actually use this argument, in practice, to construct representations for our beliefs, but rather we begin with a representation and invent beliefs about observables to be consistent with the representation.

Secondly, the formulation implies consistencies in our posterior beliefs which are unreasonably strong. Rarely are we logically certain that our beliefs will necessarily be exchangeable after we have observed a sample. (Indeed, in general statistical investigations, considerable effort is invested in questioning this assumption.) However, the standard account forces exchangeability of future beliefs, with no obvious way to reflect the actual complexities of the process of learning about future observations from the past.

In this chapter, we seek to strengthen the subjectivist account of inductive argument by constructing 'subjective judgements' to replace those features of the standard Bayes explanation for inductive reasoning that are typically granted an 'objective status', namely the requirement that a full, meaningful prior probability specification should be made (whereas, subjectively, only a limited, partial specification is possible) and the assumption that beliefs will be revised by formal conditioning (whereas, subjectively, many forms of reasoning are brought to bear).

We therefore confront the difficulties with the standard arguments by, firstly, creating the exchangeability representation from a more limited and achievable prior specification, and secondly, using less rigid tools to analyse our revisions of belief. A simplified formulation for exchangeability was given in Goldstein (1986). In this formulation, only second-order prior specifications over observables are made, and a second-order exchangeability representation is constructed directly from the reduced specification. In this chapter, we examine the logical issues that arise when we make inferences under this representation.

We proceed as follows. In section 14.2, we introduce our development by giving an informal treatment of our argument as applied to coin tossing. In section 14.3, we describe the limited exchangeability representation that we shall employ to describe our prior expectation structure. Section 14.4 introduces the posterior expectations, expressing our revisions of belief at some future time as primitives whose properties we shall investigate. In section 14.5, we consider the interpretation of posterior statements about quantities derived from exchangeability representations. In section 14.6, we consider the effect upon our beliefs of observing exchangeable data. Section 14.7 describes the systematic decomposition of beliefs for learning about exchangeable structures. In section 14.8, we introduce the basic symmetry assumption for belief revision which allows us to construct a full exchangeability representation for the revision of exchangeable beliefs. This representation is given and interpreted in section 14.9. Some final discussion of the development is contained in section 14.10.

14.2 TOSSING COINS

In this section, we introduce our general development by considering a simple case, namely learning about the probability of heads, when we toss a coin. We proceed informally, and present only the outline of the argument, leaving details to later sections.

Second-order exchangeability for tossing coins requires that we have the same probability that each individual toss lands heads and the same probability that each pair of tosses lands heads. The probability that a toss is heads is the expectation of the corresponding indicator function. Let $I_j = 1$, if toss j is heads, $I_j = 0$, otherwise. Our conditions are

$$E(I_j) = q_1, \quad \forall j, \tag{1}$$

$$E(I_j I_k) = q_2, \quad \forall j \neq k. \tag{2}$$

Let P_n be the proportion of heads in the first n tosses, so that

$$P_n = \frac{1}{n} \sum_{i=1}^{n} I_i .$$

We can verify that

$$E((P_n - P_m)^2) \to 0, \quad n, m \to \infty,$$

so that P_n is a Cauchy sequence in mean square. Therefore, we can construct the further quantity P for which

$$\lim_{n \to \infty} E(P_n - P)^2 = 0.$$

The quantity P may be interpreted informally as the 'underlying probability of heads'. For each k, we can construct the (second-order) exchangeability representation

$$I_k = P + R_k, \tag{3}$$

where

$$E(R_k) = 0, \quad \text{Var}(R_k) = q_1 - q_2 = V_R,$$

and $P, R_1, R_2, \ldots$ are mutually uncorrelated.

Now suppose that we are going to collect further evidence by tossing the coin n times, and then, at some time, T say, after the tosses, $D_n = \{I_1, \ldots, I_n\}$, have been observed and analysed, announce new probability values $E_T(I_j)$, by whichever means seem appropriate to us at that time. Given the representation (3), we would like to write, for $k > n$,

$$E_T(I_k) = E_T(P) + E_T(R_k), \tag{4}$$

and interpret our new expectation for toss k as the sum of a new belief for the 'underlying probability' plus a new belief specific to the individual for toss k. (This is analogous to the traditional Bayes formulation, where we treat the 'true probability' P of heads as a real random quantity, and deduce all of our posterior predictive belief statements about future tosses from the posterior model.) However, we must be cautious in treating P in this way. Having observed the sample, we may decide that future observations are not exchangeable: for example, we may notice some simple pattern in the sequence. In such cases, we cannot simply change our beliefs about P, for P, as part of the belief system within which the exchangeability representation was constructed, ceases to exist!

Instead, we interpret $E_T(P)$, as follows. We assume that our beliefs are conglomerable, meaning that current expectations for future expectations agree with current expectations, so that

$$E(E_T(I_j I_k)) = E(I_j I_k), \quad \forall j, k. \tag{5}$$

We may deduce that the sequence $E_T(P_m)$ is Cauchy in mean square. We may therefore interpret $E_T(P)$ as the Cauchy limit

$$E_T(P) = \lim_{m \to \infty} E_T(P_m). \tag{6}$$

We cannot suppose that our collection of expectations at time T will necessarily be exchangeable. However, in many cases we do not know now how our future beliefs will fail to be exchangeable. We suggest a weak assumption about future beliefs that is sufficient to retain the essence of the argument. We replace conditions (1), (2) for our prior expectations with similar conditions, satisfied in expectation, for our posterior beliefs. Thus we suppose that

1. $E(E_T(I_k))$ takes the same value $\forall k > n$;
2. $E((E_T(I_k))^2)$ takes the same value $\forall k > n$;
3. $E(E_T(I_j)E_T(I_k))$ takes the same value $\forall j \neq k > n$.

With the above assumptions, the sequence $\{E_T(I_k)\}$ is second-order exchangeable for $k > n$. From this, we may deduce a consistent interpretation of (4), namely that the exchangeability representation is preserved in expectation, i.e. for $k > n$,

$$E_T(I_k) = E_T(P) + E_T(R_k), \tag{7}$$

where each $E_T(R_k)$ has zero expectation, equal variance, and the sequence $E_T(P)$, $E_T(R_1)$, $E_T(R_2)$, ... is mutually uncorrelated.

In the usual exchangeability formulation, if we observe r heads in n tosses, then we can treat the quantity P as a real unknown quantity, update our prior probability measure for P, using Bayes's theorem, and derive all future predictive statements about observables from the revised mixture distribution. Within this approach, a sufficiently large number of tosses will reduce posterior variance for P to an arbitrarily small value while having no implication for the outcome of future tosses given P. Now, we describe the corresponding effect of observation of r heads in n tosses upon posterior beliefs within our formulation.

Our actual posterior beliefs will be formulated by subjective assessment at time T, using whatever further information and analysis seem appropriate to us at that time. Now, we may draw certain probabilistic inferences about these intended revisions by evaluating the Bayes linear expectation in $P_n = r/n$, the proportion of heads in the first n tosses, for P.

The Bayes linear expectation is the linear combination $E_n(P) = \alpha P_n + \beta$ which minimizes the squared distance $E((\alpha P_n + \beta - P)^2)$, over all choices of α, β. (The values of α, β and the minimal squared distance are determined by the specification (1), (2).)

Informally, we may consider that observation of $E_n(P)$ reduces our expected posterior variance for P from $Var(P)$ by an amount $Var(E_n(P))$. Because $Var(E_n(P)) \to Var(P)$, as $n \to \infty$, we deduce that our prior expectation for the variance for P that we shall specify at T tends to zero as n tends to infinity. Thus, a sufficiently large sample is currently expected to identify P with an arbitrarily high precision.

For an individual toss, I_k, $k > n$, we have $E_n(R_k) = 0$, so that no similar argument will refine our beliefs for individual tosses, given P, by observation of P_n. We may decompose our uncertainty about each I_k, $k > n$, as the sum of mutually uncorrelated components, namely

$$I_k - E(I_k) = \{P - E_T(P)\} + \{E_T(P) - E_n(P)\} + \{E_n(P) - E(P)\}$$
$$+ \{R_k - E_T(R_k)\} + \{E_T(R_k)\}. \tag{8}$$

With the exception of $\{E_n(P) - E(P)\}$, each term in the decomposition is uncorrelated with the outcomes of the first n tosses.

Informally, our beliefs concerning our planned revision of beliefs are consistent, in expectation, with the following procedure:

1. On observation of P_n, we replace the prior expectation for P, $E(P)$, by $E_n(P)$ and reduce uncertainty for P from $Var(P)$ by an amount $Var(E_n(P))$;

2. At time T, we replace $E_n(P)$ by $E_T(P)$, reducing the variance of P by an amount $\text{Var}(E_T(P) - E_n(P))$;
3. We subtract from each I_k, $k > n$, the mean $E_T(R_k)$, reducing each $\text{Var}(R_k)$ by an amount $\text{Var}(E_T(R_k))$, (which is constant over k).

All of the changes in expectation are mutually uncorrelated. The modified sequence is second-order exchangeable in expectation.

14.3 EXCHANGEABLE STRUCTURES

We make a series of measurements $C = \{X_1, X_2, \ldots\}$, finite or infinite, on a collection of individuals. Let $C_k = \{X_{1k}, X_{2k}, \ldots\}$, be the values of the measurements for person k. (For example, X_1 might be blood pressure, X_2 might be temperature, and X_{1k}, X_{2k} might be the observed blood pressure and temperature for the kth patient in some list.) We specify our prior mean, variance and covariance for each pair of quantities. These expectations are specified directly, as primitive quantities, rather than being derived through some intermediary probability specification (see, for example, the development of expectation, or prevision, as the natural primitive for the subjectivist theory in de Finetti 1974).

Definition The collection of measurements C is (second-order) exchangeable over the individuals if

1. $$E(X_{ik}) = m_i, \quad \forall k; \tag{9}$$
2. $$\text{Cov}(X_{ik}, X_{jk}) = v_{ij}, \quad \forall k; \tag{10}$$
3. $$\text{Cov}(X_{ik}, X_{jl}) = c_{ij}, \quad \forall k \neq l. \tag{11}$$

We shall suppose that the collection of individuals is, at least in principle, infinite.

Notation We say that collections of random quantities $\{A\}$ and $\{B\}$ are orthogonal, written $\{A\} \perp \{B\}$ if every element of $\{A\}$ is uncorrelated with every element of $\{B\}$.

In Goldstein (1986), the following exchangeability representation is derived.

Theorem 1 If C is exchangeable over individuals, then we may introduce the further collections of random quantities $\bar{C} = \{\bar{X}_1, \bar{X}_2, \ldots\}$, termed the *population collection* and $R_k = \{R_{1k}, R_{2k}, \ldots\}$, the *residual collection* for individual k, which satisfy the following:

1. $$X_{ik} = \bar{X}_i + R_{ik}, \quad \forall i, k; \tag{12}$$
2. $$E(\bar{X}_i) = m_i, \quad E(R_{ik}) = 0, \quad \forall i, k; \tag{13}$$

3. $\quad \text{Cov}(\bar{X}_i, \bar{X}_j) = c_{ij}, \quad \text{Cov}(R_{ik}, R_{jk}) = v_{ij} - c_{ij} \quad \forall i, j, k;$ (14)

4. $\quad \bar{C} \perp R_1 \perp R_2 \perp \ldots$ (15)

In contrast to the usual exchangeability representations, this representation is operational in practice as well as in principle. All that we must elicit are a comparatively small number of quantitative statements, namely those of (9)–(11), and from these beliefs we construct the collections $\bar{C}$, $R_1, R_2, \ldots$ These quantities are not observable. Rather, the quantities in $\bar{C}$ may be considered to be analogous to 'underlying population averages' for the various measurements, so that the collections R_i may be considered to be the individual discrepancies from the overall averages. The construction is given in Goldstein (1986). We now summarize those details of the construction that we shall require.

From the collection of measurements C, we form the linear space $\langle C \rangle$ of finite linear combinations

$$Y = \sum_u \alpha_u X_{r_u},$$

of elements of C, where $\{r_1, r_2, \ldots\}$ is a general finite subset of the positive integers. For each $Y \in \langle C \rangle$, and each individual k, we may construct the corresponding quantity Y_k defined as

$$Y_k = \sum_u \alpha_u X_{r_u k}.$$

We denote the collection of finite linear combinations $\{Y_k\}$ by $\langle C_k \rangle$. We view $\langle C_k \rangle$ as a vector space, where each element of the collection C_k is a vector, and linear combinations of the vectors are the corresponding linear combinations of the random quantities. We construct the inner product space $[C_k]$ from $\langle C_k \rangle$ by imposing the inner product and norm, for X, $Y \in \langle C_k \rangle$,

$$(X, Y) = \text{Cov}(X, Y), \quad \|X\|^2 = \text{Var}(X). \quad (16)$$

(We sometimes refer to $[C_k]$ as the (partial) belief structure with base $\{C_k\}$; see Goldstein 1988.) We restrict C_k to contain elements with finite prior variance, and identify equivalence classes of vectors whose differences are constant. We say that $[A]$ and $[B]$ are orthogonal if the underlying collections of random quantities $\{A\}$ and $\{B\}$ are orthogonal.

We now construct the composite linear space

$$\langle C_+ \rangle = \langle \cup_k C_k \rangle.$$

We construct the corresponding inner product space over $\langle C_+ \rangle$, so that a typical member of this space is a finite linear combination

$$\sum_u \alpha_u X_{r_u k_u}.$$

We denote the minimal closure of the space by $[C]$.

(Recall that any inner product space, S say, can be embedded in a minimal closure by adding, for each Cauchy sequence $\{s_j\}$ of elements of S for which the limit point of the sequence does not exist in S, a new element s^*, where

$$\lim_j \|s_j - s^*\| = 0, \tag{17}$$

and whose inner product satisfies

$$(s, s^*) = \lim_j (s, s_j), \tag{18}$$

for each $s \in S$.)

We may verify that the sequences of partial sums, $S_m(\cdot)$, defined for each $Y \in \langle C \rangle$ as

$$S_m(Y) = \frac{1}{m} \sum_{k=1}^m Y_k, \tag{19}$$

satisfy, for each $Y \in \langle C \rangle$,

$$\lim_{m, u \to \infty} \|S_m(Y) - S_u(Y)\| = 0, \tag{20}$$

and so are Cauchy sequences in $[C]$.

Therefore the limit

$$\lim_m S_m(Y) = \bar{Y} \tag{21}$$

exists in $[C]$, and for each Z in $[C]$

$$(\bar{Y}, Z) = \lim_m \frac{1}{m} \sum_{k=1}^m (Y_k, Z). \tag{22}$$

In particular,

$$(\bar{X}_i, X_{jk}) = (\bar{X}_i, \bar{X}_j) = c_{ij}, \tag{23}$$

and the representation theorem follows.

Notice that we have extended the representation theorem over $\langle C \rangle$ in the natural way. For $Y \in \langle C \rangle$ we have a representation

$$Y = \sum_u \alpha_u X_{r_u},$$

so that, for each k,

$$Y_k = \bar{Y} + R_k(Y), \tag{24}$$

where

$$\bar{Y} = \lim_m S_m(Y) = \sum_u \alpha_u \bar{X}_{r_u} \quad \text{and} \quad R_k(Y) = Y_k - \bar{Y} = \sum_u \alpha_u R_{r_u k}.$$

Note in particular that, for $Y, Z \in \langle C \rangle$,

$$\operatorname{Cov}(\bar{Y}, \bar{Z}) = c(Y, Z), \quad \operatorname{Cov}(R_k(Y), R_k(Z)) = d(Y, Z), \qquad (25)$$

where

$$c(Y, Z) = \operatorname{Cov}(Y_k, Z_l), \ k \ne l, \quad d(Y, Z) = \operatorname{Cov}(Y_k, Z_k) - c(Y, Z).$$

(The values of $c(\cdot, \cdot)$, $d(\cdot, \cdot)$ do not depend on the choice of k, l.)

14.4 POSTERIOR EXPECTATIONS

Now suppose that we are going to collect certain information and then, at future time T (for example, when we have collected and analysed all of the information) we shall make revised belief statements about various individuals in the population. For any random quantity X, we denote by $E_T(X)$ the expectation that we will declare at time T for X. Each $E_T(X)$ is currently a random quantity, and we now investigate the constraints that it is reasonable to impose for our beliefs over such quantities. We do not assume, even in principle, that there is some automatic method by which the value of $E_T(X)$ can be assessed, based, for example, on a combination of prior beliefs and data such as is provided by Bayes' theorem. Our posterior judgements are subjective assessments, to be made at time T, and can no more be forced into a predetermined form than can our prior assessments. What subjectivist theory does provide is a probabilistic description of the relationships between our assessments at the various time points, which may, as with any probabilistic description, provide guidance as to our thoughts and actions. Any further restrictions upon our revisions of belief, when imposed as logical requirements as opposed to pragmatic simplifications in particular situations, are both unwarranted and unnecessary.

In our development, we shall assume that posterior beliefs are conglomerable, defined as follows.

Definition Posterior expectation $E_T(\cdot)$ is conglomerable if, for all quantities X for which $E_T(X)$ will be specified,

$$E(E_T(X)) = E(X). \qquad (26)$$

Conglomerability would automatically follow if $E_T(\cdot)$ was determined, for example, by conditioning over a countably additive measure. If generated by conditioning on a finitely, but not countably, additive measure, then conglomerability need not hold. In practice, it is very unusual for non-conglomerable belief revisions to be constructed. However, the theoretical status of such non-conglomerable revisions is somewhat controversial. My view is that, while it is reasonable to make strictly finitely additive conditional probability statements, problems arise when we try to force a logical identification between such conditional probabilities and actual posterior probabilities. In Goldstein (1983), I offer a direct argument as to why actual posterior expectations should be conglomerable. While still finding this argument convincing, I would now argue the case for for the particular version of conglomerability defined by (26) from more general principles. As it would take us somewhat far afield to pursue this issue here, I will discuss such principles in detail elsewhere. Thus, in what follows we shall simply view conglomerability as a necessary, but plausible, condition for our various results to hold.

For conglomerable expectations, we have the following relations:

1. For every W, Z

$$E(E_T(W)Z) = E(E_T(E_T(W)Z)) = E(E_T(W)E_T(Z)) = E(WE_T(Z)), \quad (27)$$

so that

$$\mathrm{Cov}(E_T(W), Z) = \mathrm{Cov}(E_T(W), E_T(Z)) = \mathrm{Cov}(W, E_T(Z)). \quad (28)$$

2. In particular, if the value of Z will definitely be revealed by T, then

$$Z = E_T(Z),$$

so by (28)

$$\mathrm{Cov}(W - E_T(W), Z) = 0. \quad (29)$$

3. Let $\mathrm{Cov}_T(W, Z)$ be the covariance that we will assign at time T, i.e.

$$\mathrm{Cov}_T(W, Z) = E_T((W - E_T(W))(Z - E_T(Z))).$$

From (26) and (27) we have the analogue of the familiar variance decomposition

$$E(\mathrm{Var}(U \mid X)) = \mathrm{Var}(U) - \mathrm{Var}(E(U \mid X)),$$

namely

$$E(\mathrm{Cov}_T(W, Z)) = \mathrm{Cov}(W, Z) - \mathrm{Cov}(E_T(W), E_T(Z)). \quad (30)$$

4. Setting $Z = W$ in (30), gives both

$$E(\mathrm{Var}_T(W)) \leq \mathrm{Var}(W), \quad (31)$$

and

$$E((E_T(W))^2) \leq E(W^2). \quad (32)$$

5. We may strengthen inequality (31) in various ways as follows. Suppose that D is any random quantity for which

$$E((W - E_T(W))D) = 0, \quad (33)$$

(for example, by (29) any quantity whose value will certainly be known by T). We have

$$E((W - D)^2) = E((W - E_T(W) + E_T(W) - D)^2)$$
$$= E((W - E_T(W))^2) + E((E_T(W) - D)^2) \quad \text{(by (28), (33))}$$
$$\geq E((W - E_T(W))^2) = E(\text{Var}_T(W)). \tag{34}$$

14.5 POSTERIOR EXPECTATIONS FOR POPULATION QUANTITIES

We now consider the general implications of the exchangeability representation theorem for assessing posterior beliefs about collections of exchangeable individuals. In (24), we have the representation, $\forall Y \in \langle C \rangle$, that

$$Y_k = \bar{Y} + R_k(Y), \quad \forall k.$$

We would therefore like to write

$$E_T(Y_k) = E_T(\bar{Y}) + E_T(R_k(Y))$$

and explain our belief revision for Y_k in terms of our separate belief revisions for $\bar{Y}$ and for $R_k(Y)$. However, quantities such as $\bar{Y}$ are not observable random quantities but rather subjectivist constructs. For example, at the time T, we might have changed our opinion as to whether the individuals are exchangeable. In such circumstances we would not simply change our beliefs about the quantities in $\bar{C}$, but rather these quantities would cease to exist. Therefore, we cannot define $E_T(\bar{Y})$ as the expectation that we will define for $\bar{Y}$ at time T, as this may not be a well-defined quantity. We now construct a consistent interpretation for such quantities.

Corresponding to the collection of measurements $C = \{X_1, X_2, \ldots\}$, we have the collection of posterior belief measurements $E_T C = \{E_T(X_1), E_T(X_2), \ldots\}$, where the value of $E_T(X_i)$ for individual k is $E_T(X_{ik})$. We denote the collection of values of $E_T C$ for individual k as $E_T C_k = \{E_T(X_{1k}), E_T(X_{2k}), \ldots\}$. For each individual k we may construct the linear space $\langle E_T C_k \rangle$. We then construct the composite linear space

$$\langle E_T C_+ \rangle = \bigcup_k \langle E_T C_k \rangle.$$

We form the inner product space $[E_T C]$ over the closure of $\langle E_T C_+ \rangle$. (Thus a typical element of $[E_T C]$ would be a finite linear combination of quantities $E_T(X_{ik})$, with inner product given by prior covariance as in (16).) Note, from (28), that for each $X, Z \in [C]$,

$$(E_T(X), Z) = (E_T(X), E_T(Z)) = (X, E_T(Z)). \tag{35}$$

Therefore, defining the inner product over $[E_T(C)]$, automatically defines, by (35), the inner product for the composite belief structure

$$[C_T] = [C \cup E_T(C)].$$

In equation (21), we identified each $\bar{Y}$ as an element in the closure of $[C]$. We now define $E_T(\bar{Y})$ to be the unique point in the closure of $[C_T]$ which satisfies the relations (35) for $X = \bar{Y}$ and each $Z \in [C_T]$.

Theorem 2 For each $Y \in \langle C \rangle$, there exists an element $E_T(\bar{Y})$ in the closure of $[C_T]$ for which

1.
$$\lim_m E((E_T(S_m(Y)) - E_T(\bar{Y}))^2) = 0, \qquad (36)$$

where $S_m(Y)$ is as defined by (19);

2. For all $Z \in [C_T]$

$$(E_T(\bar{Y}), Z) = (E_T(\bar{Y}), E_T(Z)) = (\bar{Y}, E_T(Z)). \qquad (37)$$

Proof For each m, u

$$\|E_T(S_m(Y)) - E_T(S_u(Y))\|^2 = E(E_T(S_m(Y) - S_u(Y)))^2$$
$$\leq E(S_m(Y) - S_u(Y))^2 \quad \text{(by (32))}$$
$$= \|S_m(Y) - S_u(Y)\|^2 \to 0, m, u \to \infty \quad \text{(by (20))}$$

so that the sequence $E_T(S_m(Y))$ is Cauchy. We identify $E_T(\bar{Y})$ with the corresponding Cauchy limit

$$E_T(\bar{Y}) = \lim_m E_T(S_m(Y)), \qquad (38)$$

and (36) follows from (17). Further

$$\mathrm{Cov}(E_T(\bar{Y}), Z) = \lim_m \mathrm{Cov}(E_T(S_m(Y)), Z) \quad \text{(by 18)}$$
$$= \lim_m \frac{1}{m} \sum_{k=1}^m \mathrm{Cov}(E_T(Y_k), Z)$$
$$= \lim_m \frac{1}{m} \sum_{k=1}^m \mathrm{Cov}(Y_k, E_T(Z)) \quad \text{(by 28)}$$
$$= \lim_m \mathrm{Cov}(S_m(Y), E_T(Z))$$
$$= \mathrm{Cov}(\bar{Y}, E_T(Z)) \quad \text{(by 22)}$$

and the other equalities of (37) follow similarly. □

We now define

$$E_T(R_k(Y)) = E_T(Y_k) - E_T(\overline{Y}), \quad Y \in \langle C \rangle,$$
$$E_T(\overline{C}) = \{E_T(\overline{X}_1), E_T(\overline{X}_2), \ldots\},$$
$$E_T(R_k) = \{E_T(R_k(X_1)), E_T(R_k(X_2)), \ldots\}.$$

14.6 LEARNING FROM EXCHANGEABLE DATA

In a standard Bayes analysis, changes in beliefs over observables, given data, can be represented in terms of the change in the mixing distribution over the 'underlying population distribution' F from prior to posterior. In the limit, the sample observations identify F uniquely, while giving no information about the values of future observations conditional on F. We now develop a corresponding statement within our formulation, namely, given that we shall see a sufficiently large number of individuals we expect to reduce posterior variance over population means to an arbitrarily small value, while the corresponding bound for individual observations tends to the prior residual variance for the individual.

Suppose that by time T we will definitely have learned the values of a sample of n individuals, $C_1, C_2, \ldots, C_n$. We construct the combined collection of observed quantities

$$C_{[n]} = \bigcup_{k=1}^{n} C_k.$$

Every element in $[C_{[n]}]$ will certainly be known by time T. Therefore from (34) we have for any W and any Z in $[C_{[n]}]$ that

$$E(\text{Var}_T(W)) \leq E((W - Z)^2),$$

so that

$$E(\text{Var}_T(W)) \leq \inf_{Z \in [C_{[n]}]} E((W - Z)^2). \tag{39}$$

The best such bound is therefore given by orthogonal projection.

Notation For any closed subspace $[U] \subset [C_T]$, we denote the orthogonal projection operator into $[U]$ as $E_U(\cdot)$. We write the squared orthogonal distance $E((W - E_U(W))^2)$ as

$$\text{Var}_U(W) = E((W - E_U(W))^2) = \inf_{Z \in [U]} E((W - Z)^2).$$

Projections of the above type are often termed Bayes linear rules. Properties of such rules are given in Goldstein (1988), where orthogonal distances from vectors to subspaces, of the form $\text{Var}_U(X)$, are interpreted as adjusted variances. In particular, note that $E_U(W)$ is the unique element of $[U]$ for which

$$(W - \mathrm{E}_U(W)) \perp [U], \tag{40}$$

and also that

$$\mathrm{Var}_U(W) + \mathrm{Var}(\mathrm{E}_U(W)) = \mathrm{Var}(W). \tag{41}$$

To simplify the notation, we write $\mathrm{E}_n(W)$, $\mathrm{Var}_n(W)$ for the projection $\mathrm{E}_{C_{[n]}}(W)$ and the distance $\mathrm{Var}_{C_{[n]}}(W)$. Note that as, for each $Y \in \langle C \rangle$,

$$R_k(Y) \perp [C_{[n]}], \quad \forall k > n$$

we must have

$$\mathrm{E}_n(Y_k) = \mathrm{E}_n(\bar{Y}), \quad \forall k > n. \tag{42}$$

We have the following theorem.

Theorem 3 Suppose that C is exchangeable, and beliefs are conglomerable. Suppose that by time T we will have observed n individuals $C_1, C_2, \ldots C_n$. Then, for each $Y \in \langle C \rangle$,

1. $$\mathrm{E}(\mathrm{Var}_T(\bar{Y})) \leq \mathrm{Var}_n(\bar{Y}) \leq \frac{d(Y, Y)}{n}, \tag{43}$$

where $d(Y, Y)$ is given by (25).

2. $$\mathrm{E}(\mathrm{Var}_T(Y_k)) \leq \mathrm{Var}_n(\bar{Y}) + \mathrm{Var}(R_k(Y))$$
$$\leq (1 + 1/n)d(Y, Y), \quad \forall k > n. \tag{44}$$

Proof The inequality

$$\mathrm{E}(\mathrm{Var}_T(\bar{Y})) \leq \mathrm{Var}_n(\bar{Y})$$

follows directly from (39), with $Z = \mathrm{E}_n(\bar{Y})$, $W = \bar{Y}$.

As the partial sum

$$S_n(Y) = \frac{1}{n} \sum_{t=1}^{n} Y_t,$$

is in $[C_{[c]}]$, we must have

$$\mathrm{Var}_n(\bar{Y}) \leq \mathrm{E}((\bar{Y} - S_n(Y))^2) = \mathrm{E}\left(\left(\frac{1}{n}\sum_{k=1}^{n} R_k(Y)\right)^2\right) = \frac{d(Y, Y)}{n},$$

completing (43). Equation (44) follows similarly, using (42), as for $k > n$, $R_k(Y)$ is orthogonal to both $[C_{[n]}]$ and $\bar{Y}$. □

The bound in (43) goes to zero with n. In the limit, we expect to identify the underlying mean values to arbitrary expected precision. We now explore further the role of the Bayes linear estimators in the revision of belief.

14.7 SEPARATING POSTERIOR BELIEFS FOR THE EXCHANGEABLE MODEL

Suppose that we observe $C_{[n]}$ and evaluate the Bayes linear rules $E_n(Y)$ for each Y in some finite or infinite collection $B = \{Y_1, Y_2, \ldots\}$ of interest. Subsequently, we form our posterior expectations $E_T(Y)$, using whatever further information seems relevant to us at the time. The Bayes linear rules allow us to make various probabilistic inferences about these subjective assessments. For example, we have described in the preceding section how the linear rules give upper bounds for expected posterior variances. These bounds derive from the underlying orthogonal decomposition of each Y as

$$Y - E(Y) = Y^* + E_T^*(Y) + E_n^*(Y), \tag{45}$$

where

$$Y^* = Y - E_T(Y), \tag{46}$$

$$E_T^*(Y) = E_T(Y) - E_n(Y),$$

$$E_n^*(Y) = E_n(Y) - E(Y).$$

We have

$$E(Y^*) = E(E_T^*(Y)) = E(E_n^*(Y)) = 0,$$

and, denoting

$$B^* = \{Y_1^*, Y_2^*, \ldots\}, \quad E_T^*(B) = \{E_T^*(Y_1), E_T^*(Y_2), \ldots\},$$

$$E_n^*(B) = \{E_n^*(Y_1), E_n^*(Y_2), \ldots\},$$

we have

$$B^* \perp E_T^*(B) \perp C_{[n]}. \tag{47}$$

As $E_n^*(B)$ is in $C_{[n]}$, we have from (45), (47), for each Y, that

$$\text{Var}(Y) = \text{Var}(Y^*) + \text{Var}(E_T^*(Y)) + \text{Var}(E_n^*(Y)). \tag{48}$$

We have decomposed the variance of each Y into three pieces. One part, $\text{Var}(E_n^*(Y))$, is removed when we observe $C_{[n]}$. One part, $\text{Var}(E_T^*(Y))$, is removed when, having observed $C_{[n]}$, we make the actual posterior assessment $E_T(Y)$. The final part is the expectation of the variance remaining when we have made our best judgements at time T.

Note in particular the implication of the orthogonality between $C_{[n]}$ and $E_T^*(B)$, namely that the differences $E_T(Y) - E_n(Y)$ are not linearly predictable from observation on $C_{[n]}$. The second-order specification that we have made only specifies the linear rules. (Formally, such a specification determines only the inner product spaces, and the associated projections, that we have constructed in our account.) Therefore, if we wish to make a more detailed prior inference for $E_T(Y)$ from observation of $C_{[n]}$, then we need to enlarge the inner product space by making a more detailed prior specification.

The most detailed specification that we could possibly make is a full joint prior probability specification over all the elements of C. Such a specification is a special case of our formulation, in which the collection of measurements C contains all the products of indicator functions for all finite combinations of random quantities. In practice, this is an extremely complex specification task, for all but the simplest of problems. However, the logical argument for such a full specification proceeds exactly as described by (45). The prior and posterior probabilities are expectations for the corresponding indicator functions, and Bayes linear rules obtained by projection are the conditional expectations, given the data, evaluated using Bayes theorem. The judgement as to whether, having observed the sample, the conditional expectation is equal to our posterior expectation, is a subjective choice that, in any problem, can only be made at time T. Logically equating conditional and posterior probabilities (i.e. assessing that $\mathrm{Var}(\mathrm{E}_T^*(Y)) = 0$), corresponds to a logical certainty on our part that we shall not examine our beliefs, or the data, in any way further before forming our posterior judgements, for example by plots, residual analysis or any of the other tools of the statistician's trade. This is not appropriate as a logical benchmark for the theory. Rather the Bayesian argument is justified in the more limited, but more precise, probabilistic sense that we have described above, namely as a prior inference for the posterior values. The conditional specification removes some of the uncertainty from the posterior specification. Any differences between the conditional and the posterior probabilities are uncorrelated, in this case, with the collection of indicator functions for the data, which, given a full probability specification, is equivalent to the condition that these differences are conditionally independent of the data. Thus, while the level of detail for our prior specification is a matter of subjective choice, the probabilistic implications of any such prior specification for our actual revisions of belief are as summarised by (45) and apply equally to any Bayes or Bayes linear assessment.

Applying the decomposition to the exchangeable formulation, by substituting (24) into (45) using (42), gives the following result which summarises much of our discussion to this point:

Theorem 4 For exchangeable and conglomerable beliefs, for any $Y \in \langle C \rangle$, $k > n$, we have

$$Y_k - \mathrm{E}(Y_k) = Y_k^* + \mathrm{E}_T^*(Y_k) + \mathrm{E}_n^*(Y_k), \qquad (49)$$

with notation as in (46). Further,

$$Y_k^* = \overline{Y}^* + R_k^*(Y),$$

where
$$E_T^*(Y_k) = E_T^*(\bar{Y}) + E_T(R_k(Y)),$$
$$E_n^*(Y_k) = E_n^*(\bar{Y}),$$

We have
$$R_k^*(Y) = R_k(Y) - E_T(R_k(Y)).$$

Setting
$$R_k^*(Y) \perp E_T(R_k(Y)) \perp C_{[n]}.$$

$$\bar{C}^* = \{\bar{X}_1 - E_T(\bar{X}_1), \bar{X}_2 - E_T(\bar{X}_2), \ldots\},$$
$$E_T^*(\bar{C}) = \{E_T(\bar{X}_1) - E_n(\bar{X}_1), E_T(\bar{X}_2) - E_n(\bar{X}_2), \ldots\},$$

we have
$$\bar{C}^* \perp E_T^*(\bar{C}) \perp C_{[n]}.$$

Thus, the sample observations directly reduce our uncertainty over the population quantities, by projection. There are no predictive implications from $[C_{[n]}]$ for our revision of beliefs over the individual residual collections given the mean collection.

(General decompositions of this type for our current beliefs about our future beliefs are termed separations of beliefs in Goldstein (1986), where a central role is argued for such decompositions in subjectivist analysis.)

We now suggest a plausible strengthening of our prior assumptions under which we can identify our adjustment of belief more closely with the representation theorem.

14.8 EXCHANGEABLE POSTERIOR BELIEFS

We cannot be sure that our beliefs will be exchangeable at time T. However, it will often be the case that we do not now know in what ways the judgement of exchangeability will be broken. Therefore, our current expectations for our future beliefs will often exhibit the symmetries of exchangeability, even when we may doubt whether our actual future beliefs will be exchangeable.

Just as prior exchangeability gives a simple representation theorem for our prior beliefs, so shall the corresponding assumption of 'preserving exchangeability in expectation' give us a simple representation for our posterior beliefs. We make the following definition, replacing each of the three conditions, (9)–(11) in the definition of exchangeability for current beliefs by the expectation over that condition for our future beliefs.

Definition C is (second-order) exchangeable in expectation over individuals from n at time T, if C is currently exchangeable over individuals, expectations at T are conglomerable and

1. $\qquad E(E_T(X_{ik})) = t_i, \quad \forall k > n;$ (50)
2. $\qquad E(\text{Cov}_T(X_{ik}, X_{jk})) = u_{ij}, \quad \forall k > n;$ (51)
3. $\qquad E(\text{Cov}_T(X_{ik}, X_{jl})) = b_{ij}, \quad \forall k \neq l > n.$ (52)

The value n in the above definition corresponds to the 'maximum possible sample size' of individuals that we might observe by time T, so that the definition restricts our expected exchangeability to individuals whose values will certainly be unknown to us at time T. Conditions (50)–(52) would automatically be satisfied in a Bayes analysis for the standard exchangeable case, namely where we take a sample of n observations and update beliefs over all future individuals using Bayes' theorem, as, under such Bayes updating, actual future beliefs must be exchangeable. However, the assumption that our future beliefs will necessarily be exchangeable is far too strong to hold even approximately for most applications. Preserving exchangeability in expectation is the minimal operationally verifiable assumption under which we may identify the exchangeability representation with our revision of our beliefs.

We first offer an alternative interpretation for the preservation of exchangeability in expectation.

Theorem 5 Given conglomerability, C is exchangeable in expectation from n at T, if and only if the collection of measurements $E_T C$ is exchangeable over all individuals $k > n$.

Proof $E_T C$ is exchangeable over all individuals $k > n$ if

$$E(E_T(X_{ik})) = t_i, \quad \forall k > n; \tag{53}$$
$$\text{Cov}(E_T(X_{ik}), E_T(X_{jk})) = w_{ij}, \quad \forall k > n; \tag{54}$$
$$\text{Cov}(E_T(X_{ik}), E_T(X_{jl})) = a_{ij}, \quad \forall k \neq l > n. \tag{55}$$

Condition (53) follows from (9) by conglomerability, and $t_i = m_i$, $\forall i$.

As beliefs over the collections C_i are exchangeable, substituting from (10), (11), (54), (55) into (30), gives immediately that $E_T C$ is exchangeable for $k > n$ if and only if C is exchangeable in expectation from n at T, with

$$w_{ij} = v_{ij} - u_{ij}, \, a_{ij} = c_{ij} - b_{ij}, \quad \forall i, j, \tag{56}$$

in this case.

From this theorem, we may express our belief revision as follows.

Theorem 6 If C is exchangeable in expectation from n at T, then we have, for each i, that

where
$$E_T(X_{ik}) = E_T(\bar{X}_i) + E_T(R_k(X_i)), \quad \forall k > n, \tag{57}$$

and
$$\text{Cov}(E_T(\bar{X}_i), E_T(\bar{X}_j)) = a_{ij}, \quad \text{Cov}(E_T(R_k(X_i)), E_T(R_k(X_j)))$$
$$= w_{ij} - a_{ij}, \quad \forall k > n; \tag{58}$$

so that
$$E_T(\bar{C}) \perp E_T(R_{(n+1)}) \perp E_T(R_{(n+2)}) \perp \ldots \tag{59}$$

$$E(\text{Cov}_T(X_{ik}, X_{jk})) = E(\text{Cov}_T(\bar{X}_i, \bar{X}_j))$$
$$+ E(\text{Cov}_T(R_k(X_i), R_k(X_j))), \quad \forall i, j, k \tag{60}$$

$$E(\text{Cov}_T(X_{ik}, X_{jl})) = E(\text{Cov}_T(\bar{X}_i, \bar{X}_j)), \quad \forall i, j, \forall k \neq l. \tag{61}$$

Proof From theorem (5), $E_T C$ is exchangeable for $k > n$. Therefore, we may write
$$E_T(X_{ik}) = M_i + W_{ik}, \quad \forall k > n,$$
where the collections $\{M_i\}$, $\{W_{i(n+1)}\}$, $\{W_{i(n+2)}\}, \ldots$ satisfy the conditions of theorem 1. From (21), M_i is the Cauchy limit of the partial sums
$$U_m(E_T(X_i)) = \frac{1}{m} \sum_{k=n+1}^{n+m} E_T(X_{ik}).$$
By the linearity of $E_T(\cdot)$,
$$U_m(E_T(X_i)) = \left(1 + \frac{n}{m}\right) E_T(S_{(m+n)}(X_i)) - \frac{n}{m} E_T(S_n(X_i)).$$
It therefore follows from (38) that for each i
$$M_i = E_T(\bar{X}_i),$$
as each M_i corresponds to the same element in the closure of $[C_T]$ as does $E_T(\bar{X}_i)$. Therefore, also, $W_{ik} = E_T(R_k(X_i))$ and (57) follows. Therefore, applying theorem 1 to the collection $E_T C$, (58) follows from (14) and (59) follows from (15). Equation (60) follows by substituting the uncorrelated sum (57) into (30).

We may extend theorem 6 over $\langle C \rangle$, in a similar way to theorem 1, as the following decomposition.

Corollary 1 If C is exchangeable in expectation from n at T, then we have, for each $Y \in \langle C \rangle$, that
$$E_T(Y_k) = E_T(\bar{Y}) + E_T(R_k(Y)), \quad k > n, \tag{62}$$

where the sequence $E_T(R_{(n+1)}(Y))$, $E_T(R_{(n+2)}(Y))$, ... is mutually uncorrelated (and uncorrelated with $E_T(\bar{Y}))$, with zero means, and equal variances.

Thus, if C preserves exchangeability from n at T in expectation, then our posterior expectation for each Y_k is the sum of two uncorrelated random quantities, the first corresponding to the posterior expectation for the population mean $\bar{Y}$, and the second to the posterior expectation for the individual discrepancy $R_k(Y)$. Further, our expected covariances at time T may be constructed by separately assessing expected covariances over the population means and over the individual discrepancies (which are the same for each individual $k > n$).

14.9 EXCHANGEABLE REVISIONS OF BELIEFS

We now collect together the results of the preceding sections to give the following account of the revision of beliefs which preserve exchangeability from n at T in expectation.

We begin with the belief structure $[C]$, expressing our (second-order) beliefs about all the observables. Each individual, k, is represented as a subspace $[C_k]$. Here $[C]$ is contained in the orthogonal decomposition

$$[\bar{C}] \oplus [R_1] \oplus [R_2] \oplus \ldots .$$

Each $[R_j]$ has the same inner product, and $[C_k] \perp [R_l]$, $k \neq l$.

Each $Y_k \in [C_k]$ may be decomposed as the uncorrelated sum

$$Y_k = \bar{Y} + R_k(Y), \qquad (63)$$

or, equivalently, determined by the orthogonal projections,

$$E_{[\bar{C}]}(Y_k) = \bar{Y}, \qquad E_{[R_k]}(Y_k) = R_k(Y).$$

We now introduce two sources of information. Firstly $[C_{[n]}]$ will be revealed, and secondly posterior judgements will be made over $[C]$. The first source of information is already contained in $[C]$. We represent the second source as the belief structure $[E_T(C)]$. We combine this with $[C]$ to create the composite structure $[C_T]$.

Let $[C^n]$ be the subspace of $[C]$ constructed from all individuals $k > n$ (i.e. so that $[C^n] = [\cup_{k>n} C_k]$). $[C^n]$ is contained in the orthogonal decomposition

$$[\bar{C}^*] \oplus [E_T^*(\bar{C})] \oplus [C_{[n]}] \oplus [R_{n+1}] \oplus [R_{n+2}] \oplus \ldots, \qquad (64)$$

and each residual space may be further orthogonally decomposed as

$$[R_k] = [R_k^*] \oplus [E_T(R_k)]. \qquad (65)$$

Each $[R_k^*]$ has the same inner product as does each $[E_T(R_k)]$.

Each $Y_k \in [C_k]$ may now be decomposed as the uncorrelated sum

REVISING EXCHANGEABLE BELIEFS 221

$$Y_k - \mathrm{E}(Y) = \bar{Y}^* + \mathrm{E}_T^*(\bar{Y}) + \mathrm{E}_n^*(\bar{Y}) + R_k^*(Y) + \mathrm{E}_T(R_k(Y)), \quad (66)$$

or, equivalently, determined by the orthogonal projections, for $k > n$,

$$\mathrm{E}_{[\bar{C}^*]}(Y_k) = \bar{Y}^*, \quad \mathrm{E}_{[\mathrm{E}_T^*(\bar{C})]}(Y_k) = \mathrm{E}_T^*(\bar{Y}), \quad \mathrm{E}_{[C_{[n]}]}(Y_k) = \mathrm{E}_n^*(\bar{Y}),$$

$$\mathrm{E}_{[R_k^*]}(Y_k) = R_k^*(Y), \quad \mathrm{E}_{[\mathrm{E}_T(R_k)]}(Y_k) = \mathrm{E}_T(R_k(Y)).$$

Thus, our current beliefs about our revision of beliefs are consistent, in expectation, with the following process:

1. We observe $[C_{[n]}]$. For each $Y \in \langle C \rangle$, our beliefs for each Y_k, $k > n$, still follow the representation (63), but we replace the prior mean and variance for $\bar{Y}$, $\mathrm{E}(\bar{Y})$ and $\mathrm{Var}(\bar{Y})$, by the Bayes linear mean $\mathrm{E}_n(\bar{Y})$ and the adjusted variance $\mathrm{Var}_n(\bar{Y})$. We do not modify our beliefs about each $R_k(Y)$.
2. We now formulate our posterior beliefs $\mathrm{E}_T(\cdot)$. Our change in expectation from $\mathrm{E}_n(\bar{Y})$ to $\mathrm{E}_T(\bar{Y})$ is uncorrelated with all the observations $[C_{[n]}]$. The variance of $\bar{Y}$ is reduced to $\mathrm{Var}(\bar{Y}^*)$.
3. We change our expectations for each $R_k(Y)$ to $\mathrm{E}_T(R_k(Y))$. These changes are mutually uncorrelated and uncorrelated with the observations $[C_{[n]}]$, and with our changes in belief over $\bar{C}$.
4. Having subtracted from each Y_k the corresponding individual residual expectation $\mathrm{E}_T(R_k(Y))$, the resulting sequence is again exchangeable in expectation, a priori uncorrelated, with common expected residual variance $\mathrm{Var}(R_k^*(Y))$, constant over k. All the residual terms $R_k^*(Y)$ are uncorrelated with $[C_{[n]}]$, and with all our changes in belief from $\mathrm{E}(\cdot)$ to $\mathrm{E}_T(\cdot)$.

Thus, when exchangeability is preserved in expectation, we may draw detailed probabilistic inferences about our changes in belief, by progressively modifying our mean and variance specifications within the representation theorem.

In a sense, our analysis therefore complements the standard Bayes development which is concerned with prior inferences for posterior values and is the first stage in the general procedure that we have described. The principal methodological differences between the approaches lie partly in the simplification of the specification of prior beliefs that we have suggested, and partly in the explicit representation that we have constructed for the revision of beliefs. As this chapter is a discussion of the foundations of the inductive argument, we shall pursue these methodological issues elsewhere.

14.10 CONCLUDING COMMENTS

This chapter offers a subjectivist account of inductive inference. The subjectivist approach is an attitude before it is a collection of techniques. De Finetti (1937) writes:

We are sometimes led to make a judgement which has a purely subjective meaning, and this is perfectly legitimate; but if one seeks to replace it afterward by something objective, one does not make progress, but only an error. Rather than by seeking to bring everything back to the objective, one can attain clarity by reducing any such concept systematically to the subjective; the value of a concept would then result from the analysis of the deep and essential reasons which have made us, perhaps unconsciously, introduce it, and which furnish us with an explanation of its usefulness.

In our account of the inductive argument, we have criticized the overly rigid view as to the prior specifications which are required for an exchangeable analysis and the unrealistic assumptions as to how beliefs are to be updated for such an analysis. Making these subjective choices explicit does indeed furnish an explanation of the usefulness of the standard Bayesian argument, as a pragmatic approximation to the first stage of a full subjectivist analysis, namely in making prior inferences as to our posterior values.

Further, just as the objectivist view of probabilistic reasoning can best (and perhaps only!) be understood within the subjectivist framework advanced by de Finetti, so can de Finetti's argument best be understood within the more fully subjectivist framework that we have introduced (and so on, I would imagine, *ad infinitum*).

REFERENCES

De Finetti, B. (1937) Foresight: its logical laws, its subjective sources, *Annales de l'Institut Henri Poincaré*, **7**.

De Finetti, B. (1974) *Theory of Probability*, vol. 1, John Wiley & Sons.

Goldstein, M. (1983) The prevision of a prevision, *J. Amer. Statist. Ass.* **78**, 817–19.

Goldstein, M. (1986) Exchangeable belief structures, *J. Amer. Statist. Ass.* **81**, 971–6.

Goldstein, M. (1986) Separating beliefs, in *Bayesian Inference and Decision Techniques*, P. K. Goel and A. Zellner, eds, Amsterdam, North-Holland.

Goldstein, M. (1988) Adjusting belief structures, *J. R. Statist. Soc.* **B50**, 133–54.

Kyburg, H. E. and Smokler, H.E. (eds) (1964), *Studies in Subjective Probability*. Wiley, 1964.

Department of Mathematical Sciences
Durham University
Sciences Laboratories
South Road
DURHAM
DH1 3LE
UK

CHAPTER 15

On Steinian Shrinkage Estimators: the Finite/Infinite Problem and Formalism in Probability and Statistics

Bruce M. Hill[†]
University of Michigan, Ann Arbor

15.1 FINITE AND INFINITE

C. F. Gauss (1900, p. 216):

I protest... against the use of an infinite quantity as an actual entity, which is never allowed in mathematics. The infinite is only a façon de parler in which one really speaks of limits to which certain ratios come as near as desired, while others are allowed to increase unrestrictedly.

L. E. J. Brouwer (1920), who founded the 'intuitionist' school of mathematics, says with regard to the principle of the excluded middle, that this principle:

was caused historically by the fact that, firstly, classical logic was abstracted from the mathematics of the subsets of a definite finite set [i.e. a set given by exhibition of its elements], that, secondly, an *a priori* existence independent of mathematics was ascribed to this logic, and that, finally, on the basis of this supposititious apriority it was unjustifiably applied to the mathematics of infinite sets [author's italics].

[†] This work was supported by the National Science Foundation under grant DMS-9201056.
Aspects of Uncertainty edited by P. R. Freeman and A. F. M. Smith. © 1994 John Wiley & Sons Ltd.

F. P. Ramsey (1950, p. 81) also questions the meaningfulness of infinite partitions in his theory of subjective probability:

Thirdly, nothing has been said about degrees of belief when the number of alternatives is infinite. About this I have nothing useful to say, except that I doubt if the mind is capable of contemplating more than a finite number of alternatives.

and again Ramsey (1950, p. 79) regarding his derivation of subjective probability:

I have not worked out the mathematical logic of this in detail, because this would, I think, be rather like working out to seven places of decimals a result only valid to two.

B. Russell (1903, p. 347):

One of the most notable victims of posterity's lack of judgment is the Eleatic Zeno. Having invented four arguments, all immeasurably subtle and profound, the grossness of subsequent philosophers pronounced him to be a mere ingenious juggler, and his arguments to be one and all sophisms. After two thousand years of continual refutation, these sophisms were reinstated, and made the foundation of a mathematical renaissance, by a German professor, who probably never dreamed of any connection between himself and Zeno. Weierstrass, by strictly banishing all infinitesimals, has at last shown that we live in an unchanging world, and that the arrow, at every moment of its flight, is truly at rest.

and Russell (1903, p. x):

The next step was the abolition of classes. ... 'The symbols for classes like those for descriptions, are, in our system, incomplete symbols: their *uses* are defined, but they themselves are not assumed to mean anything at all. ... Thus classes, so far as we introduce them, are merely symbolic or linguistic conveniences, not genuine objects.'

and Russell (1903, p. ix):

Numbers were immutable and eternal, like the heavenly bodies; numbers were intelligible: the science of numbers was the key to the universe. The last of these beliefs has misled mathematicians and the Board of Education down to the present day.

A. Einstein (1921, p. 3):

As far as the propositions of mathematics refer to reality they are not certain, and in so far as they are certain they do not refer to reality.

Other distinguished mathematicians, such as H. Poincaré and H. Weyl, have supported parts of the intuitionist and/or constructivist thesis regarding the nature of mathematics and science, and have therefore been extremely careful to avoid formalism in their dealings with infinity. We are all aware that there has been an intense controversy, extending throughout most of this century, regarding the Bayesian approach to statistics. This approach is, in its essence, of an algorithmic nature. That is to say, one inputs information as deemed appropriate, does something with it for which there is some rational justification, and in this way obtains novel and interesting answers to real-world questions that had not, and have not, been answered in any other way.[†] It is perhaps not so well known that this controversy is related to older controversies in the foundations of logic and mathematics, pertaining to the finite/infinite question, the use of algorithms, and formalism.

It is argued here that the same mistakes and misconceptions that are alluded to or implied by the above quotations have permeated statistics to an enormous extent, primarily by confusing people who are so dimly aware of the history of logic and mathematics that they are blissfully ignorant that there has ever been any controversy. We believe this confusion is at the root (at least the scientific root) of the dispute regarding the Bayesian approach to statistics. We illustrate this thesis in connection with the Stein 'paradoxes' regarding shrinkage estimators. See section 15.8 for a return to discussion of the more general issues.

15.2 ADMISSIBILITY AND BOUNDEDNESS

Hill (1974, 1990a, 1993) suggested or implied that much of the work that stemmed from Stein's results on inadmissibility is largely irrelevant and even misleading with respect to the problem of inference about many means. This was based upon the fact that real-world implementations of any statistical procedure are necessarily done on a computer with finite memory, and in the finite case Bayes procedures are admissible, provided only that the *a priori* probability is positive for each possible parameter value; while all admissible procedures are necessarily Bayes procedures, since in the finite case the collection of Bayes procedures is closed. See Blackwell and Girschick (1954, p. 127). These articles also suggested an operational method, later called the 'evaluation game' by Berger (1985, p. 41), by means of which admissibility at least in the finite scenario, can be justified as a criterion. It should be recalled that the initial justification proposed for Stein-type estimators was due to the fact that the classical least-squares estimator was shown to be inadmissible in dimension 3 or more in the idealized model, by Stein, and that Stein-type estimators dominate them everywhere for squared error loss (and some other loss functions). Nowadays, this is sometimes forgotten.

[†] Indeed, as emphasized by de Finetti (1974), the subjective Bayesian approach represents an extension of classical logic to the case of uncertainty.

Hill (1990a) distinguished between bureaucratic decision procedures that must be fully spelled out in advance for one reason or another, and repeatedly implemented, as for example when the procedure is applied under varying conditions that might be represented by means of a parameter that varies randomly from occasion to occasion; as opposed to those that require creative data analysis. See Hill (1990b) and Smith (1986). Bayesian procedures are optimal in both scenarios, but here we only discuss the bureaucratic type of decision procedure. In any real-world decision problem where decision rules must be completely specified in advance, there are of necessity only a finite number of decision functions mapping the space of data values into the collection of estimates, or more generally, of terminal decisions. This is so because to hold everything requisite in the memory of a computer with finite memory, it is necessary to represent both the data values and the space of possible estimates (or more generally, of terminal decisions) by means of some finite collection. Hence both sufficiently large, sufficiently small, and sufficiently precise, values of any of these variables must of necessity be rounded after some fashion or other. In the usual formulation of the statistical decision problem, the number of non-randomized decision functions is the number of mappings of the finite data space into the finite space of possible terminal decisions. If there are N possible distinct data values, and there are J possible distinct terminal decisions, then the number of possible decision functions is at most J^N. This number can be so large as to be prohibitive for an exact analysis on even the most powerful of modern computers. None the less it is finite, and so theoretical issues *that arise only in the infinite case* would seem to be quite irrelevant, except as a mathematical exercise.

The rounding is ordinarily done in the obvious fashion. Consider for example the problem of estimating the proportion of human beings who survive x years, as in survival analysis. The parameter is either the true distribution function F for the population, or some parametric representation, say in terms of a parameter θ that identifies the true distribution, or approximates to it in some specified sense. To simplify matters suppose that in fact the parameter θ is the average age in the population, as for example when this parameter largely determines the distribution. The data consists of the ages at which patients either die or are censored. Most people would not take age seriously to too fine a level, for example microseconds. Thus ordinarily a person dies over some time period, at least seconds, and it would appear absurd to pretend to measure such an age to microseconds. As argued in Hill (1990a), in real-world problems the parameter ordinarily loses its meaning beyond a small number of digits. Similarly, the *Guinness Book of World Records* has 121 years as the maximum authenticated age, and 200 years seems to be a quite generous upper bound. An important class of statistical problems involves sampling from finite populations, and in such examples the parameters are simply functions of the empirical distribution for the entire population, and take on only a finite number of values. See Hill (1992a). Plainly, if we are to evaluate the loss when the true parameter is θ and its estimate is θ̂ in an actual play of the evaluation game, then because of the finiteness of memory of our computer, the interval of parameter values beyond *some* value must be represented by a single value, here taken to be 200 years. Sometimes more sophisticated rounding is done, as for example when one takes a less fine grid in regions given low *a priori* probability.

Next, in providing an estimate of θ, we first observe that it too must be bounded, since again one cannot hold in memory all possible values, and some finite grid must be selected to represent θ̂ as well. In practice, often the values are simply to the nearest month in the case of survival analysis. Even, however, if the chosen grid for θ̂ were to the nearest microsecond, it would require a remarkable degree of naïveté to present the results in such a fashion. In estimating the mean of a population one does not ordinarily wish to see the data to even a googul ($g = 10^{100}$) of decimal points, much less to infinitely many, even if this were possible.[†] See Scheffé (1959, p. 59) for a warning to round estimates in a sensible way, or be viewed with disdain by serious scientists or engineers. In conclusion, I would argue for a variety of reasons that cannot be gone into here, that the only scenario with genuine meaning as a real-world problem, requires a finite representation for the parameter space, as well as for the data space and space of terminal decisions. This does not, of course, preclude the use of idealized limits as a convenient device for communication and for obtaining insight.

The distinguished probabilist, W. Feller (1968, p. 7) argues against the use of a finite upper bound for age.[‡] Feller argues that the probability that a person lives longer than 1000 years is so small (with on the order of 10^{27} billions of zeroes) that it would require an immense time span seriously to test the hypothesis that age is bounded by 1000 years. He also questions the existence of a logical upper bound, on the grounds that if this bound were B in seconds, say, then the implication is that $B + 2$ seconds

[†] In recent years there has been substantial inflation, so we shall replace g by G as an upper bound. Let the monadic function $*$ operating on the positive numbers be such that $*a$ denotes the number a raised to the power a, this power operation being iterated a times starting from the right. For example $*3 = 3^{27} \approx 7.6 \times 10^{12}$. Let $G^{(1)}$ be $*g$, and define $G^{(r+1)}$ inductively as $G^{(r+1)} = *G^{(r)}$. Define G to be $*G^{(g)}$. This should suffice as an upper bound for most practical purposes.

[‡] Feller was in my opinion the last historically significant probabilist in the tradition of Pascal, Fermat, Huygens, D. Bernoulli, Laplace, Gauss, Poincaré, Markov, and a few others, who made serious contributions to the modelling of real-world phenomena. I attribute much of the débâcle in probability and statistics discussed in section 15.8 as stemming from the fact that formalistic texts, such as those of Kolmogorov, Halmos, Loéve and Doob, were more influential than those of Feller, whose books are rich in empirical content and have combinatorical underpinnings. These other texts are rigorous in the conventional formalistic sense, and can be useful to all of us for a variety of reasons, but lack the genuine flavour of probability. Unfortunately, Feller had some misconceptions about Bayesian statistics, that were perhaps due to the fact that the optimal character of Bayesian methods was not yet fully understood at the time he was writing. See Hill (1992b, 1992c).

would be impossible. To this we may answer, first, that there may well be a logical upper bound, although not known to us. Thus we must distinguish between the existence of logical bounds, and our knowledge of them. Secondly (and even more importantly) although I do not know of a logical bound for age span, I believe it is useful to separate out the practical problem, which involves ages decidedly smaller than 200 years, from the theoretical issue, which is of course unprovable one way or the other. Thus as suggested in my discussion of the Monette–Fraser example in Hill (1988), one can always deal with such issues by conditioning upon the variable being in some subset of particular interest, for example, conditional upon age being less than 200 years. One may reserve some tiny probability (that it is probably not useful to attempt to evaluate) for ages beyond this; but I expect that most people are interested primarily in the conditional problem, given that survival times are less than 200 years. (Curiously, in connection with the Kaplan–Meier estimator for survival times, when the largest observation is censored this issue becomes of some importance for conventional mathematical statisticians. Hill (1992d) offers both theoretical and empirical results to the effect that the Berliner–Hill survival estimator/predictor makes a substantial improvement over the Kaplan–Meier estimator, especially when a number of patients are censored in the upper tail of the distribution.) We also observe that if there are creatures resembling people that live more than 1000 years, they are perhaps something else than what is ordinarily meant by a human being. Thus I do not think that Feller's argument is on very solid grounds either logically or practically. It is this argument of Feller, incidentally, that opens the door to countable additivity and the *pretence* to deal seriously with infinite collections, thus leading to the errors Gauss and others warn against.[†]

When one restricts attention to finite decision problems the results are completely different from those suggested by naïve adherents of Stein-type theory. The basic mathematics is presented by Blackwell and Girschick (1954, sec. 5.2). Theorem 1 below is a strengthened version of both their Theorem 5.2.4 and of Theorem 1 of Hill (1990a). By a finite decision problem we mean one in which the parameter space, the space of possible data values, and the space of possible terminal decisions, are specified finite collections. Theorem 1 below is valid for all finite decision problems. It is

---

[†] It is not that I am arguing against infinite models per se. The ancient Greeks discovered that there was no ratio of integers whose square could be 2. This was a valuable discovery. It was even practically useful, in connection with the theorem of Pythagoras, in determining the length of the diagonal of a right triangle, both sides of which have the same length. The ratio of the length of the diagonal to that of either side was about 1.4142. However, it seems doubtful that the complete decimal expansion of $\sqrt{2}$ would be of much use in this regard, since there are no known instances of exact lines, or of exact orthogonality. Beyond a certain level of precision the 'granularity' of real-world objects becomes important, and carrying things to too many decimal points, as in the quotation by Ramsey, can be counter-productive. This is especially so when a useful criterion in the finite case, such as admissibility, is carried over to the infinite idealized case, without any attempt at justification.

assumed that a loss function has been specified, which can without loss of generality be assumed to be non-negative. See Savage (1972, p. 200), De Groot (1970, p. 125), for the relationship between loss and utility. As is customary, we shall sometimes identify a decision rule with its risk function. By a Bayes decision rule, we mean, as usual, one that minimizes the Bayes risk for some a priori distribution. In the finite case this definition is fully equivalent to the extensive form of such rules, that is, those obtained by minimizing the posterior expected loss, for each possible realizable data point x in the finite representation. We say that one procedure dominates another (or is better than another) if its risk function is everywhere less than or equal to that of the other, with strict inequality somewhere, and strictly dominates the other, if its risk function is everywhere strictly smaller. Let $\mathscr{S}$ be the convex hull of the collection of risk vectors corresponding to all available non-randomized decision functions $\delta(\cdot)$ in a specific problem. In the finite case, say with I possible values for the parameter (or more generally the state of nature), $\mathscr{S}$ is therefore representable as a closed convex set in I-dimensional Euclidean space. (For convenience, and because there is no loss in doing so, we allow all possible randomizations amongst the non-randomized decision rules.) By the Bayes boundary of $\mathscr{S}$ we mean the decision rules (represented as points in $\mathscr{S}$) for which there is no other rule in $\mathscr{S}$ that is strictly better. A decision rule is Bayes if and only if its risk vector is on the Bayes boundary, as proved in DeGroot (1970, p. 133). If a prior distribution attaches positive probability to each possible parameter value then we shall refer to it as everywhere positive.

Theorem 1 Consider any finite decision problem with any loss function. A procedure that is Bayes with respect to a prior distribution that is everywhere positive is admissible. Any decision function δ is either itself a Bayes procedure, or else is not admissible. If δ is not a Bayes procedure, then for each everywhere positive π, there exists a Bayes procedure with respect to π that strictly dominates δ in risk.

Proof Let δ be a Bayes rule for an everywhere positive prior distribution π. If δ were inadmissible, then there would exist some other decision rule δ_1 that is better than δ. Since π is everywhere positive, it follows that δ_1 has smaller Bayes risk against π than δ, which is a contradiction. Next, in finite problems the admissible class is contained in the closure of the Bayes class for prior distributions everywhere positive, and since in the finite case the collection of Bayes rules is closed, it follows that the admissible class is contained in the class of Bayes procedures. This proves the first two assertions of the theorem.

Let π be any everywhere positive prior distribution. If δ is not Bayes, then it is not on the Bayes boundary, so there exists some procedure δ_1

with all risk components strictly less than those for δ. Let $T(\delta_1)$ be all the risk vectors corresponding to points in $\mathscr{S}$ with each component less than or equal to the corresponding component for δ_1. First suppose that δ_1 is not Bayes. Then it is inadmissible, and so $T(\delta_1)$ is not empty and is itself a closed convex set. Let $\delta_1^B(\pi)$ be any decision rule that minimizes the π expectation of loss within the set $T(\delta_1)$. Because $\mathscr{S}$ is closed, such a rule exists and is admissible within the class of rules contained in $\mathscr{S}$. For there can be no better procedure in $T(\delta_1)$ because of the fact that $\delta_1^B(\pi)$ is Bayes (in the restricted problem) with respect to an *a priori* distribution that is everywhere positive. And there can be no procedure better than $\delta_1^B(\pi)$ outside of $T(\delta_1)$, since the risk function for each such rule will have a component that is larger than the corresponding component of that for δ_1, while $\delta_1^B(\pi)$ is everywhere at least as good as δ_1. In fact, $\delta_1^B(\pi)$ must be better than δ_1, since otherwise it would be identical in risk with δ_1 and the latter would then be admissible. Since $\delta_1^B(\pi)$ is admissible in the unrestricted problem, it then follows that $\delta_1^B(\pi)$ is a Bayes rule that is better than δ_1, and that is strictly better than δ. Plainly there is such an improvement over δ for each everywhere positive π. Next, suppose that δ_1 is already Bayes. In this case corresponding to δ and δ_1 take a proper convex combination of the risk vectors to obtain some δ_2 that is not Bayes, but is strictly better than δ, and proceed as before. □

It is an immediate consequence of this theorem that the implementation of any procedure, such as the James–Stein positive part estimator, to the finitized version of the problem, is either Bayes, or can be strictly dominated by a Bayes procedure. Although there are only a finite number of non-randomized Bayes procedures (in a finite problem) that dominate a particular non-Bayes procedure, there are (in principle) infinitely many randomized such Bayes procedures that do so, corresponding to hyperplanes on the Bayes boundary.[†] Thus let $\delta_{JS}^{\mathscr{S}}$ be any implementation of the James–Stein procedure for a specific finite representation of the problem, say in terms of a fixed finite grid $\mathscr{P}$ of possible values for the parameter and its estimate. According to Theorem 1, either this implementation is already Bayes, or else we can find a Bayes procedure that strictly dominates it in risk. Let $\delta_{JS}^B(\pi)$ be any Bayes procedure that so dominates the James–Stein estimator with respect to the prior distribution π. We shall refer to it as a Bayesianification of the James–Stein procedure in the direction π. It should be noted that such Bayes procedures are computable to within any desired precision with a computer of sufficiently large finite memory. According to the conventional Waldian admissibility theory, it is desirable to replace any decision procedure by one that dominates it

---

[†] The reader may note that although for convenience we sometimes use mathematical results concerning the non-finite case, we have attempted to restrict such to those with clear operational meaning.

everywhere, so hereafter when we refer to the James–Stein estimator we will mean its Bayesianification for some specified π unless explicitly stated otherwise, in which case we will speak of the traditional James–Stein estimator. When we omit the π we mean any member of the class of Bayesianifications.

In his rejoinder, Brown apparently agrees that the finite scenario is most meaningful, and does not question the truth of the original Theorem 1, but argues that the Bayes procedure suggested by myself, based upon a uniform prior distribution on a finite grid, although admissible, would only perform well on the boundary of the set of possible parameter values. He introduces an example in which the parameter vector is the vector of true log masses of four beehives. He takes the upper and lower bounds to be B and b, respectively, and states in Brown (1990, p. 537):

These James–Stein style estimators will dominate the uniform prior Bayes estimator in risk except when some μ_i is near b or B. But values of μ_i near either extreme are unrealistic. No one a priori expects the beehive to weigh as much as an elephant, and if it did, then we might seriously question our assumption that $X_i \sim N(\mu_i, \sigma_i^2)$, as well as our sanity and safety!

His claim will be critically examined in the next section. To do so we shall use the 'evaluation game' of Hill (1974, 1990a) employing a referee, who selects the parameter θ using some a priori distribution π_0, and from the perspective of this referee we examine the performance of the estimator proposed by myself and of the James–Stein estimator. The statistician or decision-maker does not know the *a priori* distribution $\pi_0(\cdot)$ of the referee, and instead uses some prior distribution of his own choice. If one decision procedure dominates another, then from the perspective of the referee the former is at least as good as the latter, with the degree of improvement being the π_0 expectation of the non-negative difference in risk functions. This method provides an objective operational criterion for choice of decision rules.

15.3 USING THE TRUE SPHERE

We wish to separate Brown's argument into two distinct cases. The first case is that in which there is a known finite least upper bound for possible $\|\theta\|$. The second case is where either a finite bound is not known to exist, or even if it does exist, it is unknown to the statistician. Brown plainly has in mind the second case, but let us consider the first case first.

Let S be the solid sphere of radius s centred at the origin, including the boundary, in k-dimensional Euclidean space. Suppose that the parameter vector θ is known certainly to lie in S, and let X be a measurement of θ, with $X \sim N(\theta, \sigma^2 \cdot I)$. We begin with the simplest case of (conditionally) independent normally distributed errors with known variance, discussed by Brown, so σ^2 is assumed known. Brown argues that only for θ near the boundary of S does the admissible Bayes estimator that 1 proposed (from now on referred to as the Hill estimator) based upon a uniform *a priori* distribution on a finite set contained in S perform better. The set S is not assumed to be the smallest sphere in which the parameter necessarily lies, but merely one that is logically certain to contain θ, as for example (if one wants to split hairs), through conditioning upon the requirement $\theta \in S$. Brown in fact considers the case where one uses an S that seems rather excessive. The example in his rejoinder has for parameter θ the logarithm of the masses of four beehives, and he takes as his upper bound the mass of a herd of elephants, while for his lower bound (which he does not specify) he might perhaps suggest the log-mass of a hydrogen atom. Both seem more than adequate.

Suppose, however, that there is a smallest sphere S_0 centred at a specified point (which we call the origin 0) in which θ is necessarily logically contained, so that the true θ necessarily lies inside S_0, and there is no smaller sphere centred at 0 with this property. In the usual development of the mathematics of the continuum, as soon as one has assumed that there is some finite sphere S centred at 0 in which θ must lie, it then follows that there is a greatest lower bound (possibly 0) for the radii of such spheres, by the complete ordering postulate for the real numbers as applied to the radii. See Birkhoff and Mac Lane (1953, p. 92). Let s_0 be this greatest lower bound, so that it is in fact the radius of the sphere S_0.[†]

Even without such mathematics one can sometimes argue that there is such a sphere, at least up to trifles. The type of example we have in mind is the following. Ignore the fact that the earth is not an exact sphere, and suppose that the parameter θ can be any point on or inside the earth, with all such points possible. We will be given a Gaussian measurement X of θ, with known variance. The 'sphere' S_0 is the earth itself, and the problem is such that it meets our requirement that it is the smallest sphere in which the parameter is necessarily logically contained. In the more general case, such as that discussed by Brown, where knowledge of an upper bound is less solid, we use the conditioning device mentioned in the discussion of age earlier. That is, we take a sphere that is sufficiently large for our purposes, condition upon the parameter lying in that sphere and then discuss only the conditional problem, since the unconditional problem is not of much interest. In this case we do have a logical upper bound in the conditional problem. In the remainder of this article, whether by conditioning, or by assuming explicit knowledge, as in the earth example, we suppose that there is a true S_0 which may or may not be known.

[†] Note that this is the type of mathematics of the non-finite that Brouwer argues against. As remarked earlier, it is different to discuss the existence of a bound than to discuss knowledge of the value of such a bound. Apart from trifles, θ must lie in S_0 under the postulate.

In this context we now examine the performance of the traditional James–Stein estimator relative to the Hill estimator based upon a uniform a priori distribution on the true S_0, in connection with a specified finite grid to represent the parameter and its estimate. From a pre-data point of view we can suppose that the referee must select a parameter value in S_0, as for example, when he uses an *a priori* distribution with support a finite subset of S_0. The radius s_0 of our true sphere is then the largest value of $\|\theta\|$ given positive mass by the *a priori* distribution of the referee. Note that it is not necessary that the selected θ necessarily lies on the boundary of S_0, which would be quite a different matter. An alternative formulation is where the referee must select the parameter so that each component is bounded in absolute value, as in Brown's example. Since this leads to essentially the same conclusions as with spheres, we shall not discuss the case of such boxes. It should be remarked that although most of our discussion in this section concerns the finite case, many of the conclusions would also be valid using the continuum as parameter space, as in section 15.4.

Would Brown think that the traditional James–Stein estimator, or even its Bayesianification, is necessarily better in terms of risk than the Bayes procedure for a uniform *a priori* distribution on the finite grid of points in S_0, when S_0 is known? Plainly it need not be from the point of view of our referee. If the referees' prior distribution happens to place sufficiently high weight near the boundary, then the Hill estimator is better, as in the quotation by Brown. On the other hand, if his prior distribution places relatively more weight near the origin towards which the James–Stein estimator shrinks, then the Bayesianification of the James–Stein procedure is better. There does not seem to be any objective sense in which the James–Stein estimator is better, but merely a subjective judgement as to where the referee has put his mass. For Brown to argue against the Bayes estimator based upon a uniform distribution over the true S_0 is to fly in the face of almost all serious theories of statistics, including the Fisherian fiducial approach, the Neyman–Pearson approach, and the Waldian minimax approach, all of which (sometimes implicitly) emphasize the uniform case as being both robust and relatively 'objective'. If he objects in dimension 3 or more, would he also object in the case $k = 1$ on the line? Is there more reason to object to a uniform distribution in higher dimensions than in the case of one dimension? As explained later, we regard the dimensionality argument in which one only argues against the uniform *a priori* distribution when the dimension is 3 or more, as yet another red herring of the Stein theory.

The Steinian literature, including the previous quotation from Brown, implies that the traditional James–Stein procedure is objectively better in some sense irrespective of the empirical context of the problem. So if the James–Stein estimator were better in some objectivistic sense, then it would have to improve upon the Bayes decision procedure for a uniform prior distribution, even if this were the referees' true prior distribution. This is plainly impossible. Furthermore, how is one even to decide in a particular example which Bayesianification $\delta_{JS}^B(\pi)$, that is, which π, to use to dominate the traditional James–Stein procedure amongs all (positive) prior distributions with support S_0, any of which could be that of the referee, without specific consideration of the empirical context of the problem, including the upper bound?[†] In the special case where the true sphere is known it is clear that just as with any other two Bayes procedures, neither is objectively better than the other, and it is a question of where one allocates his *a priori* mass (even if only to decrease risk in the appropriate places) based upon subjective *a priori* knowledge.[‡]

Now let us consider the case where a true sphere exists but is unknown. In this case the centre of the true sphere may or may not be known. Suppose that the centre is known and called 0, and define S_0 using the previous arguments, but with the radius s_0 now unknown. Suppose then that I use the Hill procedure for some S with an excessively large radius s, as in Brown's example, but to begin with, centred at the same place as the true S_0. It would seem that such an S can easily be constructed in real-world problems, as Brown himself has done. How much will my ignorance of the true s_0 then cost me? Again, it is plain that if my sphere is only slightly too large, then I will still do better (in the sense of lowering risk) near the boundary of the true sphere. On the other hand, if my S is sufficiently large, then I can do everywhere worse than the traditional James–Stein estimator within the true sphere S_0 (although not within S) since at the boundary of the true sphere (when s_0 is not too small) his risk is a little less than $k \cdot \sigma^2$, while if s is sufficiently large then my risk approaches $k \cdot \sigma^2$ throughout the true sphere. So we can assert, as Brown implies, that a Bayesian who uses an absurdly large value for the upper bound does less well than either James–Stein, any Bayesianification of James–Stein, or than a Bayesian who uses a uniform prior distribution on the true sphere. This is not entirely surprising. A very poor prior distribution can lead to very poor decisions. Note, however, even here that the use of a uniform *a priori* distribution on S with s very large is not really so awful, being nearly a minimax procedure, as discussed below. It is merely the fact that the risk could be improved upon everywhere in the true sphere S_0 (if it is sufficiently small relative to S) that is used to castigate our procedure when used with too large S.

[†] As we shall see below, the subjective knowledge regarding the point towards which to shrink is even more critical than subjective knowledge as to the upper bound for θ.
[‡] It is Theorem 1 that justifies this conclusion, since from this theorem we know that the traditional James–Stein estimator is dominated by any of its Bayesianifications, and we know that no Bayesian procedure can dominate any other if each is Bayes for an everywhere positive *a priori* distribution.

But how much can our procedure be improved upon in risk over the true sphere? Very close to the origin towards which the James–Stein procedure shrinks, here the assumed known centre of the true sphere S_0, the improvement can be substantial. On the other hand, it is apparent from the form of the James–Stein estimator that this estimator is only likely to be appreciably different from X itself when the true θ is quite close to the origin towards which Stein shrinks. Here 'closeness' is measured in terms of the known standard deviation σ of the measuring instrument. Plainly $1/\|\theta\|$ is negligible, and therefore $1/\|X\|$ is very likely to be negligible, if $\|\theta\|$ is sufficiently large. Thus, if we again look at matters from the perspective of our referee, while it is true that with sufficiently large s he sees myself as doing less well in the true sphere than James–Stein, on the other hand the magnitude of the improvement that he sees is trifling, unless in fact the referee's *a priori* distribution is quite closely concentrated near the origin towards which they shrink. To seriously discuss this issue we must now address an extremely interesting question, namely, the choice of origin for the James–Stein estimator.

In the traditional Steinian theory it is implicitly assumed (though not always explicitly stated) that the origin towards which one shrinks is either objectively known in some way or other, or is a matter of subjective judgement, or as in random effects models, is itself estimated from the data. We consider now the first two cases, as the third is best discussed in the context of the random effects model in section 15.6. Our basic assumption is that there is a true value for θ, say the value selected by the referee; and that we get to observe $X = \theta + \varepsilon$, where $\varepsilon \sim N(0, \sigma^2 \cdot I_k)$. Of course observations from a normal distribution are not really possible, but we assume, for the purpose of insight, that the approximation is excellent.

Now the origin has no objective significance. For example, we might measure X and/or θ relative to any origin whatsoever, say θ_0, at least in principle. In fact we could use a different reference point for X than for θ. Here θ_0 also need not have any particular significance, other than as a reference point. Physical measurements, such as lengths and weights, are typically ratios, relative to some standard, such as an official kilogram. Suppose that θ is measured relative to a fixed and specified reference point, θ_0, for example relative to the centre of the earth in our example; and that X of our previous discussion is originally measured relative to any other reference point. Now let the reference point for X be altered to become θ_0, also. Thus the observation is now $X_0 = X - \theta_0$, where X is still measured relative to its original reference point. In the classical least-squares approach, it makes no difference how we chose θ_0, since our estimate of the original θ would still be $\theta_0 + X_0 = X$. But this is not true for the traditional James–Stein estimator. If X_0^{JS} is the James–Stein estimator based upon the data X_0 shrunken towards the origin (the vector of zeroes), then the obvious James–Steinian estimator of θ would be (with $\sigma^2 = 1$)

$$Y_0 = \theta_0 + X_0^{JS}$$
$$= \theta_0 + (X - \theta_0)\left[1 - \frac{k-2}{\|X - \theta_0\|^2}\right]$$
$$= X\left[1 - \frac{k-2}{\|X - \theta_0\|^2}\right] + \frac{(k-2)\theta_0}{\|X - \theta_0\|^2}. \quad (1)$$

See Anderson (1984, p. 86).

Apart from the fact that the 'weight' given to X can be negative, this exhibits the original Stein estimator as a weighted average of the original observation X and the arbitrary reference point θ_0, in that the coefficients sum to 1. (For the positive part estimator this deficiency was removed, but see section 15.6 for a new difficulty with the positive part estimator.) Such estimators are familiar to Bayesians, apart from the negative weight, as for example when the *a priori* distribution for θ is of the Gaussian form centred at θ_0. However, there is an important distinction that may not have been noticed or given much attention in the Steinian literature. In the Bayesian scenario the value θ_0 is given special significance, as representing genuine a priori information. It is for example, the prior expectation of θ. By contrast, in the Stein scenario, θ_0 can be any reference point whatsoever, and particularly in Stein's anti-Bayesian context, should presumably be chosen 'objectively' without reference to *a priori* information.

But the *numerical value* of his estimate of θ actually depends quite sensitively upon the choice of reference point θ_0! For example, suppose he behaved like a foolish Bayesian and in Brown's example he used for his arbitrary reference point θ_0, the log mass of a neutron star. Then our referee (who thinks that beehives weigh considerably less) does not expect to see such a large X, and judges that the weight given to X in the above equation is virtually certain to be essentially 1, so that the James–Stein estimator is very likely to be nearly identical with the classical least-squares estimator X. Even with an absurd choice of reference point the James–Stein estimator dominates X as an estimator of θ in the idealized scenario, but the domination is only non-trifling if θ_0 is sufficiently close to the true θ. It is understood by sophisticated users of the Stein machinery, that θ_0 should be taken as the *a priori* expectation of θ, just as in the Bayesian approach, although this point is not emphasized, perhaps because it begins to destroy the 'objective' character of the Stein approach. Even with this best possible interpretation, however, somehow or other the *a priori* expectation must in fact be very close to the true θ for there to be any non-trivial improvment. This is even true in my earth example if the point θ_0 selected by Stein, say the centre of the earth, is far from where the referee has concentrated his mass.

Thus in Brown's example suppose one were to shrink instead towards a vector with all components equal to the log mass of a neutron star, say $\theta_0 = N$ as origin, using the (traditional) James–Stein procedure. Then the risk function for the James–Stein estimator of the original θ, using this origin, would have the value approximately $k \cdot \sigma^2$ even for θ in the sphere S based upon Brown's upper bound in terms of the log mass of a herd of elephants. My procedure, based upon the same S with Brown's upper bound, would have a slightly smaller risk than that of the James–Stein estimator shrunk towards N, for every value of θ in S. The referee, who perhaps has some much smaller S_0 in mind than S, would think that we are all mad, but would certainly find my procedure a little better than the James–Stein estimator shrunk towards the log mass of a neutron star. Plainly, however, we are talking about trifles, and we could all do much better than any of these procedures by using realistic prior knowledge.

We can now see that there is a curious special character to the Steinian type of analysis. To give a non-trivial improvement over X the *a priori* reference point, θ_0, must be quite close to the true θ. So in order to recommend the use of the James–Stein procedure as a non-trivial improvement over the classical least-squares estimator, one must be rather confident that his *a priori* information for choice of the centre is on a solid footing. On the other hand, as in Brown's example, the Steinians would like also to pretend to be extremely sceptical about *a priori* information with respect to the magnitude of the departure of θ from their reference point θ_0. Presumably this is the basis for Brown's choice of the mass of elephants. In Bayesian terms, they are comfortable with the *a priori* expectation, but dubious about the prior variances. This is a possible state of mind, although in either Brown's example or my earth example, the magnitude of departure is plainly not so vaguely known as they pretend. At any rate my use of perhaps a far too large sphere, that is my choice of an unfortunate prior distribution, only sets me back to a risk of about $k \cdot \sigma^2$, just as with the least-squares estimator; and this is also precisely where the Steinians are set back, if they make a serious misjudgment in the selection of θ_0. Furthermore, precisely in the case (such as Brown's elephant) in which s is chosen far too large, then such people must be quite lucky to get a reference point θ_0 that is only a few multiples of σ from the true θ,[†] since this would represent a negligible fraction of the volume of the large sphere S that Brown would have us use. Of course, in practice often the same background knowledge that leads to a good guess at the reference point suggests a reasonable discrepancy as well. Even when this is not the case, once we recognize that any implementation of the James–Stein estimator can always be dominated by a Bayes estimator, we see that the real issue is just the usual one of a choice between two Bayes decision procedures. It was only the way in which the Stein paradoxes were presented, using the infinite sphere as a straw man, that made this seem otherwise. Whether

---

[†] And it is only in this case that their improvement would be non-trivial.

with regard to my choice of s or Stein's choice of a reference point, the issue comes down to the appropriateness of the *a priori* information being employed. And what I think is especially unfortunate is that while the reasons for the importance of making careful a priori assessments are entirely clear in the Bayesian analysis, this is hidden by the 'objectivistic' manner in which Stein's results were originally presented.

15.4 THE RISK FUNCTIONS

The various risk functions are now discussed. Here we shall pretend to be dealing with the continuum, as in the standard Steinian literature. To begin with we consider the special case where an origin is given, as in the earth example, and is called θ_0. The referee then selects a parameter θ, which is any point in the sphere S_0 of radius s_0 centred at θ_0. As explained earlier, S_0 can be defined as the true 'smallest' sphere in which the referee *necessarily* selects θ. The observation is $X \sim N(\theta, \sigma^2 . I)$ as before. In the special case most favourable to Stein, θ_0 is somehow or other known to him, and so he shrinks towards this θ_0 as reference point, yielding (1) as the Steinian shrinkage estimator.

We shall compare the risk functions for the James–Stein estimator shrunk towards the origin θ_0, the Bayes estimator for a uniform prior distribution on S_0, and the Bayes estimator for a uniform *a priori* distribution on any larger sphere S centred at θ_0, such as the one that a Bayesian might use if he knows the centre of the true sphere, and chooses $s \geqslant s_0$ without knowing s_0 itself. Thus here the first issue to be discussed only concerns the radius of the sphere S to be used, and not the choice of reference point. Let $\theta_s(X)$ be the Bayes estimator based upon a uniform prior distribution over the sphere of radius s centred at the origin, and let $R_s(\cdot)$ be its risk function as a function of θ. As $s \to \infty$ the risk function $R_s(\theta) \to k . \sigma^2$ for each θ.

We first consider the risk function based upon the uniform *a priori* distribution over S, as a function defined for all θ, and later examine it from the perspective of the referee who knows that $\theta \in S_0$. It is easy to verify that $R_s(\theta)$ depends only upon $\|\theta - \theta_0\|$. In this discussion it is convenient to take $\theta_0 = 0$, so that the risk function can be written as

$$R_s(\theta) = r_s(\|\theta\|),$$

and is a function of a single non-negative real variable. The function $R_s(\theta)$ has a stationary value at 0, and this value is a relative maximum whenever s/σ is not too small. Under this condition the value at 0 is typically only slightly less than $k . \sigma^2$, and $R_s(\theta)$ then decreases slowly until we approach to some distance, say $c_k . \sigma$, from the boundary of S. For as proved in greater generality in Hill (1974, p. 562), for s/σ not too small, the posterior distribution given the data $X = x$, based upon the uniform prior distribution in S, is essentially

$$\theta \sim N(x, \sigma^2 . I),$$

unless x is within a fixed distance $c_k . \sigma$ from the boundary of S. The constant c_k is here obtained from the chi-square distribution with k degrees of freedom, and is such that the sphere of radius $c_k . \sigma$ about the origin contains most of the mass of the error distribution. It follows that for $\|x\| \leq s - c_k . \sigma$ the truncation to S has negligible effect on the posterior distribution of θ. Furthermore, for values of θ with

$$\|\theta\| \leq s - 2 . c_k . \sigma,$$

the random variable X has high probability of being in the sphere of radius $s - c_k . \sigma$ centred at the origin, and so the risk function $R_s(\cdot)$ in the sphere

$$\|\theta\| \leq s - 2 . c_k . \sigma$$

is very nearly that of the classical least-squares estimator X. The risk function $R_s(\theta)$ does not have its minimum at the boundary, but rather a little before the boundary (whenever s/σ is not too small), and the magnitude of the minimum achievable risk can be substantially smaller than that for X.

If we go outside the sphere S then somewhere beyond s the risk function $r_s(\cdot)$ begins to increase (essentially) quadratically, and tends to infinity as $\|\theta\| \to \infty$. This is because our Bayes estimator shrinks into S, so if θ is far outside of S then with high probability so also will be X, and the Bayes estimator is a shrinkage of X to essentially its perpendicular projection on to the surface of S. Hence the risk function for $\theta_s(X)$ is then close to $\|\theta\|^2 - s^2$ for large $\|\theta\|$. The risk function of the Bayes estimator for a uniform prior distribution on the true S_0 is, of course, of exactly the same nature, merely replacing s by s_0.

For any S with $s < \infty$ the Bayes estimator $\theta_s(X)$ is admissible with respect to squared error loss for the restricted problem in which $\theta \in S$, and so all three risk functions, $R_{JS}(\cdot)$, $R_s(\cdot)$, and $R_{s_0}(\cdot)$ must cross within S. Plainly, in comparing R_s with R_{s_0}, the latter is better for θ within S_0, while the former is substantially better well outside of S_0. For small s_0 this still holds, with the improvement of the latter now only for θ very close to the origin. When s_0/σ is not too small, then the risk function $R_{s_0}(\theta)$ in the sphere

$$\|\theta\| \leq s_0 - 2 . c_k . \sigma$$

is close to $R_s(\theta)$ for any $s > s_0$, with both being approximately $k . \sigma^2$, for the reasons given earlier. It follows that the major reduction in risk derived from knowing the true radius s_0 as opposed to using some larger value s, occurs for θ within the annulus of width $2 . c_k . \sigma$ from the boundary of the true sphere of radius s_0, since this is where the risk is minimum in the true sphere. In the way we have set up the problem, with a true sphere; there is of course never any positive advantage to using a larger sphere than necessary, but one loses very little by using too large a sphere unless θ is quite close to the boundary of the true sphere. Of course, if one is mistaken about the existence of a true finite sphere, then it may be advantageous to use a very large value for s, so that one does not shrink an X that comes from a huge θ that is truly there and larger than ordinarily thought possible down to a preconveived size. In any case, in applications one cannot shrink towards the sphere of radius s_0 without having very strong *a priori* knowledge as to the existence and value of that upper bound. Although it is better to know the value of s_0 should it exist, the above discussion suggests a strong robustness property for the Bayes procedure based upon a uniform prior distribution over the sphere of radius s for even much too large s. See Hill (1980a, 1992c) for a general formalization of robustness in Bayesian statistics.

For $s > s_0$ we shall now compare the traditional James–Stein estimator θ_{JS}, the Bayes estimator θ_s and the Bayes estimator θ_{s_0} based upon the true s_0. The estimator θ_s is admissible for any fixed finite s, and typically makes a substantial improvement over the others for θ within S and sufficiently near the boundary of S. However, such admissibility does not hold *in the limit* as $s \to \infty$. By sleight of hand, ignoring the real-world boundedness of parameters and data, Stein implicitly lets the band in which R_s improves upon the other risk functions, drift off to infinity (that is, to nowhere). As we let $s \to \infty$ the risk function for θ_s tends to $k \cdot \sigma^2$ at every fixed θ, while the James–Stein estimator has risk only slightly less than $k \cdot \sigma^2$, except very close to the origin, where the improvement is substantial.

It is precisely here that Stein flies in the face of the warnings of first-rate mathematicians such as Gauss. He passes to a limit and then ignores how he got there, as well as the meaning of this limit as an approximation.[†] Brown correctly states that the Hill estimator only makes a substantial improvement upon the James–Stein estimator near the boundary of S, and that if my S is excessively large (as in his example) I will do everywhere worse in the true sphere S_0 than the James–Stein estimator. In the same way I can say that if my sphere is $S = S_0$ as in the earth example, or is only slightly too large, then it improves upon the James–Stein-type estimator (including its Bayesianifications) except near the origin, i.e. when θ_0 is very close to the true θ. The fact is that the risk functions must cross within S, and so if my s is sufficiently close to s_0 then my estimator is substantially better in some solid annulus near the boundary of S, while the James–Stein estimator is substantially better very close to the centre.

† This caused some confusion in the rejoinder by Brown, who thought that the estimator Hill was proposing based upon a bounded sphere, was different from the classical least-squares estimator in a non-trival sense. Hill did not take the trouble to distinguish between the classical estimator and that based upon a uniform distribution over S with s sufficiently large, as perhaps Laplace and Gauss did not also, since the classical estimator does not really exist as an implementable procedure. A computer would have to have infinite memory to deal with arbitrarily large $\|x\|$.

The next question is precisely how to measure the closeness of θ to the boundary, say in the case where $S = S_0$, in order that my estimator improve substantially upon that of James–Stein? Brown implies that θ must be very close in order for this to be the case, as though such were highly unlikely. There has been substantial confusion on this issue due to the dimensionality question. This is discussed in Hill (1974, p. 562). For a sphere of radius R in k-dimensional Euclidean space, the volume is proportional to R^k. Hence the volume contained in the annulus extending from the surface of the sphere of radius $R - r$ to the surface of the sphere of radius R, with $0 < R - r < R$, as a fraction of the entire volume of the sphere of radius R, is given by $1 - [1 - r/R]^k$. For fixed $0 < r/R < 1$ this fraction of the total volume tends to 1 as $k \to \infty$, so the mass becomes relatively more concentrated near the boundary as the dimensionality increases. On the other hand, for fixed dimension k, as $R \to \infty$ with fixed r, plainly the relative mass near the boundary goes to 0, which is why for sufficiently large (but finite) s a Bayesian who uses the uniform distribution over S behaves very nearly as though he were using the improper Jeffreys uniform *a priori* distribution.

The limit with respect to k has given rise to the following confusion on the dimensionality question. Suppose k is fixed but large, say $k = 100$. There will be a relatively thin annulus extending to the surface of S, in which, say, half the volume of the sphere of radius R is contained, and let us say the values in this set have a certain property. Then one might deceive oneself into thinking that only for parameter values in the 'small' set corresponding to the region between the spheres of radius $R - r$ and R does the property hold. Returning to the quotation by Brown, while it is true that the radius of the sphere in which the James–Stein estimator is better in risk than my estimator is larger than r, this is misleading, since the relative volume of the annulus extending from the sphere of radius $R - r$ to the boundary, may be quite large. Indeed, because of the first dimensionality limit, if the annulus where my estimator is better were not 'small' when measured in terms of the radius, then the volume where such improvement occurs would be an enormous percentage of the entire volume. For example, if $k = 100$, it turns out that when $r/R = 0.007$, half the volume of the sphere of radius R lies between the sphere of radius $R - r = 0.993R$ and the boundary of the sphere of radius R. While in a certain sense it is true that the Hill estimator improves only 'near' the boundary, it is incorrect to imply that therefore this is a small set in the sense of volume! The comparison between the risk functions for the James–Stein estimator and the Hill estimator is like the comparison of the risk functions for any two real-world Bayesian estimators. Each being admissible, must do better somewhere. By sleight of hand one can make the set where mine does better look small, for example by considering only a narrow band of 'size' $0.007R$ in terms of the radius, and suggesting that therefore over 99% of parameter values have the property that they lie in the region where the James–Stein procedure is better. But this would not change the fact that my set still would have half the volume.

Although over-simplified, I believe that the qualitative nature of the traditional James–Stein procedure in real-world problems is as follows. Suppose that the parameter is known to lie in a huge but finite specified set, and you have a choice between two estimators (or their associated risk functions). For the first, corresponding to the James–Stein estimator, you arbitrarily select a point θ_0 in the set; then if in fact the true value is $\theta = \theta_0$, the risk is 0 near θ_0, and is everywhere else the constant $k \cdot \sigma^2 > 0$. The second risk function, corresponding to the Hill estimator, is nearly constant (except close to the boundary where it is smaller) with a value slightly less than $k \cdot \sigma^2$, say $k \cdot \sigma^2 - \varepsilon$. Then the naïve argument for the James–Stein estimator (or risk function) is that if you guess correctly then you do extremely well, and even if you are far off the mark, you (mostly) only lose ε as compared to the Hill estimator. (This naïve argument can be likened to an argument for playing a lottery game where one tries to guess a 12-digit number, on the grounds that one wins an enormous prize if one guesses correctly, and only loses the price of a ticket if not.) What this argument ignores is that if you have no real basis for the selection of θ_0, as for example when your *a priori* distribution is uniform over the large set, then you must think that your chance of being correct is negligible. Since you do prefer $k \cdot \sigma^2 - \varepsilon$ to $k \cdot \sigma^2$, therefore you may prefer the Hill procedure unless you have a reason, as represented by your *a priori* knowledge, to think that you can choose θ_0 wisely. Note that the Bayes risk for my procedure will be smaller than that for the James–Stein procedure when the referees' *a priori* distribution is in fact uniform over S, and by continuity, also for distributions sufficiently close to uniform. This implies that the Hill procedure is generally preferable for procedures that respect volume. Even the slight improvement of ε throughout most of S would be picked up in repeated plays of the evaluation game, to lead to a substantial difference in favour of the Hill procedure when the *a priori* distribution is nearly uniform. This analysis is an over-simplification primarily in that we have replaced values of the risk functions that are constant only up to trifles, by constants, and if anything, errs on the side of being overly generous to the James–Stein procedure. It may be noted that while choice of risk functions is very nearly equivalent to choice of prior distributions, in examples such as this it is much easier to be misled if the problem is viewed purely in terms of the risk function.

15.5 EXTENDED ADMISSIBILITY

If one is to deal with infinite parameter spaces at all, then, as argued in Hill (1992a), extended admissibility is a better concept than admissibility. In the finite case, which we have argued is the only case of importance in real-world problems, the criterion of extended admissibility is mathematically equivalent to the restriction to Bayes procedures based upon *a priori* distributions that are everywhere positive, and which are therefore necessarily admissible. There are pros and cons in the finite case for thus restricting to the class of extended admissible procedures. For example, a prior probability of 0 in this context, would mean 0 to infinitely many decimal points. Excluding events known to be logically impossible, such as $A \wedge (\sim A)$, I doubt that anyone has such probabilities of exactly 0. In this case the distinction is not of great importance.

However, the purpose of various authors in proposing extended admissibility, is instead to provide a handle on the infinite case. With this criterion, when one procedure dominates another it cannot be only at a single point, such as at an isolated transcendental number that has never even been thought of, but must be everywhere. In this case the Stein paradox disappears for yet another reason, since Heath and Sudderth (1978) have proved that the classical least-squares estimator is admissible in the extended sense. Note that as $\|\theta\| \to \infty$, the risk functions for the various Stein-type estimators approach that of the least-squares estimator. The Stein procedures then cannot improve uniformly in a neighbourhood of infinity. There are prior distributions that give substantial weight outside any finite sphere, i.e. finitely additive priors, for which the classical least-squares estimator is optimal. Thus in a certain sense the classical least-squares estimator, if taken literally, can be interpreted as a Bayes estimator that shrinks towards infinity, i.e. it does not shrink at all. I do not know of any examples, however, where (taken literally) this prior distribution is appropriate. I do know of many examples where the least-squares procedure provides an excellent approximation, since one knows that one is virtually certain not to see extreme data, and in this case the stable estimation argument of L. J. Savage (1962, p. 20) gives the required justification for the uniform *a priori* distribution. See Hill (1974, 1990b) for discussion of the extreme data problem.

Although better than ordinary admissibility as a criterion, I do not think that even the criterion of extended admissibility allows us to deal in a satisfactory way with non-finite sets. Thus as an operational procedure, one would have to be able to record which of a non-finite number of possible observations is realized, and also to choose a decision from a non-finite set, with all such choices being possible. But as we argued earlier, in connection with human age, it is necessary at some point to replace the infinite collection by a finite set, if a procedure is to be implementable. Whether this is justified on the grounds of the finite memory of a computer, or of common sense, or by replacing the original problem by a finite conditional problem, is not important. But I think that it must be done in order to avoid complications that in my opinion are almost totally irrelevant to real-world problems of inference, prediction or decision-making.

And as soon as this is done all the known paradoxes disappear, including those discussed in Hill (1980b) concerning non-conglomerability.

15.6 RANDOM EFFECTS MODELS

The importance of the random effects model was emphasized in Hill (1977, 1980a), both with respect to modelling of real-world phenomena, robustness of Bayesian methods and in regard to the admissibility question. Such models are now widely used in many real-world problems to represent situations where the parameter is chosen randomly according to a fixed (but possibly unknown) *a priori* distribution, as for example when it is generated by a mechanism under conditions that vary in some random fashion. As will now be shown the (Bayesian) random effects model, which is of course a hierarchical model, can incorporate highly informative *a priori* distributions, and constitutes an important alternative to the type of analysis based upon weak *a priori* knowledge presented above. See Lindley and Smith (1972) for a general presentation of the Bayesian random effects model, and Bhat (1988) for an important extension of the results of Hill (1977) to principal components analysis.

The simplest random effects model is equivalent to the one-way random model that arises in the analysis of variance, as in Scheffé (1959, Ch. 7). Suppose that the data to be observed have the structure

$$y_{ij} = \mu + \alpha_i + \varepsilon_{ij}$$

for $i = 1, \ldots, I$, $j = 1, \ldots, J_i$. Here we assume that the random quantities $\varepsilon_{ij} \sim N(0, \sigma^2)$, and $\alpha_i \sim N(0, \sigma_a^2)$, are conditionally independent, given $\theta = (\mu, \sigma^2, \sigma_a^2)$, where the components of θ are unknown constants, and have some *a priori* distribution. Note that the model is mathematically the same whether interpreted in conventional ANOVA terms, where the α_i are random in the usual sense, or whether they are fixed but unknown constants, just as with θ. Hence we shall not attempt to distinguish between these two cases.[†]

Let $\tau^2 = \sigma_a^2/\sigma^2$. It is easily seen that inference about all other parameters, given τ^2, is quite simple, and that the difficulties in the analysis of this model reside only in the determination of the posterior distribution of τ^2. Define $\mu_i = \mu + \alpha_i$, for $i = 1, \ldots, I$. Then the structure of the data is $y_{ij} = \mu_i + \varepsilon_{ij}$, which is plainly a generalization of the conventional Stein model for inference about an unknown mean in an I-dimensional Euclidean space. Our model is more general both in that σ^2 is not assumed known, and also in allowing for unequal numbers of observations in the different rows. It is immaterial whether the parameter is taken to be θ as in conventional random effects ANOVA, or whether it is taken to be (μ, σ^2) as

---

[†] It may be observed that such a distinction is also not meaningful in theories in which 'randomness' arises through instability, as in conventional chaos theory.

in the Stein model, with $\boldsymbol{\mu}$ the I-component vector consisting of the μ_i. In the Bayesian framework, a prior distribution for θ generates a prior distribution for, and therefore a logically consistent inference about, $\boldsymbol{\mu}$. This issue has caused some confusion in the non-Bayesian scenario, since in the conventional random effects model α_i is thought of as the realization of a random variable, rather than as an unknown constant. But in the Bayesian approach, or from a purely logical/mathematical point of view, there is no inherent difference between the case where a quantity is unknown, and the case where it is modelled as the realization of a conventional random variable. In both cases probability is used to represent our *uncertainty* about the quantity in question.

In the following analysis we use the standard sufficiency properties of the normal distribution, as for example in Scheffé (1959, p. 225) and Hill (1967). The random variable

$$y_{i.} = \frac{\sum_{j=1}^{J_i} y_{ij}}{J_i}$$

has the distribution $N(\mu_i, \sigma^2/J_i)$, conditional upon μ_i, σ^2, with the $y_{i.}$ conditionally independent, given these parameters. Next, conditional upon θ, we have $\mu_i \sim N(\mu, \sigma_\alpha^2)$, with the μ_i conditionally independent, given θ. Thus we are in the most familiar of Bayesian problems, namely, the measurement of means μ_i by normal observations $y_{i.}$ in the case where the μ_i are themselves jointly normal and (conditionally) independent. Hence the partial posterior distribution of $\boldsymbol{\mu}$ given only the data $y_{i.}$, $i = 1, \ldots I$, and the parameter θ, is

$$\mu_i \sim N(\gamma_i[y_{i.}] + (1 - \gamma_i)[\mu], [1/\sigma_\alpha^2 + J_i/\sigma^2]^{-1}), \tag{2}$$

with the μ_i conditionally independent, given the data $y_{i.}$ and the parameter θ. Here

$$\gamma_i = \gamma_i(\tau^2) = \frac{J_i \cdot \tau^2}{1 + J_i \cdot \tau^2}.$$

Such a posterior distribution for the μ_i is what is desired, except that we want it conditional upon all the data, and unconditional upon all parameters. Our next step is to uncondition upon μ.

If either μ is unknown in conventional terms, or if μ has a sufficiently large variance in Bayesian terms, then the optimal conditional estimator for μ, given τ^2, is

$$\hat{\mu}(\tau^2) = \frac{\left[\sum_{i=1}^{I} \frac{J_i \cdot y_{i.}}{1 + J_i \cdot \tau^2}\right]}{\sum_{i=1}^{I} \frac{J_i}{1 + J_i \cdot \tau^2}}. \tag{3}$$

This estimator is in fact the posterior expectation of μ, given τ^2 and all the data. It follows that the posterior expectation of μ_i, given all the data, and conditional only upon τ^2, is

$$\mathscr{E}(\mu_i \mid \text{Data}, \tau^2) \equiv \hat{\mu}_i(\tau^2) = \gamma_i[y_{i.}] + (1 - \gamma_i) \cdot \hat{\mu}(\tau^2). \tag{4}$$

Finally, conditioning only upon the data gives the desired posterior expectation:

$$\mathscr{E}(\mu_i \mid \text{Data}) = [\mathscr{E}\gamma_i(\tau^2)] \times [y_{i.}] + \mathscr{E}[1 - \gamma_i(\tau^2)] \times \hat{\mu}(\tau^2), \tag{5}$$

where in the last equation the expectations are taken with respect to the posterior distribution of τ^2.

In the balanced case $J_i \equiv J$

$$\gamma_i(\tau^2) = \frac{J \cdot \tau^2}{1 + J \cdot \tau^2}, \tag{6}$$

and

$$\hat{\mu}(\tau^2) = \bar{y} = \frac{\sum_{i,j} y_{ij}}{IJ}. \tag{7}$$

Thus in the balanced case the only quantity that needs to be evaluated explicitly to obtain the Bayes estimator for the μ_i is the posterior expectation of

$$\frac{J \cdot \tau^2}{1 + J \cdot \tau^2}.$$

From here on we shall confine attention to the balanced case. As discussed in Hill (1977, 1980a) there is a lack of robustness of the Bayesian analysis in this problem. There are two important scenarios. In the first the variance components σ_α^2 and σ^2 are *a priori* independent, while in the second scenario the ratio τ^2 is independent of the within-variance component σ^2. The second scenario comes closest to conventional non-Bayesian analysis of the random effects model. *Within* the framework of the second scenario the Bayesian analysis is sometimes robust, and except for one important difference, the Bayes estimator more or less agrees with the James–Stein estimator.

Consider the new parameter

$$\gamma_1 = \ln(1 + J \cdot \tau^2).$$

The Bayesian analysis of the problem in terms of this new parameter is then extremely simple. With the usual notation for the between and within mean squares, we have conditional upon τ^2,

$$\frac{MSB}{MSW} \sim (1 + J \cdot \tau^2) \cdot F_{m,n},$$

where $m = I - 1$, $n = I(J - 1)$, and $F_{m,n}$ is a random variable having the F-distribution with m and n degrees of freedom.

Taking logarithms yields

$$\hat{\gamma}_1 = \gamma_1 + \delta_{m,n}$$

where $\hat{\gamma}_1 = \ln MSB/MSW$, and $\delta_{m,n}$ has the distribution of $\ln F_{m,n}$. This is a simple location model, but note that while $\hat{\gamma}_1$ can be anywhere on the real line, it is necessarily the case that $\gamma_1 \geq 0$. Now suppose that we are given only the data MSB/MSW, and using this data are to obtain the posterior distribution for γ_1. If $p_{n,m}(\cdot)$ is the density of $\delta_{n,m}$ then the likelihood function for γ_1 is

$$l(\gamma_1) = p_{n,m}(\gamma_1 - \hat{\gamma}_1),$$

defined only for $\gamma_1 \geq 0$. However, since the *a priori* density for γ_1 is 0 when $\gamma_1 < 0$, we can extend the definition of the likelihood function $l(\gamma_1)$ to the entire real line, and then utilize the *a priori* density to rule out negative values. The mode of $p_{n,m}(\cdot)$ is 0, and the (extended) likelihood function for γ_1 is then simply the density $p_{n,m}(\cdot)$ translated to have a mode at $\hat{\gamma}_1$, irrespective of whether the latter is positive or negative.

As shown in Hill (1980a, sect. 2) this method of analysis is robust within the second basic scenario, in the sense that the results do not depend very much on the choice of *a priori* distribution for γ_1, provided that MSB/MSW is sufficiently large. In the case $\hat{\gamma}_1 \gg 0$, so that the part of the lower tail cut off by the truncation from below at 0 is negligible, we obtain the approximation

$$\mathscr{E}[\gamma(\tau^2) | \hat{\gamma}_1] = 1 - [n/(n-2)][MSW/MSB]. \tag{8}$$

Here $\mathscr{E}[\gamma(\tau^2) | \hat{\gamma}_1]$ is the desired numerical weight to be attached to $y_{i.}$, in equations (5, 6), and the expectation is with respect to the posterior distribution of τ^2, given the data MSB/MSW.

This approximation is for the easy case in which MSB/MSW is very large, and is then virtually identical with the original Stein estimator, except that we have an unknown σ^2 and so (in effect) use MSW as its estimate. The mean square MSB here corresponds to Stein's $\|X\|^2/I$, since in our scenario the $y_{i.}$ correspond to the elements of Stein's X, and the reference point to which we shrink is $\bar{y}$. Note that in this case the reference point has a relatively objective character, and need not be specified *a priori*.

On the other hand, when MSB/MSW is small, which is the more interesting case and corresponds to a negative conventional estimate for the between-variance component, we obtain for our estimate

$$\mathscr{E}[\gamma(\tau^2) | \hat{\gamma}_1] \approx 2/(I + 1). \tag{9}$$

In our analysis this is obtained by taking the limit as $MSB/MSW \to 0$. There is a resemblance to the James–Stein positive part estimator, but with the important difference that in the exact Bayesian analysis the weight attached to $y_{i.}$ does not become zero (or even negligible), as for the positive part estimator, when X is very close to the origin. There is in fact a lower bound of approximately $2/(I+1)$ that is determined by the number of rows, or dimensionality of the problem, and this lower bound can be quite substantial when I is small. Thus from our point of view, when the Steinian weight is negative, and is then replaced by 0, this is a poor approximation. The correct conclusion, as discussed in Hill (1965, 1967) in regard to the variance components problem, is rather that when MSB/MSW is small the data are very uninformative about certain functions of the variance components, so that in real-world problems some of the posterior distributions are closely approximated by the corresponding *a priori* distributions. Thus although we have a robust case in the second scenario when MSB/MSW is sufficiently large, this is not the case otherwise, and then the James–Stein positive part estimator systematically shrinks too much unless I is very large.

Next, it may be observed from Hill (1977, 1980a) that it is always optimal to do some shrinkage towards $\bar{y}$, provided that the model is appropriate. However, it is apparent from the form of the weight $\gamma(\tau^2)$, or from the usual variance versus bias considerations, that the optimal shrinkage will ordinarily be small (i.e. one will use nearly, $y_{i.}$ itself) when J is large, unless one knows the value of τ^2 to be sufficiently small, that is, unless one has very solid a priori knowledge to this effect. Technically, this would only have to be true of the *posterior* distribution for τ^2, but often the information in the data regarding τ^2 will not be sufficient to lead to substantial shrinkage in and of itself. This also is in qualitative agreement with our analysis of the earth example. We saw that the improvement of the James–Stein estimator over the least-squares estimator was quite trivial, and would be so regarded by both ourselves and the referee, unless the parameter θ is known to be quite close to the origin. Precisely in the case where s is chosen too large, one would have little reason to regard this as the case, and so one would regard the improvement as trivial; and indeed so it would be in typical applications. For example, if the point is measured with a standard deviation of an inch, and could be anywhere within the earth, then it is plainly absurd to pretend that the James–Stein estimator is an improvement over X. In the earth example, $\sum \mu_i^2/I$ plays the role of $\sigma_\alpha^2 + \mathscr{E}(\mu^2)$ in the Bayesian analysis of the random effects model, as discussed in Hill (1974, sect. 3). Here also when the relevant *a priori* knowledge is weak, as for example when the ratio τ^2 has an *a priori* distribution not concentrated near 0, then there is ordinarily only minor shrinkage. And this is true even in the best case for Stein, since the origin $\bar{y}$ towards which we shrink is known.

An extremely important issue regarding objectivity arises in connection with shrinkage in applications of the random effects model, such as to census estimations. Here we are dealing with substantially more complicated problems than the ones considered up to now, although of the same basic nature. A non-trivial amount of shrinkage is not only not objectively justified, but has been quite controversial, as in some recent discussions of shrinkage estimators in connection with census data. See Freedman and Navidi (1986). Thus under the pretence of objectivity, we have non-Bayesians like Stein implicitly recommending Bayes shrinkage estimators, which are not at all objective. Here I take the liberty of presuming that Stein would agree that the Bayesianified versions of his estimators are superior to the ones that he proposed, at least in the real-world finitized problems with which the census is concerned.

15.7 THE STEINIAN PARADOXES AND ASSORTED RED HERRINGS

The admissibility criterion is, in our opinion, an important one, but only in finite spaces. The evaluation game provides a justification in such spaces for this criterion in terms of the additional risk, as perceived by a referee, if one uses an inadmissible procedure as compared with one that dominates it. However, naïve adherents of the admissibility criterion pass to the idealized limit without realizing that the admissibility criterion is meaningful only in the finite case. At least I am not aware of any attempt to justify admissibility in the non-finite case, and such an attempt would fly in the face of the warnings by Gauss and Brouwer and others that the idealized version is meaningless except as an approximation. This is much like the error made by beginning students of mathematics who mistake the limit of integrals as being necessarily the integral of the limit. Perhaps Stein was relying upon some verbal non-Bayesian argument that suggested to him that the admissibility principle was of some real-world importance in non-finite cases. If he has such a demonstration that would be most interesting.

I do not mean to imply that there is anything necessarily wrong with idealizations. They suggest interesting and nearly indispensable insights, as in the Bayesian analysis of the random effects model. My criticism is rather that proofs of inadmissibility in the non-finite case prove nothing whatsoever. The worst part is the silly theorizing against Bayes procedures, or the ancillarity principle, that such proofs have given rise to. A serious problem (to which Stein's results indirectly make a contribution) has been transformed into a mathematical exercise. As with the quotations at the outset, historically first-rate mathematicians may view with scepticism such exercises.

As a consequence of such theorizing numerous so-called paradoxes were then discovered. Articles discussed such profound questions as how can it be that the inadmissibility result applies no matter where one shrinks to, when it is plainly absurd to shrink towards an absurd reference point. And how can it be that when the parameters are quite sensibly viewed as independent (I presume, for example, that most scientists would regard the temperature at the centre of Venus as independent of the parameters estimated in the last census) that the Stein theory is not able to distinguish 'objectivistically' this from the case where they are not, as for example when they are dependent in the random effects model, or when the parameters are exchangeable. Again in my opinion this is because his theory has been largely liberated from the possession of empirical context or content, and is largely a mathematical exercise. Some of his results and calculations were useful, but in my opinion his theory was awful, and was in its essence a theoretical argument as to why one should use suboptimal procedures. One must, of course, in practice use approximations to Bayesian methods because of real-world constraints, but it is not necessary to build a theory of suboptimality to justify it. To their credit, at a time when the Bayesian approach was considered scandalous, the admissibility people did manage successfully to disguise the approximate Bayesian nature of their shrinkage procedures, and thus opened the door to some interesting research. Some of the Steinian results, although misguided from a Bayesian viewpoint, can be of importance for other reasons. The use of Bayesian arguments, if only for proofs, did in part lead to the resurgence of the Bayesian approach, at least among some conventional mathematical statisticians (for better or for worse).

We now list and discuss the various 'paradoxes' in detail in light of our previous results.

1. The first paradox was that the classical least-squares estimator could be beaten everywhere by various Steinian shrinkage estimators. Many found this difficult to accept, and people questioned the loss function, and even the entire decision formulation of the problem, because it led to such a surprising conclusion. See Hill (1974, p. 560). This paradox is, however, easily resolved, and relates to the abuse of infinite idealized models. The classical least-squares estimator cannot even be implemented, since among other things, our computer will have to truncate the data when its norm is sufficiently large. If we replace the classical least-squares estimator by its practical equivalent, namely the Bayes estimator for some large but finite sphere of radius s, then the inadmissibility disappears. The latter is an admissible Bayes estimator relative to the finite representation being used, and it is clear, as Brown himself states, that it substantially dominates the James–Stein estimator at least for θ near the boundary of the sphere of radius s, and to a lesser extent, throughout most of the true sphere if s is not chosen too large. Hence the inadmissibility pertained only to the comparison between the James–Stein estimator and the idealized limit of such proper Bayes estimators, as $s \to \infty$. If we agree that in any real world problem

a sufficiently large bound can be found (even pretending to work with the continuum within the sphere), then the Bayes estimator based upon the too large bound cannot be dominated by the traditional James–Stein estimator or even by a Bayesianification of the latter, everywhere in the sphere of radius s. If there is a true sphere S_0 that is known to all concerned, then use of a uniform *a priori* distribution would be just fine, and typically better than the James–Stein estimator, except in the quite special case where by luck the true θ is very close to the arbitrarily chosen reference point θ_0. In any case a Bayesianification of the James–Stein estimator along the lines of Theorem 1 is strictly better than the James–Stein estimator, and makes it clear that the real issue concerns the comparison of two Bayes estimators. As is always the case in real-world problems, neither can dominate the other in any logical sense.

2. The next paradox concerns the reference point for shrinkage. The Stein results caused consternation here because his theory suggests that an 'objective' real improvement is possible no matter to what point one shrinks. This is of course absurd and would be rejected by anyone doing serious practical work. Imagine, for example, someone suggesting that one should shrink census data towards the log mass of a neutron star because this procedure dominates the classical least-squares estimator. This paradox is another artefact of improper dealings with infinity. For his 'improvement' to be non-trivial, even in his idealized formulation of the problem, he must make a selection of the origin θ_0 towards which to shrink X, and he must be lucky enough for this choice to be extremely close to the true θ. Any trivial improvement in risk, even in the idealized version of the problem, would be swamped by the various respects in which the model is only an approximation. And note that he does not have to miss by much in order for this to be the case.

3. Paradox of lack of empirical context. The 'objectivistic' nature of Stein's work suggests that there are purely logical reasons why the Stein shrinkage procedures improve upon the classical least-squares estimator. Hence empirical context and content can be ignored. If this were the case, then it would be true in all examples, including my example of the earth, where the true sphere is known. Yet I hope both Brown and Stein, as well as the referee, will at least agree that in this example there is no objective reason why the James–Stein procedure is better. Even when the centre θ_0 is known, and even in the idealized version of the problem, the risk functions for the James–Stein estimator and for the Bayes estimator based upon a uniform *a priori* distribution over S_0, must cross within this known true sphere. If they do agree with this, then I think they must also agree that the real-world issue concerns subjective knowledge regarding both the appropriate reference point and possible upper

bounds (or *a priori* variances) for θ. In this case, in a specific problem one must give careful consideration to *a priori* knowledge regarding these quantities, rather than use a mechanical procedure such as that of James–Stein that ignores the empirical context of the problem.

4. Paradox of independent parameters. It was argued in Hill (1974, p. 566) that the Stein formulation of the problem could not differentiate between the case where the parameters are judged to be *a priori* independent, and the case where they are sensibly judged to be dependent, as in the random effects model. This is again because his theory does not allow such subjective considerations. I proved that if one restricts attention to decision procedures in which the estimate of θ_i is a function only of X_i, then the classical least-squares estimator is admissible (in all dimensions) for squared error loss, just as in dimension 1 or 2, even in the idealized formulation. The class of appropriate decision procedures is never just handed to one in real-world problems. The statistician must make a decision with respect to which class is relevant and justifiable for a particular problem, and I think that many will agree that in problems concerning the temperature of Venus and census data, it is quite reasonable to make such a restriction. It is interesting that Stein's apparent emphasis on 'objectivity' opens the door to highly subjective procedures, such as making adjustments to estimates of census data on the basis of any variable that one chooses.

5. Paradox of dimensionality and structure. We have already explained how the elementary fact that the relative volumes behave as they do, with most of the volume concentrated near the boundary of a sphere in high dimensions, has been used to confuse the issues, and to imply that there is some magic reason separating dimensions 2 and 3. From our perspective, with all problems necessarily finitized, the role of dimension is diminished. Indeed, we view the essential (qualitative) reasons for shrinkage to be the same in dimension 1 as in all other dimensions, although of greater importance when the dimension is high, because of the possibility of more complex structure to the data. One then obtains modifications of the type of analysis presented in connection with the random effects model, and the reader may note that the nature of our analysis there did not depend upon the choice of I, although we require $I > 1$ if we are to obtain much information about σ_α^2 from the data. Structure is quite compatible with finite models. When there is structure, such as in terms of interactions between variables in a random effects analysis of variance, then the *a priori* information is all the more important, and mechanical shrinkage methods such as those of Stein are even less appropriate. See Hill (1969) for Bayesian shrinkage with interactions in complex arrays.

Now I do not believe any of the above are particularly paradoxical, and certainly do not have the status of the paradoxes of Zeno, let alone those of Russell. Although there have been some beneficial developments stemming from the work of Stein, the introduction of such pseudo-paradoxes and commentary on the inadequacy of Bayes rules, is not one of them.

We consider that we have now established, even in the most idealized version of the problem, that there is little loss in using a too large sphere S as opposed to the true S_0, unless the referee's true *a priori* distribution gives a substantial mass to the near vicinity of the arbitrary reference point that Stein has chosen, in which case the referee is putting much mass on a set of relatively small volume. Assuming conditions under which the James–Stein estimator is better than the Hill estimator is little more than a hidden way to make assumptions under which Stein's *a priori* information is better than that of our own. Such assumptions may or may not be true in a real problem. In a game-theory context where the referee is an intelligent opponent, note that it would be especially dangerous to choose such a θ_0 from a huge sphere S_0 even when the true sphere is known. Without such an intelligent opponent, Stein's procedure is generally inappropriate when the parameter is determined by a mechanism that respects volume. It might be argued that the James–Stein estimator loses very little even with a poor choice of θ_0, and gains much if the guess is good. This is not correct, since in repeated applications of the evaluation game the suboptimality of the James–Stein procedure will be picked up by the referee. Even more importantly, however, one must also consider the loss of respect that such procedures will give rise to among serious scientists and decision-makers, once it is fully understood that they can, with equal justification in the Steinian theory, shrink towards the log mass of neutron stars as towards any other point.

It seems curious that Brown, who has made many interesting contributions to both minimax and admissibility theory, in his example rejects the uniform estimator based upon a finite sphere. For this estimator is admissible, is minimax except for trifles when s_0/σ is not too small, and respects the real-world issue of finitization that any practising statistician must deal with. Furthermore, he pretends to some objective superiority of the James–Stein estimator, which by now I hope the reader will see can hardly be the case. Also, Brown apparently thought that Hill was recommending the uniform prior distribution over solid spheres as though it were somehow uniquely appropriate. Rather, Hill was merely arguing that there is nothing wrong with it, and that conventional Neyman–Pearson theory, the minimax theory and the principle of maximal entropy in information theory, all lend support for the use of a uniform distribution on S_0. However, Hill's approach does not necessitate uniformity, and one can use any prior distribution thought appropriate. The random effects model provides an example of a non-uniform distribution.

L. Dubins has suggested that the idea for Stein's admissibility result is at least partly due to L. J. Savage. If true, perhaps Savage did not pursue this topic himself, because like the other first-rate mathematicians quoted above, he was not interested in formalistic mathematical exercises. He may have regarded mathematics as the art of not making computations, through understanding. See de Finetti (1975, p. 6), an author whose own work exemplifies this higher form of mathematics. I think the real argument for the James–Stein positive part estimator, at least as a crude approximation, is in connection with the Bayesian analysis of random effects model, where for data with large MSB/MSW it more or less agrees with the result based upon equation (8). Unfortunately here too mechanical Steinian shrinkage rules go astray when MSB/MSW is not large, even in the more robust of the two basic scenarios for the one-way model, and cannot even begin to provide an approximation in the scenario in which the variance components are *a priori* independent.

15.8 CONCLUSIONS

The Steinian results have some valuable aspects. Our objection is as to the way they were presented, in which the reliance upon formalistic infinite idealized models obscured their actual real-world significance. Indeed, they were sometimes used to argue against the Bayesian approach because it was claimed that use of this approach would lead to inadmissibility. They ignored the fact that inadmissibility in the non-finite case has no operational meaning. There is no simulation game, such as I challenged frequentists to in Hill (1990a), that could possibly lead to the importance of admissibility in non-finite problems, since the referee in human games will have to do everything on a computer with finite memory. There are other deeper reasons to reject the idealized formulation, such as the meaninglessness of parameters beyond a certain precision, due to time changes in parameters, to the uncertainty principle of Heisenberg, etc. In our earth example we should perhaps bear in mind that the thing we call earth changes constantly, as it is a plastic, non-rigid structure, that distorts constantly. From my perspective, although some of their results are interesting and useful, the arguments of Brown and Stein are essentially theoretical arguments as to why one should use suboptimal procedures in real-world problems. This was obscured by a constant confusion of the logical issues with the practical and pragmatic issues involving Bayes procedures.

A real-world event of possible significance with respect to the Bayesian controversy may have occurred recently in connection with some articles by Marilyn vos Savant in *Parade Magazine* in 1990–91. Marilyn gave the correct frequentist solution to one of the most elementary problems in conditional probability, now known as the game-show problem, and was vigorously and rudely attacked by large numbers of professional mathematicians, physicists, engineers, biostatisticians, strategic defence analysts, etc. She did not present the Bayesian solution in terms of conditional probability, but rather an argument that agrees with that solution for the single most appropriate *a priori* distribution. What was astonishing is that her attackers could neither understand her original crystal-clear frequentistic argument, nor solve the problem themselves, nor look up the correct analysis in first-rate textbooks such as Feller (1968); nor could they for the most part take the trouble even to simulate the simple experiment. Eventually she challenged grade school classes to do so, and third graders along with Los Alamos and MIT engineers were able to reproduce, to many decimal points, a result which is intellectually barely above the level of $2 + 1 = 3$.

But many of us who have been seriously involved with the Bayesian controversy should not find this so astonishing after all. In fact, it is my contention that it is precisely the same confusion regarding conditional probability that has been the stumbling block in probability and statistics; and that this confusion in turn has its origin in connection with the finite/infinite problem and formalism. Although the professions have largely mastered and made good use of the basic ideas of probability, this has not been the case for conditional probability, and I believe this is the root cause for both the game-show fiasco, and also for the primitive level at which the debates regarding Bayesian statistics have traditionally taken place. It is my opinion that initially the confusion arose because of some fundamental misconceptions about both mathematics and science held by a politically powerful segment of the mathematics profession, namely the formalistic *Bourbaki* school, the leading members of which held that mathematics consists of 'making meaningless marks on paper', in which enterprise many of that school undoubtedly succeeded. See Hersh (1979). Gleick (1987, p. 88) offers in partial explication of the Bourbaki débâcle in mathematics and science: 'In part, Bourbaki began in reaction to Poincaré, the great man of the late nineteenth century, a phenomenally prolific thinker and writer who cared less than some for rigor.... The group stressed the primacy of mathematics among sciences, and also insisted upon a detachment from other sciences'.

And Gleick (1987, p. 52) quoting the mathematician R. Abraham: 'The romance between mathematicians and physicists had ended in divorce in the 1930s. These people were no longer speaking. They simply despised each other. Mathematical physicists refused their graduate students permission to take math courses from mathematicians.'[†]

[†] It is not widely known that Poincaré, in addition to being a historically significant mathematician, made several important contributions to Bayesian statistics, and like only a few other first-rate thinkers, penetrated through to qualitative aspects of Bayesian procedures rather than merely perform mechanical mathematical exercises.

This 'rigorous' attitude was initially in part an understandable reaction to the various paradoxes with which mathematics and logic were beset at the end of the last century, such as serious paradoxes of the infinite, and Russell's paradox of self-reference. Nowadays many products of this school are, however, blissfully ignorant of the reasons why it developed, or even that they are its product and victim, just as many purported 'frequentists' have little comprehension of the frequentistic argument to which they pay lip service, and merely repeat meaningless phrases.[‡] First-rate logicians/ mathematicians, such as Russell and Ramsey, have long been aware that the attempt to reduce mathematics (let alone probability and statistics) to classical logic was a failure. For example it has been argued that the Dedekind cut principle in analysis requires consideration of higher types in Russell's theory of types. But lesser mathematicians have continually ignored these issues and proceeded with the type of naïve analysis that has been known to be meaningless formalism for over a century. The Bourbaki viewpoint eventually drifted down to the various professions, until we reached the abysmal state of affairs represented by the Marilyn episode.

In the field of probability and statistics the 'rigorization' of probability by means of the reduction to measure theory by Kolmogorov (1950), was a product of the Bourbaki line of thought. Kolmogorov, who was himself a mathematician of the first rank, eventually adopted a less formalistic approach in his theory of algorithmic complexity, which emphasized information and finite combinatorical arguments, as in Kolmogorov (1963, 1983). I am delighted to report also that some other segments of the mathematics community appear to have made basic alterations in their opinions about the nature of mathematics and science. A respectable mathematician, L. Gillman (1992), gives the correct *Bayesian* solution to the game-show problem in *The American Mathematical Monthly*. There is in fact an older and deeper strain of mathematics, represented for example, by B. Pascal, D. Bernoulli, P. Laplace, C. F. Gauss, H. Poincaré, E. Borel, F. P. Ramsey, B. de Finetti, and others, for whom mathematics was an integral part of science and thought, rather than a formalistic exercise or game, and which may be reasserting itself. It is no secret that such individuals produced a rather large proportion of the mathematics that has survived to the present, and which is usefully employed in the sciences and decision-making. Let us hope that the probability and statistics professions have the integrity, intelligence, and courage to make similar alterations. Real-world problems of statistical inference and decision-making are quite subtle, and require serious knowledge of conditional probability and utility theory. If the professions cannot yet understand and solve the game-show

---

[‡] One of de Finetti's greatest contributions was his resolution of the meaningful aspects of the frequentist paradigm with the subjectivist theory, but this is not understood by many self-proclaimed frequentists.

problem, let alone inference for the normal distribution in dimension 3, what hope is there to deal with any serious issue?

Although as T. S. Kuhn argues, scientific paradigms are difficult to replace (or hackneyed ideas never die, and only very very slowly fade away), the controversy about Bayesian statistics seems to me to be of an even more embarrassing nature than for example those that recently occurred in geology and physics, prior to the introduction of the then scandalous theories of plate tectonics in geology and of chaos in physics. B. de Finetti found the situation in probability and statistics unbelievable. He writes in his University of Rome farewell speech, (de Finetti 1977, p. 349), regarding the contrast between himself and L. J. Savage (another historically significant mathematician who arrived at the same conclusions by a different route): 'I, of course, was in quite the opposite situation—a layman or barbarian who had the impression that the others were talking nonsense but who virtually does not know where to begin as he cannot imagine the meaning that the others are trying to give to the absurdities that it is imposing as dogma.'

Fortunately, the Bayesian controversy appears at long last to be drawing to a close. This is partly due to the development of the digital computer, which is requisite for the solution of real-world Bayesian problems, and to the closely related fact that important segments of the community of serious users of statistics are beginning to understand the nature of the issues. In addition to the paradoxes of Stein, we have had sequential analysis, called a hoax by F. Anscombe (1963), one of the major contributors to the subject; the Dempster–Shafer theory of upper and lower probabilities, which as shown in Hill (1991a) is subject to a Dutch book (or sure loss) and provides a 'justification' for the incorrect answer of 1/2 in the game-show problem; and numerous other 'great breakthroughs' that replace each other with some regularity. Generally speaking there is some element of content in each of these developments (for example in Bayesian sequential design, and in the Bayesian version of Dempster's theory), and the authors of these approaches believed they were making a serious contribution. But in the attempt to avoid its subjective Bayesian nature, the proponents twist and turn to the point where this content is almost totally hidden. In each case it has taken a number of years before such things could be seen clearly for what they are. The Bayesian approach, in conjunction with finite mathematics, is the easiest way to penetrate such gems of reasoning, and this may be one of the reasons why both have been so controversial.

Whether intentional or not, every attempt to depart from the underlying subjective Bayesian theory has eventually led to nonsense, although fortunately not without leaving some residue of truth. As de Finetti (1977, p. 370) says regarding 'objective' probability: 'The intrinsic contradictions in claims to 'objective' definition or explanation can always be shifted elsewhere but never eliminated; it can be said of them (to use the apt image suggested by Bernard O. Koopman) that: "unlike Napoleon's Guard they always retreat but never die".'

REFERENCES

Anderson, T. W., (1984) *An Introduction to Multivariate Statistical Analysis*, 2nd ed, New York, John Wiley & Sons.
Anscombe, F. J. (1963) Sequential medical trials, *J. Amer. Statist. Assoc.* **58**, 365–83.
Berger, J. (1985) In defense of the likelihood principle: axiomatics and coherency (with discussion), in *Bayesian Statistics 2*, J. M. Bernardo, M. H. DeGroot, D. V. Lindley, and A. F. M. Smith, eds. Amsterdam, North-Holland, Elsevier Science Publishers B. V., Valencia University Press, pp. 33–65.
Bhat, A. (1988) Applications of Bayesian statistics in econometrics, doctoral dissertation, the University of Michigan.
Birkhoff, G. and Mac Lane, S. (1953) *A Survey of Modern Algebra*, rev. edn, Macmillan.
Blackwell, D. and Girshick, M. A. (1954) *Theory of Games and Statistical Decisions*, New York, John Wiley & Sons.
Brouwer, L. E. J. (1920) *Jahresberichte der Deutschen Mathematiker-Vereinigung* **28**.
Brown, L. D. (1990) An ancillarity paradox which appears in multiple linear regression, *The Annals of Statistics*, pp. 471–538 (with discussion).
De Finetti, B. (1937) La prévision: ses lois logiques, ses sources subjectives. *Annales de l'Institut Henri Poincaré*, **7**, 1–68.
De Finetti, B. (1974) *Theory of Probability*, vol. 1, London, John Wiley & Sons.
De Finetti, B. (1975) *Theory of Probability*, vol. 2, London, John Wiley & Sons.
De Finetti, B. (1977) Probability: beware of falsifications! In *New Developments in the Application of Bayesian Methods*, A. Aykac and C. Brumat, eds, Amsterdam, North-Holland, pp. 347–85.
DeGroot, M. (1970) *Optimal Statistical Decisions*, New York, McGraw-Hill.
Einstein, A. (1921) *Geometrie und Erfahrung*, Berlin, J. Springer.
Feller, W. (1968) *An Introduction to Probability Theory and its Applications*, 3rd edn, rev. printing, New York, John Wiley & Sons.
Freedman, D. A., and Navidi, W. C. (1986) Models for adjusting the census (with discussion), *Statistical Science* **1**, 3–39.
Gauss, C. F. (1990) *Werke*, vol. 8, Leipzig.
Gillman, L. (1992) The car and the goats, *The American Mathematical Monthly*, **9**, 37.
Gleick, J. (1987) *Chaos*, Harmonds Worth, Penguin.

Heath, D. and Sudderth, W. (1978) On finitely additive priors, coherence, and extended admissibility, *The Annals of Statistics*, **6**, 333–45.

Hersh, R. (1979) Some proposals for reviving the philosophy of mathematics, reprinted in *New Directions in the Philosophy of Mathematics*, T. Tymoczko, ed., Birkhäuser.

Hill, B. M. (1965) Inference about variance components in the one-way model, *Journal of the American Statistical Association*, **58**, 918–32.

Hill, B. M. (1967) Correlated errors in the random model, *Journal of the American Statistical Association*, **62**, 1387–1400.

Hill, B. M. (1969) Foundations for the theory of least squares, *Journal of the Royal Statistical Society*, **B31**, 89–97.

Hill, B. M. (1974) On coherence, inadmissibility and inference about many parameters in the theory of least squares, in *Studies in Bayesian Econometrics and Statistics in Honor of L. J. Savage*, S. Fienberg and A. Zellner, eds, North-Holland, Amsterdam, 555–84.

Hill, B. M. (1977) Exact and approximate Bayesian solutions for inference about variance components and multivariate inadmissibility, in *New Developments in the Application of Bayesian Methods*, A. Aykac and C. Brumat, eds, Amsterdam, North-Holland, pp. 129–52.

Hill, B. M. (1978) Decision theory, in *Studies in Statistics*, vol. 19, R. V. Hogg, ed., The Mathematical Association of America, pp. 168–209.

Hill, B. M. (1980a) Robust analysis of the random model and weighted least squares regression, in *Evaluation of Econometric Models*, J. Kmenta and J. Ramsey, eds, New York, Academic Press, pp. 197–217.

Hill, B. M. (1980b) On finite additivity, non-conglomerability, and statistical paradoxes (with discussion), in *Bayesian Statistics*, J. M. Bernardo, M. H. DeGroot, D. V. Lindley, A. F. M. Smith, eds, Valencia, Spain, University Press, pp. 39–66.

Hill, B. M. (1985–86) Some subjective Bayesian considerations in the selection of models (with discussion), *Econometric Reviews*, **4** (2), 191–288.

Hill, B. M. (1988) Comment on *The Likelihood Principle*, 2nd edn, by J. Berger and R. Wolpert, *IMS Lecture Notes—Monograph Series*.

Hill, B. M. (199a) Comment on 'An ancillarity paradox that appears in multiple linear regression', by L. D. Brown, *The Annals of Statistics*.

Hill, B. M. (1990b) A theory of Bayesian data analysis, in *Bayesian and Likelihood Methods in Statistics and Econometrics: Essays in Honor of George A. Barnard*, S. Geisser, J. S. Hodges, S. J. Press, A. Zellner, eds, Amsterdam, North-Holland, pp. 49–73.

Hill, B. M. (1991a) Coherency, upper and lower probabilities, and Bayesian data analysis, unpublished manuscript, Department of Statistics, the University of Michigan.

Hill, B. M. (1992a) Bayesian nonparametric prediction and statistical inference (with discussion), in *Bayesian Analysis in Statistics and Econometrics*,

P. K. Goel and N. S. Iyengar, eds, Springer-Verlag Lecture Notes Series No. 75. New York, Springer-Verlag.

Hill, B. M. (1992b) Bayesian statistics, in *Encyclopedia of Physical Science and Technology*, vol 2. Academic Press.

Hill, B. M. (1992c) Statistics, robustness, *Encyclopedia of Physical Science and Technology*, vol 15. Academic Press.

Hill, B. M. (1992d) Bayesian nonparametric survival analysis: a comparison of the Kaplan–Meier and Berliner–Hill estimators (with discussion), in *Survival Analysis: State of the Art*, J. P. Klein and P. K. Goel, eds, NATO ASI Series, Kluwer Academic, pp. 25–46.

Hill, B. M. (1993) Dutch books, the Jeffreys-Savage theory of hypothesis testing and Bayesian reliability, in *Reliability and Decision Making*, R. Barlow, C. Clarotti and F. Spizzichino, eds, London, Chapman and Hall, pp. 31–85.

Kolmogorov, A. N. (1963) On tables of random numbers, *Sankhya* **A25**, 369–76.

Kolmogorov, A. N. (1983) On logical foundations of probability theory, *Probability Theory and Mathematical Statistics, Lecture Notes in Mathematics 1021*, pp. 1–5.

Lindley, D., and Smith, A. F. M. (1972) Bayes estimates for the linear model, *Journal of the Royal Statistical Society*, **B34**, 1–41.

Ramsey, F. P. (1950) *The Foundations of Mathematics and Other Logical Essays*, R. B. Braithwaite, ed., New York, The Humanities Press.

Russell, B. (1903) *The Principles of Mathematics*, 2nd edn, W. W. Norton.

Savage, L. J. et al. (1962) *The Foundations of Statistical Inference*, London, Methuen.

Savage, L. J. (1972) *The Foundations of Statistics*, 2nd rev. edn, Dover, New York.

Scheffé, H. (1959) *The Analysis of Variance*, New York, John Wiley & Sons.

Smith, A. F. M. (1986) Some Bayesian thoughts on modelling and model choice, *The Statistician*, **35**, 97–102.

Department of Statistics
University of Michigan
419 S State
Ann Arbor
MI 48109–1027
USA

CHAPTER 16

Bayesian Decision Theory and the Legal Structure

Joseph B. Kadane,[†]
Carnegie Mellon University, Pittsburgh

16.1 INTRODUCTION

Dennis Lindley is among the ablest defenders, expositors and scholars of the principle of maximizing expected utility. He had the courage to take up and defend an unpopular opinion, and he did so with great clarity. Statisticians in general, and Bayesians in particular, are vastly in his debt. His papers are full of important ideas, and define a view of Bayesian decision theory and statistics to which I fully subscribe. I cannot think of a single matter of decision-making principle about which we disagree.

Dennis Lindley is also a man of strong social views, some inkling of which may be gained from Lindley (1992). These opinions are also deeply held and forcefully articulated by him. The very qualities of courage and inconoclasm that make him so effective as a Bayesian advocate are fully in play in his expression of his social views.

Because expected utility theory is so much a part of his thinking, it is unsurprising that his social views are expressed in terms of it. Lindley raises the question of whether current legal procedure might also be found to be a consequence of expected utility theory, writing: 'It is easy to err in applying a mathematical model, usually through the omission of some important feature, so that some of the disagreements between the present

---

[†] Research was supported in part by NSF grants DMS-9005858, DMS-8705646, DMS-8701770, SES-8900025 and ONR contract N00014-89-J-1851. I thank Jim Press and Teddy Seidenfeld for helpful comments.

Aspects of Uncertainty edited by P. R. Freeman and A. F. M. Smith. © 1994 John Wiley & Sons Ltd.

application of the model and legal procedure, which has a much longer pedigree, will show practice to be sound.' (Lindley, 1991) My intent in this chapter is to accept Lindley's challenge of trying to reconcile current legal practice with Bayesian ideas.

Lindley's proposals for changing the legal procedure are as follows:

1. Judgement of guilt (or liability) should be abolished, and replaced by fines or imprisonment based on a probability of guilt (or liability).
2. All cost-free evidence should be admitted.
3. The adversarial approach should be replaced by a cooperative one.

In the succeeding sections, I examine Lindley's argument for each of these conclusions, and propose alternatives.

16.2 ABOLITION OF JUDGEMENTS OF GUILT

Lindley, following Cohen (1977), discusses the gatecrasher problem, in which 501 of 1000 spectators at an event did not pay. Should each be held civilly liable, since each has probability $501/1000 > 1/2$ of having crashed the gate? Lindley prefers to discuss the fine to which each should be subject.

If the (court's) utility of fine f for an innocent person I is

$$u(f; I) = 1 - f \tag{1}$$

and for a guilty person G is

$$u(f; G) = 1 - (2 - f)^2, \tag{2}$$

then Lindley shows the optimal fine is

$$f = 2 - (1 - p)/2p, \tag{3}$$

where p is the probability of guilt, here, $p = 501/1000$. This yields a fine of about 3/2, less than the fine of 2 that would be imposed if one were sure of guilt (every spectator crashed the gate). Lindley does not mention negative fines, so I take him to mean, strictly, the larger of zero and f as given in (3). Lindley thus has found the optimal fine f in a decision structure that allows the court only the option of deciding the fine.

Perhaps it would be useful for me to offer an analysis based on my understanding of the current system. At present, in a civil case such as the gatecrasher problem, there are two decisions to be made: should the defendant be found guilty, and, if so, what fine should be imposed. I take the current system to suggest that if the defendant is found not guilty, no penalty will be imposed. It is also my understanding that a finding of guilt, in a criminal case even more strongly than in a civil one, is a moral burden to the person found guilty, or liable, and hence in itself is a kind of punishment.

BAYESIAN DECISION THEORY AND THE LEGAL STRUCTURE 263

To establish some notation, let C be the action of the court in convicting (or finding liable), and let R stand for releasing, or failing to convict or find liable. Thus this notation distinguishes between C and G on the one hand, and between R and I on the other, as Lindley (rightly, I think) insists it should. My understanding of the social utilities involved suggest that they satisfy

$$u(R; I) > u(C, f; I) \qquad (4)$$

for all fines f.

Suppose $u(C, f; I)$ and $u(C, f; G)$ are at least once differentiable in f. With these and other assumptions made along the way, I shall show that there is a p^* such that if the court's probability p of the defendant's guilt or liability exceeds p^* it is optimal to convict and impose a penalty, and if p is less than p^* it is optimal to release.

The expected utility of convicting and imposing fine f is

$$pu(C, f; G) + (1-p)u(C, f; I) \qquad (5)$$

while the expected utility of releasing is

$$pu(R; G) + (1-p)u(R; I). \qquad (6)$$

The optimal fine f^* if the person is convicted is a function of the probability that the person is guilty, so I write it as $f^*(p)$. There may be some $\bar{p} > 0$, such that $f^*(\bar{p}) = 0$ for all $p \leq \bar{p}$. For $p > \bar{p}$, $f^*(p) \geq 0$. Suppose that the maximum fine possible for the offence in question is $\bar{f}$, and that an optimal fine $f^*(p)$, if the person is convicted, exists in $(0, \bar{f})$ for each $p > \bar{p}$. Then $f^*(p)$ satisfies

$$pu'(C, f^*; G) + (1-p)u'(C, f^*; I) = 0. \qquad (7)$$

Suppose that the optimal fine $f^*(p)$ satisfies

$$u(C, f^*(p); G) > u(R; G) \quad \text{for all } p > p^*. \qquad (8)$$

This means that the optimal punishment has greater utility than releasing.

Let $\Delta(p)$ be the advantage (positive or negative) of finding guilty over releasing. Then

$$\Delta(p) = pu(C, f^*(p); G) + (1-p)u(C, f^*(p); I)$$
$$- pu(R; G) - (1-p)u(R; I). \qquad (9)$$

Now

$$\Delta(1) = u(C, f^*(1); G) - u(R; G) > 0,$$

by (8), while

$$\Delta(\bar{p}) = u(C, f^*(\bar{p}); I) - u(R; I) < 0$$

by (4). Furthermore,

$$\frac{\partial \Delta(p)}{\partial p} = u(C, f^*(p); G) + pu'(C, f^*(p); G)\mathrm{d}f^*/\mathrm{d}p$$
$$- u(C, f^*(p); I) + (1-p)u'(C, f^*(p); I)\mathrm{d}f^*/\mathrm{d}p$$
$$- u(R; G) + u(R; I)$$
$$= u(C, f^*(p); G) - u(R, G) + u(R, I) - u(C, f^*(p); I) > 0 \quad (10)$$

using (7), (4) and (8). Thus $\Delta(p)$ is monotone increasing in $(\bar{p}, 1)$, and differentiable, so there is a unique p^*, $\bar{p} < p^* < 1$ such that $\Delta(p^*) = 0$. Furthermore, for all $p > p^*$, $\Delta(p) > 0$, so finding guilty is optimal, while if $p < p^*$, $\Delta(p) < 0$, so releasing is optimal. If another derivative of u (in f) is assumed to exist, differentiation of (7) yields $\mathrm{d}f^*(p)/\mathrm{d}p$ so $f^*(p)$ is continuous. Whether under this circumstance the optimal fine is continuous as a function of p, i.e. whether $f^*(p^*) = 0$, depends on the nature of the utility functions assumed. There is some case law supporting such a shape, as when CBS was found guilty of having libelled General Westmoreland, and was fined $1.

The point of the analysis above is to show that it is possible to obtain an expected utility characterization of current practice. What distinguishes this analysis from Lindley's is the decision space and utilities assumed. These are appropriate questions for public discussion and legislative consideration. Neither version should be regarded as having an advantage because of its closer adherence to the principle of maximizing expected utility.

16.3 ADMISSION OF ALL COST-FREE EVIDENCE

Lindley proposes that one consequence of Bayesian principles is the admission into evidence of all cost-free evidence, and cites as examples he thinks should be changed, the right of a defendant to remain silent, and the right not to testify against a spouse. Another example, not mentioned by Lindley, is the admission of evidence illegally obtained. His analysis does not clarify for whom the evidence is thought of as cost-free.

The issue here is what is to be regarded as cost-free. My understanding is that the right to remain silent rises from the desire to avoid coerced confessions, which at one time were all too common. Similarly, the right not to testify against one's spouse has to do, in my understanding, with the social value placed on protecting marriages and marital communication. Finally, the exclusion of illegally obtained evidence is an attempt to give some teeth to the rules restraining the police. While it might be advocated that illegal police actions should be prosecuted, as a practical fact those prosecutions tend not to be brought.

In each instance, the reasons for exclusion of evidence is that inclusion is not 'cost-free' when viewed in the greater social context. Again it is reasonable to debate whether the costs of inclusion are greater or less than the costs of exclusion, but again, neither side should be able to use as an argument greater compatibility with subjective expected utility principles.

16.4 ABOLITION OF THE ADVERSARIAL APPROACH

Lindley also argues against an adversarial approach to the law. His argument here is not entirely clear to me, but the principal elements of it, in my understanding, are as follows:

1. Game theory models competition while expected utility theory models cooperation.
2. There should not be conflict in the courtroom: 'In the courtroom, the question is not one of conflict but whether a law has been broken' (Lindley, 1991, p. 90).

Taking each point in turn, it seems to me that expected utility theory models the optimal decisions that a single party should make, given that party's utilities and probabilities. Several decision-makers, depending on the sequence of decisions and the various utilities and probabilities involved, may optimally act in either a cooperative or a competitive way. Situations involving simultaneous moves by several decision-makers have traditionally been modelled using game theory, but I have written elsewhere (Kadane and Larkey 1982) about how these decisions could be made within an expected utility theory framework. I believe that both conflict and cooperation may be modeled in either game theory or expected utility theory, but I prefer the latter as a normative standard for optimal decisions.

With regard to the second issue, I believe that it is an incorrect view of the current legal practice to think of it as totally competitive. There are many rules of fair play to which attorneys must adhere. For example, in a criminal case, the prosecution must make available to the defence any evidence it has tending to show the innocence of the defendant. Documents inadvertently revealed must be returned unread. Given that the different sides in a courtroom have genuinely different interests (whether civil or criminal), a certain degree of competition is inevitable. So I suppose what is really at stake here is the extent of the moves available to attorneys for the two sides, and perhaps whether judges should be more active in posing questions, etc. These again are matters of legitimate social debate and legislative action. The expected utility idea suggests how each side should make those decisions it is permitted to make, given its beliefs and what it is trying to accomplish, but it does not directly address what decisions each side should be permitted to make.

16.5 CONCLUSION

In all three instances, Lindley's proposals would increase the power of the courts. To remove the finding of guilt (or liability) would permit the court to fine people with rather low probabilities of guilt. Using Lindley's loss functions in the gatecrasher problem, the optimal fine is positive as long as $p > 1/5$. Lindley's second proposal would permit courts to compel testimony from defendants and their spouses. His third proposal, ending the adversary system, would reduce the vigour of complaints about high-handed court action. Whether it would be wise public policy so to increase the court's power and decrease the ability of the parties to do anything about it, is a very important question of public policy. It would be possible to discuss each of these in terms of social utilities and disutilities, although neither Lindley nor I have attempted to do so. But it is likely that if such a model were written, both the current practice and Lindley's proposals would be optimal for some choices of utilities and decision spaces.

Notwithstanding the modesty of Lindley's reservation, I do not think that Lindley erred nor that current practice is necessarily sound. He has one view of the way the law should operate, and expressed his argument in expected utility terms. I have taken my understanding of current practice, and done the same. Doing this can clarify the values, expressed in utility terms, behind the social choice of a legal structure, but in and of itself, cannot recommend either choice to a reader.

Dennis Lindley's contributions to expected utility theory are many and fundamental. One of them is his discussion of the legal structure in Bayesian terms. As we pursue that debate, it is well to remember that various positions have legitimate claim to justification under the theory of maximizing expected utility, and hence the debate must proceèd on other grounds.

REFERENCES

Cohen, L. J. (1977) *The Probable and the Provable*, Oxford, Clarendon.
Kadane, J. B. and P. D. Larkey (1982) 'Subjective probability and the theory of games', *Management Science* **28**, 113–20.
Lindley, D. V. (1991) 'Subjective probability, decision analysis, and their legal consequences', *JRSS*, **A154**, Part 1, 83–92.
Lindley, D. V. (1992) 'Is our view of Bayesian statistics too narrow', *Bayesian Statistics IV*, J. M. Bernardo, J. O. Berger, A. P. Dawid, A. F. M. Smith, eds, Oxford University Press, pp. 1–15.

Department of Statistics
Carnegie-Mellon University
Pittsburgh
PA 15213
USA

CHAPTER 17

Experimental Design from a Subjective Utilitarian Viewpoint

Frank Lad and John Deely
University of Canterbury, Christchurch, New Zealand

17.1 INTRODUCTION

A unifying aspect of Dennis Lindley's research career has been his commitment to the recognition of personal value-judgements and beliefs in the conduct of scientific inference. From his important article on the decision theoretic approach to statistical inference (Lindley 1953) through to his synthetic text on *Making Decisions* (Lindley 1971) and many articles examining intricate aspects of specific details, he has been a thoughtful leader in applying the subjective utilitarian viewpoint to practical problems in all realms of statistical analysis. An academician of great character, Dennis has taken seriously the challenges and problems that have arisen in the development of the theory, and has continually developed his ideas through the give and take of professional discussion. In his honour we propose the following consideration of issues in experimental design as seen from this viewpoint, which we share.

The problem of the 'design of experiments' has long been an area of intense interest and distinguished publications. Its beginning is commonly attributed to Fisher (1925, 1935) for whom it no doubt had a specific meanings, but over time it has come to mean many things to many people. Broadly speaking, it deals with the question of what is the optimal number of observations to take from each of a collection of possible experiments in order to learn something desirable about the response of a specific measured quantity to different experimental conditions.

Aspects of Uncertainty edited by P. R. Freeman and A. F. M. Smith.© 1994 John Wiley & Sons Ltd.

A problem which has been examined extensively in the literature on experimental design concerns the classical regression set-up. In the utilitarian formulation of the problem, our interest focuses on the values of uncontrollable measurements Y that correspond to various controllable experimental conditions, summarized by a measurement, X. Typically, scientists' previsions (expectations) for Y values conditioned on different experimental settings of X are asserted as functions of X. In order to forecast Y more precisely and more accurately for any specific settings of X, information can be gained by observing some experimental outcomes of Y under various conditions. In the simplest case, it is presumed that a fixed total number of experiments can be performed due to financial or physical constraints. The design of experiments problem in this context is to decide how many of the total should be performed at what levels of the measurement X in order to learn most advantageously from limited resources. The resolution of this problem requires that we assert our opinions about values of X and Y that occur in nature, and that we assess our utility for knowing their values to any degree of precision.

Due to foundational differences, there is no unique universally accepted formulation of this problem, and there are both frequentist and Bayesian answers which do not agree. There have been attempts to reconcile these differences by appropriate use of an information matrix to represent utility. See, for example, the works by Chaloner (1984), Pilz (1991), Verdinelli (1992) and Dette and Studden (1992). Regardless of the exact problem being addressed under the label 'design of experiments', ultimately, whether or not explicitly stated, some kind of utility function must be used. But opinions differ on what utility function is appropriate. Following early suggestions of Lindley (1956), recent work has focused on technical details of the choice of functions associated with Shannon's information function. In discussion of these developments, Deely (1992) remarked that work still needed to be done to construct a utility function that is appropriate to practical purposes of experimental design. He then posed the simplest of problems which could be used to clarify the issues involved in determining a utility function and a corresponding optimal design. In this present chapter, we bring these ideas to fruition.

To fix a context, we address a specific historical problem of the experimental measurement of patients' urine sugar and blood sugar levels, when the former was inexpensive to measure relative to the latter. Since basic information regarding diabetes is widely disseminated, we believe the reader will have sufficient knowledge to appreciate and to participate in the type of judgements required to determine an optimal experimental design. The problem resolves to the structure of the regression problem in its simplest context, when the only possible values of X and of Y are 0 and 1. We solve the design question, using detailed application of the judgement of partial exchangeability as it applies to the experimental set-up.

In so doing, we construct a utility function that we believe is realistic for the motivating problem, and that yields an optimal feasible solution. We compare the optimal decision based on this utility to those based on two common forms of an information-based utility function.

We believe that the framework of this practical simple problem involves issues that are widely relevant to general problems of experimental design. Moreover, the analytic methods we apply can easily be generalized to more complicated problems. Our analysis proceeds in a completely operational programme, thus eliminating any need for estimation of or even reference to unobservable parameters, as encouraged by both Lindley and de Finetti.

17.2 THE PROBLEM

We are concerned with two experimental measurement of a person's capacity to process carbohydrates, sugars, and starches via pancreas-secreted insulin. One is a urine sugar test which is inexpensive to perform but is only mildly informative, yielding false positive and false negative diagnostics of diabetes with some regularity. The test procedure amounts to producing a chemical reaction in a urine sample. Standard categories of the test result, based on the colour of the solution (from blue through green to orange), are labelled nil, trace, medium or dark. The other is a blood sugar test, called a 'glucose tolerance test', which is more expensive to perform but is extemely informative. The test amounts to an assay of the level of sugar concentration in the blood over three hours at 30-minute intervals after the ingestion of 75 g of glucose. This test has been considered so informative that until recently, the presence of diabetes in a person was defined operationally by the person's registering at least 11.1 mmol glucose per litre of blood after two hours.

To simplify details in our problem, we abridge the measurements under consideration to events: low urine sugar will be represented by $(X = 0)$ if the result of the urine test is nil or trace, and high urine sugar by $(X = 1)$ otherwise; similarly, low blood sugar will be represented by $(Y = 0)$ if the measured blood sugar level is less than 11.1, and high blood sugar by $(Y = 1)$ otherwise.

Now suppose that the database of a particular medical clinic contains the records of 1000 patients with the specific information that L of them have low urine sugar, and the remaining $1000 - L$ have high urine sugar. The numerical value of L has been left variable in the problem formulation, since we shall exhibit how the utility maximizing design decision differs under various scenarios. The blood sugar levels have not been measured for any of these patients. You have resources to measure the blood sugar for just two patients from these 1000, and you must decide whether none, one or both of them should be selected from among the high urine sugar group. After you observe the outcome of the tests on your chosen patients, a third patient is to be selected by lot from among all patients in the database. Your goal in performing the two expensive tests is to allow you to assert your informed probability that this third patient exhibits high blood sugar, conditioned only on the results of your designed test, and on the knowledge of whether the selected person is one who has exhibited low or high urine sugar.

Aside from the simplification of the measurement scale for urine sugar and blood sugar, our problem has been slightly contrived in this way so as to preclude the consideration of a randomized design, which might select the experimental X values by sampling. But this would only prove to be a red herring with respect to the issues we want to discuss. A randomized design would possibly be useful in a context wherein you are unsure about the incidence of $(X = 0)$ and $(X = 1)$ measurements among the target population, since it could allow you to learn something about that from the experiment as well.

17.2.1 Analytic set-up

Analytically, this problem of experimental design yields to the following description. You can afford to observe two *pairs* of events, denoted by (X_1, Y_1) and (X_2, Y_2), for which you can choose whether the events X_1 and X_2 will equal 1 or 0. Once the experiments are completed, you will use the results to assert a probability for $(Y_3 = 1)$ given the test results and the urine sugar status of the third patient. But your *design decision* must be made without knowing which condition will be relevant, since you cannot control the results of the experiment, nor determine the value of X_3.

17.2.2 Partial exchangeability

The three design options available are to select none, one or two of the patients to be given the blood test come from the group of high urine sugar patients. We denote this decision space by $\mathscr{D} = \{d_0, d_1, d_2\}$. The design is to be determined in the context of your presumed judgement to regard the event vector Y_3 partially exchangeably, in the sense of conditional exchangeability. That is, you regard

Y_1, Y_2 and Y_3 exchangeably given $X_1 = X_2 = X_3 = 0$,

and you regard

Y_1, Y_2 and Y_3 exchangeably given $X_1 = X_2 = X_3 = 1$.

This is to say that Y_j values are regarded exchangeably whenever their corresponding X_j values are equal.

17.2.3 Sufficient statistics

Your regarding a sequence of events exchangeably is equivalent to your asserting the same probability for any two possible outcome sequences that yield the same sum and the same total number of events. These are the two sufficient statistics for your inference from any particular observed events about any further events that you regard exchangeably with them. The simplest extension of the structure of exchangeable judgements to partial exchangeability involves increasing the number of sufficient statistics whose equality over two possible sequences would motivate you to accord them identical probabilities. See Diaconis and Freedman (1981). In the context of our present problem, the judgement of partial exchangeability regarding the components of an event vector Y_3 amounts to the constancy of your probability assertions for any event $(Y_3 = y_3)$ as long as the event vector pairs (X_3, Y_3) yield the same vector of four sufficient statistics, denoted by $S(X_3, Y_3) = (s_{00}, s_{01}, s_{10}, s_{11})$, where

s_{ab} = the number of event pairs (X_j, Y_j) for which $X_j = a$ and $Y_j = b$.

For example, s_{01} is the number of the three pairs (X_j, Y_j) that equal $(0,1)$. Clearly, the sum of the components of the vector $S(X_3, Y_3)$ must be 3, since the statistic vector S summarizes the structure of three such pairs.

For our problem the sufficient statistic vector $S(X_3, Y_3)$ can attain only 20 possible vector values, even though the event vector (X_3, Y_3) can yield 64 possibilities. These 20 possible vectors are listed in columns below:

S	$\sum X_3 = 0$				$\sum X_3 = 1$						$\sum X_3 = 2$						$\sum X_3 = 3$			
s_{00}	3	2	1	0	2	2	1	1	0	0	1	1	1	0	0	0	0	0	0	0
s_{01}	0	1	2	3	0	0	1	1	2	2	0	0	0	1	1	1	0	0	0	0
s_{10}	0	0	0	0	1	0	1	0	1	0	2	1	0	2	1	0	3	2	1	0
s_{11}	0	0	0	0	0	1	0	1	0	1	0	1	2	0	1	2	0	1	2	3

The 20 columns are displayed in groups of four, six, six and four, corresponding to the sum of the X_3 components, which derives from the sum of s_{10} and s_{11} in each column. The choice of an experimental design in $\mathscr{D}$ would relegate the possibilities for $S(X_3, Y_3)$ still further: d_0 would allow only columns for which $\sum X_3$ equal either 0 or 1; d_1 would allow only $\sum X_3$ equal 1 or 2; and d_2 would allow only $\sum X_3$ equal 2 or 3. For each design, only the value of X_3 is not controlled, and it may equal either 0 or 1.

17.3 MIXING DISTRIBUTIONS

The three design structures each limit the possible values of $\sum X_3$ in a different way. Thus, the mixing distribution for $S(X_3, Y_3)$ characterizing your partially exchangeable judgements will depend on which design decision you make. You can specify your mixture for all designs by your asserting four conditional distributions over S, each conditioned upon a different bank of columns corresponding to a fixed value of $\sum X_3$. This will be clarified now as we consider an example of an appropriate elicitation procedure.

17.4 ELICITATION

We are ready to elicit relevant distributions for $S(X_3, Y_3)$ vectors conditioned on each of the four groupings defined by $\sum X_3$ in the array displayed above. Assessments of the probabilities within the first and last groups of four columns are relatively easy. Among the first four columns you are simply asserting your probabilities that none, one, two or three patients who show low urine sugar will show high blood sugar. Among the final four columns you are simply asserting your probabilities that none, one, two or three patients who show high urine sugar will also show high blood sugar. For a numerical example, we propose the following two conditional distributions:

$P(S = 3,0,0,0 \mid \sum X_3 = 0) = 0.8$ $P(S = 0,0,3,0 \mid \sum X_3 = 3) = 0.03$
$P(S = 2,1,0,0 \mid \sum X_3 = 0) = 0.12$ $P(S = 0,0,2,1 \mid \sum X_3 = 3) = 0.07$
$P(S = 1,2,0,0 \mid \sum X_3 = 0) = 0.06$ and $P(S = 0,0,1,2 \mid \sum X_3 = 3) = 0.3$
$P(S = 0,3,0,0 \mid \sum X_3 = 0) = 0.02$ $P(S = 0,0,0,3 \mid \sum X_3 = 3) = 0.6$.

The other two groups of columns of possible $S(X_3, Y_3)$ values are typically rather more difficult to assess. To begin, let us focus on just one of them, the second group of six possibilities, for which $\sum X_3 = 1$. So we are considering the problem of knowing that one high urine sugar patient and two low urine sugar patients are having their blood sugar checked. First, consider merely how you would rank your relative probabilities that

    (the two low urine sugars show low blood sugar
    and the high urine sugar shows high blood sugar)

with

    (the two low urine sugars show high blood sugar
    and the high urine sugar shows low blood sugar).

We choose these as the first two exemplars of an ordering decision, since we imagine most readers would agree at least in asserting their P for the first of these exceeds their P for the second. Symbolically, this judgement would be represented by the inequality

$P(S(X_3, Y_3) = (0,2,1,0) \mid \sum X_3 = 1) < P(S(X_3, Y_3) = (2,0,0,1) \mid \sum X_3 = 1).$

EXPERIMENTAL DESIGN   273

Simplifying the notation for the moment, let us propose a completed ordering of the six possibilities which might well agree with yours:

$$P(0,2,1,0) < P(1,1,1,0) \approx P(0,2,0,1) < P(1,1,0,1) \approx P(2,0,1,0) < P(2,0,0,1).$$

Tildes placed above two of these inequalities denote that you might easily want them to be nearly equal, or perhaps reversed. Of course, many orderings are possible. At the base, your ordering will be affected by your assessment of the reliability of the low urine classification and the high urine classification in identifying low and high blood sugar levels. In addition, your sense of the information provided by results on low urine sugar patients that is relevant to high urine sugar patients (and vice versa) will affect your ordering.

For a numerical example, we propose the probability distribution below left as one reasonable assessment, without detailing all the considerations involved in our assessing it. Without further discussion, we also merely list (below right) an assessed conditional distribution for the third group of six columns of possible $S(X_3, Y_3)$ values, which identify $\sum X_3 = 1$. The operational processes involved in assessing it are similar:

$P(S = 2,0,0,1 \mid \sum X_3 = 1) = 0.6804$ $\quad\quad\quad P(S = 1,0,1,1 \mid \sum X_3 = 2) = 0.2243$
$P(S = 1,1,1,0 \mid \sum X_3 = 1) = 0.0096$ $\quad\quad\quad P(S = 1,0,0,2 \mid \sum X_3 = 2) = 0.6235$
$P(S = 1,1,0,1 \mid \sum X_3 = 1) = 0.1104$ and $P(S = 0,1,2,0 \mid \sum X_3 = 2) = 0.0011$
$P(S = 0,2,1,0 \mid \sum X_3 = 1) = 0.0075$ $\quad\quad\quad P(S = 0,1,1,1 \mid \sum X_3 = 2) = 0.0224$
$P(S = 0,2,0,1 \mid \sum X_3 = 1) = 0.0325$ $\quad\quad\quad P(S = 0,1,0,2 \mid \sum X_3 = 2) = 0.0765.$

Some care must be taken in assessing probabilities such as these, to ensure that they cohere with the assertions specified conditional on $\sum X_3 = 0$ and on $\sum X_3 = 3$. In larger design problems the elicitation proceeds much more easily using integral representations of infinitely exchangeably extendible distributions. We shall expand this remark in our conclusion.

17.4.1 Sufficient statistics from the designed experiment

Just as the judgement to regard N events exchangeably requires via coherency that any subgroup of them must be regarded exchangeably, so with the judgement of partial exchangeability. Your regarding the three pairs of (X, Y) measures partially exchangeably in the way we have described requires that you regard the first two pairs partially exchangeably in the same way. The sufficient statistic values $S(X_2, Y_2)$ that can derive from the experimental evidence are relegated to only ten possibilities. Each available experimental design choices delineates exclusive possibilities for the vector of sufficient statistics. Let $S(d_j)$ denote the value of the vector $S(X_2, Y_2)$ that arises from following design d_j. The possibilities for $S(d_0)$

are (2, 0, 0, 0), (1, 1, 0, 0) and (0, 2, 0, 0); those for $S(d_1)$ are (1, 0, 1, 0), (1, 0, 0, 1), (0, 1, 1, 0) and (0, 1, 0, 1); and the possibilities for $S(d_2)$ are (0, 0, 2, 0), (0, 0, 1, 1), and (0, 0, 0, 2).

17.4.2 Implied conditional probabilities

Via coherency requirements whose algebraic detail we shall not review, your elicited conditional probability distributions over $S(X_3, Y_3)$ yield implied distributions for the experimental outcomes $S(d_0)$, $S(d_1)$, and $S(d_2)$, and they also determine the forecasting probabilities, $P(Y_3 | S(d_j), X_3 = x)$, where j may be 0, 1, or 2, and x may be 0 or 1. These implied distributions are shown in Table 17.1

Table 17.1. Probabilities $P(S(d_j))$ and conditional probabilities $P(Y_3 | S(d_j), X_3 = x)$ implied by your conditional distributions on $S(X_3, Y_3)$ given $\sum X_3$.

	$S(d_j)$	$P(S(d_j))$	$P(Y_3 \mid S(d_j), X_3 = 0)$	$P(Y_3 \mid S(d_j), X_3 = 1)$
d_0	2, 0, 0, 0	0.8400	0.0476	0.8100
	1, 1, 0, 0	0.1200	0.3333	0.9200
	0, 2, 0, 0	0.0400	0.5000	0.8133
d_1	1, 0, 1, 0	0.1644	0.0292	0.6821
	1, 0, 0, 1	0.7356	0.0750	0.8476
	0, 1, 1, 0	0.0123	0.6087	0.9130
	0, 1, 0, 1	0.0877	0.3708	0.8723
d_2	0, 0, 2, 0	0.0533	0.0200	0.4375
	0, 0, 1, 1	0.2467	0.0908	0.8108
	0, 0, 0, 2	0.7000	0.1093	0.8571

In addition, your probabilities for Y_3 given $X_3 = x$ without conditioning on any experimental results are

$$P(Y_3 | X_3 = 0) = 0.1 \quad \text{and} \quad P(Y_3 | X_3 = 0) = 0.8233.$$

To the extent that some of the predictive probabilities in Table 17.1 may appear anomalous, these could be adjusted in a full interactive elicitation procedure that would iterate between your assessments of

$$P(S(X_3, Y_3) | \sum X_3) \quad \text{and} \quad P(Y_3 | S(d_j), X_3).$$

But at the end of the day, after you have assessed your probabilities and selected your chosen design, you will perform your experiment. You will then be in the position of asserting some conditional probability of the form

$$P(Y_3 | (S(d_j) = s), (X_3 = x)) \quad \text{for } x = 0 \text{ or } 1$$

according to the value of X_3 determined by selection from the clinic's population base as we have described. Furthermore, this conditional probability, whichever one it is, will be asserted in the face of one of two actual possibilities—that Y_3 equals 0 or 1, whose conditional probabilities you assert.

17.5 EXPERIMENTAL DESIGN AS A DECISION PROBLEM

Our analysis of the design problem has characterized it as a decision problem only mildly different from a very standard one. Each possible design amounts to a lottery ticket defined by a probability distribution over a finite list of prizes whose descriptions are such as the following example:

You decide to take decision d_1 to test one low and one high urine sugar patient; the result of your experiment is that both the low and the high urine sugar patients exhibit high blood sugar; you find that the next patient shows low urine sugar; you assert your probability that this low urine sugar patient shows high blood sugar at some value, p; and it turns out that this patient has low blood sugar.

Such a description summarizes completely one of the possibilities you are facing when you choose design d_1. Symbolically, we denote such a description by the five expressions

$$d_1, \quad (S(d_1) = (0,1,0,1)), \quad X_3 = 0,$$
$$P(Y_3 \mid (S(d_1) = (0,1,0,1), X_3 = 0) = p, \quad Y_3 = 0.$$

Any such description can be denoted merely by the vector $(S(d_j) =$ s, X_3, p, $Y_3)$. The value of j may be 0, 1 or 2; $S(d_j)$ may be one of its several possibilities; you assert whatever relevant conditional probability p that you do; and X_3 and Y_3 may each be either 0 or 1. (The mild difference of this decision problem from a standard problem is that the possibilities in the 'state of nature' space depend on which decision you take. And thus your probability distribution over rewards in the state space depends on your decision.)

17.6 A BENCHMARK UTILITY VALUATION

To begin, how would you value the outcome deriving from your decision to follow d_1 if it actually turned out to be the detailed description proposed in the paragraph above? Of course you cannot be sure that this description would follow from your choice of d_1. (For example, you might get different experimental results, and the third patient coming in may be a high urine sugar patient. In such a case you would be asserting your conditional probability for the high blood sugar event under these conditions; and the blood sugar of this third patient may show up to be high.) None the less, the final aspect of your judgement required to resolve your design decision problem is to assert your expected utility valuations for the various possible outcomes that could derive from each of the three design choices.

The crux of your utility valuations revolves upon your assessing your utility for asserting your $P(Y_3 \mid \cdot)$ as some specific number in the context of Y_3 being what it actually is, either 0 or 1. We denote this symbolically by

$$U[Y_3, P(Y_3 = 1 \mid S(d_j) = s, X_3)].$$

For the specific conditional probability value $P(Y_3 = 1 \mid S(d_j) = s, X_3) = 1$, this utility would amount to your valuation for a definitive positive diagnosis of diabetes. In the case of $Y_3 = 1$ it would be a correct positive diagnosis. If $Y_3 = 0$ it would be a false positive. Similarly, for the specific conditional probability assertion

$$P(Y_3 = 1 \mid S(d_j) = s, X_3) = 0,$$

your valuations would concern a false negative and a correct negative diagnosis. Your assessing your utilities for these easily understandable scenarios will serve as benchmarks for the entire utility function over all possible values of P. Denoting these benchmark values of $U(Y_3, P(Y_3 \mid \cdot))$ appropriately, you might evaluate these values as $U(0, 0) = 10$, $U(0, 1) = 9$, $U(1, 0) = 0$, $U(1, 1) = 8$ if you were a patient, or as $U(0, 0) = 10$, $U(0, 1) = 8$, $U(1, 0) = 0$, $U(1, 1) = 10$, if your were a doctor. (Of course, different patients, doctors or concerned family members could espouse different value numbers. You might easily imagine a patient who would support U values of 10, 8, 0, 2; or a doctor who would support 10, 1, 0, 10.) The important thing here is to realize that some such value-judgement must be made if a design decision is to be concluded, and different people may make this evaluation in different respectable ways, depending on their personal concerns in the matter. It is in this way that the utilities become relevant to the particular problem being addressed.

Now for intermediate values of $P(Y_3 \mid \cdot)$ within the interval $(0, 1)$, how would you evaluate $U(Y_3 \mid P(Y_3 \mid \cdot))$ relative to your four benchmark utility values of $U(0, 0)$, $U(0, 1)$, $U(1, 0)$ and $U(1, 1)$? It seems that, at the very least, most anyone would avow that $U(Y_3 = 1, P)$ increases monotically from $U(1, 0)$ to $U(1, 1)$ as P increases from 0 to 1; and similarly that $U(Y_3 = 0, P)$ decreases monotically from $U(0, 0)$ to $U(0, 1)$ as P increases from 0 to 1. If these monotonic functions were linear in P, your required utility function would have the form

$$U(Y_3 = 1, P(Y_3 \mid \cdot) = p) = pU(1, 1) + (1 - p)U(1, 0),$$

and

$$U(Y_3 = 0, P(Y_3 \mid \cdot) = p) = pU(0, 1) + (1 - p)U(0, 0).$$

Of course, these functions need not be linear, so in general this function would look like

$$U(Y_3 = y, P(Y_3|\cdot) = p) = g_y(p)U(Y_3, 1) + g_y(1-p)U(Y_3, 0),$$

for appropriate monotonic functions $g_1(\cdot)$ and $g_0(\cdot)$. All functions of this form satisfy the special cases of

$$U(Y_3, P(Y_3|\cdot) = 1) = U(1, 1) \quad \text{and} \quad U(Y_3, P(Y_3|\cdot) = 0) = U(0, 0).$$

To keep formalities brief, we report the calculations which follow in the context of presuming that the utility functions $U(Y_3, P(Y_3|\cdot) = p)$ are linear in p as depicted above. (In practice, such a specification may prove widely appropriate for a doctor's utility function, but perhaps not for a patient.) Your expected utility in asserting $P(Y_3|\cdot) = p$ would then equal

$$P[U(Y_3, P(Y_3|\cdot) = p)] = p^2 U(1, 1) + p(1-p)[U(1, 0) + U(0, 1)]$$
$$+ (1-p)^2 U(0, 0).$$

17.7 RELATED CONCEPTIONS OF UTILITY AS INFORMATION

An alternative formulation of a utility function $U(Y_3, P(Y_3|\cdot))$ has been proposed and studied in the context of developments in the theory of proper scoring rules. (See, for example, de Finetti 1962; Good 1957; Buehler 1971; Lindley 1982; DeGroot 1984.) Among proper scoring rules, one important proposal is the function

$$U(Y_3 = 1, P(Y_3|\cdot) = p) = \log(p),$$

and

$$U(Y_3 = 0, P(Y_3|\cdot) = p) = \log(1-p).$$

This form of utility function derives from a concept of utility associated with a design decision wherein your expected utility in asserting $P(Y_3|\cdot)$ equals the entropy of your probability assessment:

$$P[U(Y_3, P(Y_3|\cdot) = p)] = p \log(p) + (1-p) \log(1-p).$$

A number of formal properties of this utility function have contributed to general interest in it. For example, the function would constitute a proper scoring rule for your assertion of $P(Y_3|\cdot)$ as p. That is, your expected utility (prevision for your score) would be the greatest if you assert the level of p that is your honest probability valuation rather than some other value. And among all proper scoring functions $U(Y_3, P(Y_3|\cdot))$, this function is the only one that depends exclusively of the value of Y_3 that is actually observed, as shown by Bernardo (1979).

A related logarithmic utility function receiving attention in recent discussion specifies the information gain from the experimental evidence as the utility indicator:

$$U(Y_3 = 1, P(Y_3 | S(d), X_3) = p, P(Y_3 | X_3) = p_0) = \log(p/p_0),$$

and

$$U(Y_3 = 0, P(Y_3 | S(d), X_3) = p, P(Y_3 | X_3) = p_0) = \log((1 - p)p/(1 - p_0)).$$

For this form of utility function, your expected utility in asserting $P(Y_3 | S(d), X_3)$ equals the gain in the entropy of your probability conditioned on the experimental evidence when compared to the entropy of your 'prior' probability:

$$P[U(Y_3, P(Y_3 | S, X_3) = p, P(Y_3 | X_3) = p_0] = p \log(p/p_0)$$

$$+ (1 - p)\log((1 - p)/(1 - p_0)).$$

But the logarithmic utility function has a striking feature that appears to make it inappropriate for the particulars of the blood sugar problem we are addressing. Its symmetric form requires that your $U(Y_3 = 1, P(Y_3 | \cdot) = p)$ must be identical to your $U(Y_3 = 0, P(Y_3 | \cdot) = 1 - p)$, that is, $\log(p)$ in both cases. However, your valuation of your probability assessments would probably not exhibit this feature in the face of either of the possible values of Y_3. For example, your valuation of asserting $P(Y_3 | \cdot) = 0.95$ when $Y_3 = 0$ would surely be quite different from your valuation of asserting $P(Y_3 | \cdot) = 0.05$ in the context of $Y_3 = 1$. (This would be true for both the patient's and the doctor's utility functions as we have characterized them.) Similarly, your valuation of asserting $P(Y_3 | \cdot) = 0.95$ when $Y_3 = 1$ could well be different from your valuation of asserting $P(Y_3 | \cdot) = 0.05$ when $Y_3 = 0$. (These are mildly different in the case of our patient, though these two are identical for our doctor's assessments.) The utility based on information gain exhibits a variation on this symmetry.

The important aspect of the practical problem we have been considering is that utility valuations of probabilistic assessments of a patient's blood sugar status depend not only on the entropy associated with the probability assertion, but on the content of the probability assertion in the face of reality as well. We suspect that our problem is not an isolated instance in which this holds, and we believe forms of our benchmark utility elicitation strategy will prove to be widely useful.

In our report on the design decision conclusions which follow, we exhibit how design decisions would differ depending on the form of the utility function, showing how our benchmark utility functions compare with the information functions.

17.8 THE EXPECTED UTILITY VALUATION OF EACH DESIGN

We can now resolve your expected utility valuation for each of the available experimental design choices. The logic of the computation is most easily understood using a formalism which de Finetti (1974) would call a 'partitioned utility'. To firm ideas, focus on the decision d_0. The formulation is similar for the other two decisions in $\mathscr{D}$. Denote by $U(d_0 \mid \mathbf{R})$ the unknown quantity representing your current assessed utility valuation for whatever unknown situation you would actually arrive at by following d_0. The event of this possible situation is a member of a partition, $\mathbf{R}(d_0)$, generated by the result of the experiment, the urine sugar status of the third patient and the patient's actual blood sugar status. Symbolically, this partitioned utility is defined as

$$U(d_0 \mid \mathbf{R}) \equiv \sum (\mathbf{S}(d_0) = \mathbf{s}, X_3 = x, Y_3 = y) U(Y_3 = y, P(Y_3 \mid \mathbf{S}(d_0) = \mathbf{s}, X_3 = x)),$$

where the summation runs over the partition $\mathbf{R}$ generated by the three possible $\mathbf{s}$ outcomes for $\mathbf{S}(d_0)$, and the two possibilities for each of X_3 and Y_3.

Mathematically, this partitioned utility is a linear combination of events, since it is the sum of 12 terms, each the product of two multiplicands. The first multiplicand is an event: $(\mathbf{S}(d_0) = \mathbf{s}, X_3 = x, Y_3 = y)$. It is one of the partition of events that can result from your design choice. Each such partition member equals either 1 or 0, but one and only one of them equals 1. The second multiplicand is your utility valuation for asserting your conditional probability for Y_3 in the situation so described,

$$U(\mathbf{S}(d_0) = \mathbf{s}, X_3 = x, Y_3 = y, P(Y_3 \mid \mathbf{S}(d_0) = \mathbf{s}, X_3 = x)).$$

In words, the partitioned utility equals your utility assessment of the situation you will experience, whatever that unknown experience might be. It is easy then to conclude that your expected utility assessment of d_0 is your prevision $P(U(d_0 \mid \mathbf{R}))$, which equals your probability weighted linear combination of these utilities:

$$U(d_0) = P(U(d_0 \mid \mathbf{R})) \equiv \sum P(\mathbf{S}(d_0) = \mathbf{s}, X_3 = x, Y_3 = y)$$
$$U(Y_3 = y, P(Y_3 \mid \mathbf{S}(d_0) = \mathbf{s}, X_3 = x)).$$

17.8.1 Expected utility calculations

Using the assessed probability distribution we have presented, and using the array of utility functions we have described as those of a doctor, a patient, the information function and the information gain, we computed expected utilities for each design under a variety of possibilities for

$P(X_3 = 0)$, which equals $L/1000$. A peculiarity of this particular example results in the utility orderings of the doctor and the patient agreeing with one another in each instance of $P(X_3 = 0)$, and similarly, the two information based utilities agreeing. This need not always happen, as we shall notice in the concluding discussion. Table 17.2 displays the orderings of the designs according to their utilities, in descending order.

When $P(X_3 = 0)$ ranges for 1 down to 0.4, d_0 is preferred no matter which of the utility functions is espoused. However as $P(X_3 = 0)$ declines further, the choice of decision switches to d_1 and then to d_0, with the switching occuring at different points for the different utilities. At the lowest values of $P(X_3 = 0)$, below 0.21, again all the orderings agree on d_2 as preferred.

Table 17.2 Descending expected utility orderings for designs in $\mathscr{D} = \{d_0, d_1, d_2\}$ under a spectrum of possible specifications of $P(X_3 = 0)$, for benchmark and Information utility functions.

$P(X_3 = 0)$ interval	Doctor and patient benchmark utility	Info and info gain information utility
[0.0, 0.21]	$d_2 > d_1 > d_0$	$d_2 > d_1 > d_0$
[0.22, 0.23]	$d_2 > d_1 > d_0$	$d_1 > d_2 > d_0$
[0.24, 0.27]	$d_2 > d_1 > d_0$	$d_1 > d_0 > d_2$
[0.28, 0.31]	$d_2 > d_1 > d_0$	$d_0 > d_1 > d_2$
[0.32, 0.33]	$d_1 > d_2 > d_0$	$d_0 > d_1 > d_2$
[0.34, 0.38]	$d_1 > d_0 > d_2$	$d_0 > d_1 > d_2$
[0.39, 1.0]	$d_0 > d_1 > d_2$	$d_0 > d_1 > d_2$

It is pleasing to find that decision orderings in an applied problem differ between the benchmark and the information utilities. We believe we have identified a procedure by which appropriate utility valuations can be introduced into the solution of experimental design problems.

17.9 CONCLUSIONS AND REMARKS

In a descriptive manner we have introduced a procedure for assessing utility functions that are relevant to a specific problem being addressed. Space precludes a complete discussion of every aspect of the calculations that were conducted. In these concluding remarks we mention some salient features of the our analysis.

Our study of the various utility functions has shown how different utilities choose different designs as optimal. This is perhaps to be expected. It then becomes a question as to which utility seems to be more appropriate. The analysis here has shown that utilities based on an information function may not always be suitable, and that a practical and intuitively appealing formulation of a utility function is possible for similar problems. Further work on what kinds of other competitive utility functions are more appropriate is in progress and will be reported elsewhere. This includes the analytic study of the various functions through their Taylor series expansions, which can identify rather precisely just when and how they are different.

Extending the ideas of this chapter to the case of designs with more than two observations is straightforward, although the probability elicitation procedure we have followed would become much more complicated. However, the recognition of the infinite exchangeable extendibility of judgements of the experiments would be one method of simplifying the procedure, and thus allowing the computation of optimal designs. Moreover, the possible measurements for X and Y can be enlarged to cover more than two categories. Again, this would involve a more complicated utility function through the required specification of more benchmarks. But the procedure is still tractable. Work on these larger designs is in progress.

An unsuspected distracting result of our analysis of this very small problem was that the expected utilities of the three possible designs differed very little from one another in absolute value, no matter which utility function was studied. This seems to indicate that the various designs offer about the same information from an experimental point of view. How this phenomenon persists in larger problems is under study.

There are many intricate aspects to the overall problem of experimental design, as is attested by its long history of investigation in the statistical literature. We offer this chapter as a contribution to its further understanding.

REFERENCES

Bernardo, J. (1979) Expected information as expected utility, *The Annals of Statistics*, **7**, 686–90.

Buehler, R. (1971) Measuring information and uncertainty, in *Foundations of Statistical Inference*, V. P. Godambe and D. A. Sprott, eds., Toronto, Holt, Rinehart and Winston.

Chaloner, K. (1984) Optimal Bayesian experimental design for linear models, *Annals of Statistics* **12**, 283–300.

Deely, J. (1992) Discussion of 'Advances in Bayesian experimental design', in *Bayesian Statistics 4*, J. Bernardo, J. O. Berger, A. P. Dawid and A. F. M. Smith, eds, Oxford, Oxford University Press.

Dette, H. and Studden, W. J. (1992) A geometric solution of the Bayesian E-optimal design problem, *Fifth Purdue Symposium on Statistical Decision Theory and Related Topics*, S. S. Gupta and J. O. Berger, eds.

DeGroot, M. (1984) Changes in utility as information, *Theory and Decision*, **17**, 287–303. See errata noted in **18** (1985), 319.

Diaconis, P. and Freedman, D. (1981) Partial exchangeability and sufficiency. *Proceedings of the Indian Statistical Institute Golden Jubilee International Conference*, Calcutta, pp. 205–36.

de Finetti, B. (1962) Does it make sense to speak of good probability appraisers? *The Scientist Speculates: An Anthology of Partially Baked Ideas*, I. J. Good, ed., London, Methuen.

de Finetti B (1974) *Theory of Probability*, vol. 1, London, John Wiley & Sons, pp. 143–6.

Fisher, R. (1925) *Statistical Methods for Research Workers*, Edinburgh, Oliver and Boyd.

Fisher, R. (1935) *The Design of Experiments*, Edinburgh, Oliver and Boyd.

Good, I. J. (1957) The appropriate mathematical tools for describing and measuring uncertainty, reprinted in I. J. Good, *Good Thinking: The Foundations of Probability and its Applications*, 1983, Minneapolis, University of Minnesota Press.

Lindley, D. (1953) Statistical Inference, *Journal of the Royal Statistical Society*, **B15**, 30–76.

Lindley, D. (1956) On a measure of the information provided by an experiment *Annals of Mathematical Statistics*, **27**, 986–1005.

Lindley, D. (1971) *Making Decisions*, London, John Wiley & Sons.

Lindley, D. (1982) Scoring rules and the inevitability of probability, *International Statistical Review*, **50**, 1–26.

Pilz, J. (1991) *Bayesian Estimation and Experimental Design in Linear Regression Models*, New York, John Wiley & Sons.

Verdinelli, I. (1992) Advances in Bayesian experimental design, *Bayesian Statistics 4*, J. O. Berger, J. Bernardo, A. P. Dawid, and A. F. M. Smith, eds, Oxford, Oxford University Press.

Department of Mathematics
University of Canterbury
Christchurch
NEW ZEALAND

CHAPTER 18

The Bayesian Analysis of Categorical Data—a Selective Review

Tom Leonard[†] and John S. J. Hsu[‡]
[†]*University of Wisconsin-Madison,*
[‡]*University of California at Santa Barbara*

18.1 THE FOUNDATIONS OF THE 1960s

Lindley (1964) proposed a Bayesian analysis for $r \times s$ contingency tables. Suppose that cell frequencies $\{y_{ij}; i = 1, \ldots, r; j = 1, \ldots, s\}$ are taken to possess a multinomial sampling distribution, with respective unconditional cell probabilities $\{\theta_{ij}; i = 1, \ldots, r; j = 1, \ldots, s\}$, satisfying

$$\sum_{rs} \theta_{kg} = 1, \quad \text{and samples size} \quad n = \sum_{kg} y_{kg}.$$

Consider multivariate logits $\{\gamma_{ij}; i = 1, \ldots, r; j = 1, \ldots, s\}$, satisfying

$$\theta_{ij} = \exp(\gamma_{ij}) / \sum_{kg} \exp(\gamma_{kg}), \qquad (1.1)$$

and any log contrast, taking the form

$$\lambda = \sum_{ij} a_{ij} \log \theta_{ij} = \sum_{ij} a_{ij} \gamma_{ij}, \qquad (1.2)$$

where $\sum_{ij} a_{ij} = 0$.

Suppose that the prior distribution of the θ_{ij} is Dirichlet, with parameters $\{\alpha_{ij}; i = 1, \ldots, r; j = 1, \ldots, s\}$, so that the posterior distribution is also Dirichlet, but with updated parameters

---

Aspects of Uncertainty edited by P. R. Freeman and A. F. M. Smith. © 1994 John Wiley & Sons Ltd.

$$\{\alpha_{ij}^* = \alpha_{ij} + y_{ij}; i = 1, \ldots, r; j = 1, \ldots, s\}.$$

Then Lindley proved that λ is distributed, in the posterior assessment, as a linear combination of independent log-Gamma variates. Equivalently

$$\lambda = \sum_{ij} a_{ij} \log u_{ij}, \qquad (1.3)$$

where the u_{ij} are independent chi-squared variates, with respective degrees of freedom $2\alpha_{ij}^* = 2(\alpha_{ij} + y_{ij})$. If $\alpha_{ij}^* \geq 5$, $\log u_{ij}$ may be taken to be approximately normally distributed with mean $\log 2 + \log(\alpha_{ij} + y_{ij})$, and variance $(\alpha_{ij} + y_{ij})^{-1}$. Hence, in many practical situations, the posterior distribution of λ will be approximately normal with mean

$$\lambda^* = \sum_{ij} a_{ij} \log(\alpha_{ij} + y_{ij}), \qquad (1.4)$$

and variance

$$v^* = \sum_{ij} a_{ij}(\alpha_{ij} + y_{ij})^{-1}. \qquad (1.5)$$

Bloch and Watson (1967) proposed some refinements to this approximation. However, Leonard *et al.* (1989) and Hsu (1990) show that Lindley's original approximation is remarkably accurate, when compared with the exact result. Next consider a 2×2 table, set $r = s = 2$, and let

$$\lambda = \log \theta_{11} + \log \theta_{22} - \log \theta_{12} - \log \theta_{21} = \gamma_{11} + \gamma_{22} - \gamma_{12} - \gamma_{21}, \qquad (1.6)$$

denote the log measure of association. Then (e.g. Lindley 1965, p. 150) λ is exactly distributed in the posterior assessment as

$$\lambda = \log \alpha_{11}^* + \log \alpha_{22}^* - \log \alpha_{12}^* - \log \alpha_{21}^* + \log F_{2\alpha_{12}^*}^{2\alpha_{11}^*} + \log F_{2\alpha_{21}^*}^{2\alpha_{22}^*}, \qquad (1.7)$$

where $F_{2\alpha_{12}^*}^{2\alpha_{11}^*}$ and $F_{2\alpha_{21}^*}^{2\alpha_{22}^*}$ are independent F-variates, with the notationally obvious degrees of freedom. Hence the posterior distribution of λ is approximately normal, with mean

$$\lambda^* = \log \alpha_{11}^* + \log \alpha_{22}^* - \log \alpha_{12}^* - \log \alpha_{21}^*,$$

and variance

$$v^* = (\alpha_{11}^*)^{-1} + (\alpha_{22}^*)^{-1} + (\alpha_{12}^*)^{-1} + (\alpha_{211}^*)^{-1}. \qquad (1.8)$$

In situations involving vague prior information, Lindley (1964) suggests setting $\alpha_{ij} = 0$ for $i = 1, 2$ and $j = 1, 2$, and proposes

$$B^2 = (\lambda^*)^2 / v^*, \qquad (1.9)$$

as an excellent alternative to the usual chi-squared statistic, for testing $\lambda = 0$, on one degree of freedom. The choices $\alpha_{ij} = 0.5$, for $i = 1, 2$ and $j = 1, 2$, however, provide the 'Jeffreys prior' and may be preferable.

Let $\xi_1 = \theta_{11}/(\theta_{11} + \theta_{12})$ and $\xi_2 = \theta_{21}/(\theta_{21} + \theta_{22})$, denote the conditional probabilities given the rows. Then the log-measure of association λ in (1.6) satisfies

$$\lambda = \text{logit}(\xi_1) - \text{logit}(\xi_2), \tag{1.10}$$

where logit $(\xi) = \log(\xi) - \log(1 - \xi)$. Note that, in the posterior assessment, ξ_1 and ξ_2 possess independent beta distributions with

$$\xi_j \sim \text{Beta}(\alpha_{1j}^*, \alpha_{2j}^*) \quad (j = 1, 2). \tag{1.11}$$

Altham (1969) exploits this representation to develop an explicit expression, involving a hypergeometric series, for the posterior probability, that $\lambda > 0$. For fixed p, let

$$I_p(\alpha_1, \alpha_2) = \text{prob}(\xi \leq p \mid \xi \sim \text{Beta}(\alpha_1, \alpha_2)). \tag{1.12}$$

Then, if a is an integer, the cumulative distribution function (1.12) satisfies

$$I_p(a, n - a + 1) = \sum_{j=a}^{n} {}^nC_j p^j (1 - p)^{n-j} \tag{1.13}$$

$$= p(Y \geq a \mid Y \text{ possesses a binomial distribution}$$

with probability p and sample size n). (1.14)

Altham uses (1.14) to obtain a representation for the posterior distribution of ξ_2 in (1.10), and then develops an expression for the posterior probability that $\xi_1 > \xi_1$, via an integration with respect to ξ_1. In the special case, when $\alpha_{11} = 0$, $\alpha_{12} = 1$, $\alpha_{21} = 1$, and $\alpha_{22} = 0$, her posterior probability reduces to a significance probability for Fisher's exact test. However, Jeffreys' choices $\alpha_{ij} = 0.5$ for and $i = 1, 2$ and $j = 1, 2$ also possess some frequency justification, since, whenever a is an integer

$$\sum_{j=a+1}^{n} {}^nC_j p^j (1-p)^{n-j} \leq I_p(a + 1/2, n - a + 1/2) \leq \sum_{j=a}^{n} {}^nC_j p^j (1-p)^{n-j}. \tag{1.15}$$

For example, the posterior probability that ξ_j is less than some hypothesized value $\xi_j^{(0)}$ (i.e. a 'Bayesian significance probability') will, under the Jeffreys prior, always lie between two frequency-based significance probabilities i.e., (1) the probability that y_{1j} is strictly greater than the value of y_{1j} actually observed, if indeed $\xi_j = \xi_j^{(0)}$, (2) a similar probability, but with weak inequality. It is possible to develop extensions of this argument, as a frequency-based justification for the Jeffreys prior ($\alpha_{ij} \equiv 0.5$), for a general $r \times s$ contingency table. For a 2×2 table, the choices $\alpha_{11} = 1$, $\alpha_{12} = 0$, $\alpha_{21} = 0$ and $\alpha_{22} = 1$, would also lead to a significance probability for Fisher's exact test. The Jeffreys prior provides a compromise between the two frequency-based significance probabilities, involving weak and strong inequalities.

For an $r \times s$ table, Lindley's (1964) Bayesian approach can be regarded as formalizing a non-Bayesian approach developed by Goodman (1964). Goodman's full rank interaction model, assumes that the multivariate logits γ_{ij} satisfy

$$\gamma_{ij} = \lambda_i^A + \lambda_j^B + \lambda_{ij}^{AB} \quad (i = 1, \ldots, r; j = 1, \ldots, s), \quad (1.16)$$

where λ_i^A denotes the ith row effect, λ_j^B denotes the jth column effect, and λ_{ij}^{AB} denotes the (i, j)th interaction effect. Subject to the constraints $\lambda_.^A = \lambda_.^B = \lambda_{i.}^{AB} = \lambda_{.j}^{AB} = 0$, where the dot notation denotes average with respect to that subscript, we have

$$\lambda_i^A = \gamma_{i.} - \gamma_{..}, \quad (1.17)$$

$$\lambda_j^B = \gamma_{.j} - \gamma_{..}, \quad (1.18)$$

and

$$\lambda_{ij}^{AB} = \gamma_{ij} - \gamma_{i.} - \gamma_{.j} + \gamma_{..}. \quad (1.19)$$

Since each expression in (1.17)–(1.19) is a linear contrast, the preceding developments (e.g. (1.4) and (1.5)) may be used to provide a normal approximation to the posterior distribution of any parameter of interest. It is, for example, straightforward to calculate an approximate Bayesian significance probability to investigate whether any particular interaction effect can reasonably be set equal to zero. Non-zero interactions can be used to indicate cells of possible interest in the contingency table. One possible choice of an overall test statistic, for independence of rows and columns, is

$$B^2 = \frac{(r-1)(s-1)}{rs} \sum_{ij} (\tilde{\lambda}_{ij}^{AB})^2 / v_{ij}^{AB}, \quad (1.20)$$

where $\tilde{\lambda}_{ij}^{AB}$ and v_{ij}^{AB} are respectively the approximate posterior mean and variance of λ_{ij}, under a Jeffreys prior. The observed value of B^2 may be compared, as a first-order asymptotic approximation, with appropriate upper percentage points of the chi-squared distribution with $(r-1)(s-1)$ degrees of freedom.

Irving Jack Good (1965, 1967) also developed the analysis of contingency tables, using Dirichlet priors. For example, under our above specification the posterior mean of θ_{ij} may be represented in the form

$$\theta_{ij}^* = \frac{np_{ij} + \alpha \mu_{ij}}{n + \alpha} \quad (1.21)$$

where $p_{ij} = y_{ij}/n$ denotes the corresponding cell proportion, $\mu_{ij} = \alpha_{ij}/\alpha$ is the prior mean of θ_{ij}, and $\alpha = \sum_{kg} \alpha_{kg}$ denotes the 'prior sample size'. Good argued that the choice of α should be based upon the data, either via a hierarchical Bayes, or bootstrap Bayes approach. This is with the intention

BAYESIAN ANALYSIS OF CATEGORICAL DATA 287

of representing possible smoothness of the contingency table, and purposefully violates Johnson's sufficientness postulate (Johnson and Braithwaite 1932). The genesis of Good's ideas can be found in his historical collaboration with Alan Turing, the father of machine intelligence, during the Second World War, when the problem of breaking the Nazi codes was solved via a cryptanalysis based on the estimation of letter frequencies in the German language.

18.2 NUMERICAL EXAMPLES

Consider, firstly the gene frequency data, reported by Lindley (1965, p. 180). Out of $n = 106$ individuals, $y_{11} = 6$ possess both gene A and gene B, $y_{22} = 61$ possess neither gene, $y_{12} = 13$ possess gene A, but not gene B, and $y_{21} = 16$ possess gene B, but not gene A.

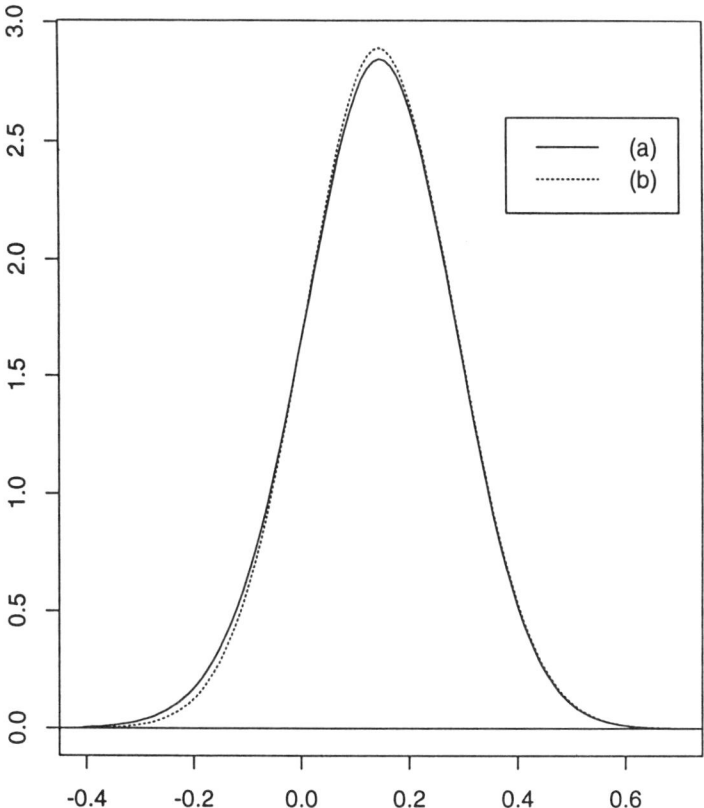

Figure 18.1 Posterior density of log measure of association gene data: (a) exact and BCM; (b) normal approximation.

Curve (a) of Figure 18.1 discribes the exact posterior density of the log measure of association (1.6), using a Jeffreys prior for the unconditional cell frequencies, and curve (b) describes Lindley's normal approximation, which is remarkably accurate. The exact Bayesian significance probability i.e. posterior probability that $\lambda < 0$, is 0.153, while the normal approximation gives 0.140. The test statistic in (1.10) is evaluated as $B^2 = 1.12$, on one degree of freedom, while the usual chi-squared statistic for testing $\lambda = 0$, gives $X^2 = 0.87$.

Table 18.1. The engineering apprentice data.

Section head's assessment	Written test result			
	A	B	C	D
Excellent	26	29	21	11
	(0.034)	(0.311)	(0.721)	(0.916)
	[0.036]	[0.301]	[0.726]	[0.907]
Very good	33	43	35	20
	(0.251)	(0.402)	(0.560)	(0.743)
	[0.248]	[0.390]	[0.548]	[0.752]
Average	47	71	72	45
	(0.860)	(0.703)	(0.197)	(0.231)
	[0.860]	[0.699]	[0.204]	[0.248]
Needs to improve	7	12	11	9
	(0.851)	(0.586)	(0.460)	(0.109)
	[0.842]	[0.612]	[0.484]	[0.108]

The 4×4 contingency table in Table 18.1 was reported by Lindley (1965, p. 180), and cross-classifies $n = 492$ engineering apprentices, according to their section head's assessment, and their grade on a written test. Under Goodman's full rank interaction model (1.17), histogram (a), of Figure 2, represents the exact posterior density of the interaction effect λ_{11}^{AB} for the (1, 1)th cell, obtained by Monte Carlo simulation, and curve (c) gives the normal approximation. The bracketed entries of Table 18.1 describe the exact Bayesian significance probabilities (i.e. posterior probabilities that the interaction effect is less than zero under a Jeffreys prior), for each of the 16 cells; the figures in square brackets were instead obtained using the normal approximation, and they are very close to the exact values. The B^2 statistic (1.21) becomes $B^2 = 6.76$, while $X^2 = 7.51$, on nine degrees of freedom.

When analysing any two-way contingency table, it is always important to consider a summary of the cells with significant interaction effects, since this is comparable with the analysis of residuals, in regression analysis. By highlighting the important cells in the table, and considering patterns of important cells, the statistician is able to infer the main conclusions from the data. In this case, cells (1, 1) and (4.4) give low significance prob-

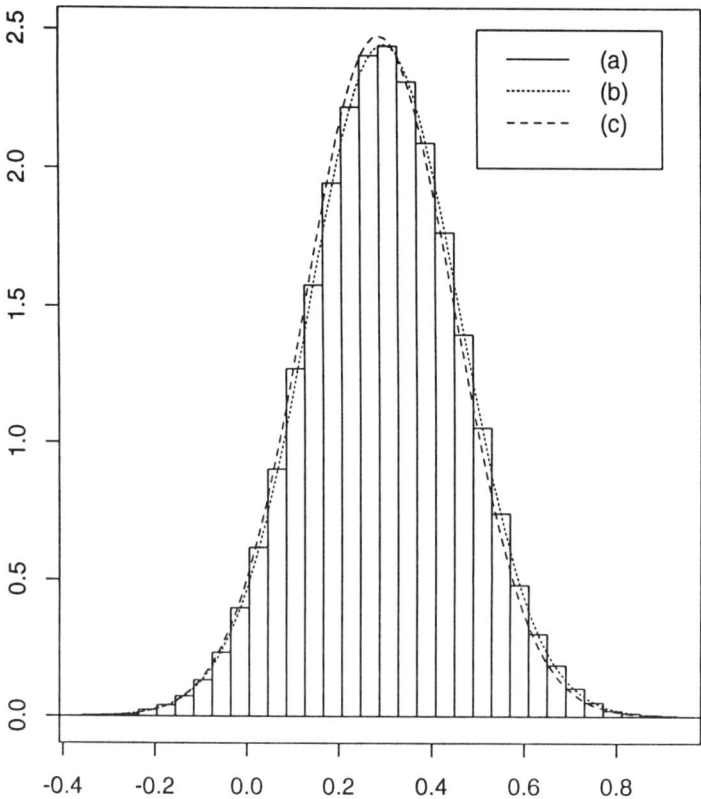

Figure 18.2 Posterior density of (1, 1)th interaction effect (engineering apprentice data): (a) exactly simulated histogram; (b) BCM; (c) normal approximation.

abilities, while cells (1, 4) and (4.4) give high significance probabilities. This enables us to infer the (in this case obvious) conclusion that there is some, but not overwhelming, positive association between the section head's assessment and the written test result. While it would be possible to perfectly fit the significant cells using Goodman's (1968) quasi-independence procedures, this can frequently give a model which overfits, but with less practical interpretability. A full rank interaction analysis of a 12×8 table (the Marine Corps data) is proposed by Leonard (1985) and Leonard and Novick (1986), using a more informative prior.

18.3 BAYES–STEIN METHODS OF THE 1970s

When the first co-author commenced graduate studies at University College London, in September 1970, as a 22-year-old Masters student, Dennis

Lindley gave him a preprint of Lindley (1971) and asked him to 'do the same thing for the binomial distribution'. Lindley and Smith were at the time developing their research on Bayes–Stein estimators for several normal means, and the linear model, (e.g. Lindley and Smith 1972). The corresponding analysis of categorical data turned out to be a question of finding a convenient transformation, and then handling the non-normality in the posterior distribution. This led to choices of non-conjugate prior distributions, other than the Dirichlet, and a general approach to the Bayesian analysis of categorical data (Leonard 1972, 1973a, b, 1975, 1976, 1977a; Laird 1978) using either multivariate normal prior distributions, or mixtures thereof, for logistic transformations of the parameters. The basic idea was simple though very unconventional at the time, as convention required conjugacy: if everybody is interested in normal approximations to the posterior distribution of the logits, then why not assume normality in the prior?

To illustrate this approach, we describe some hitherto unpublished research, for a generalization of Leonard's original thesis problem, i.e. the simultaneous estimation of the parameters of the product multinomial model (e.g. simultaneously smoothing several histograms). For $i = 1, \ldots, m$, consider cell frequencies $y_{i1}, \ldots, y_{is}$, which are taken to possess a multinomial distribution, with respective cell probabilities, $\theta_{i1}, \ldots, \theta_{is}$ satisfying $\sum_g \theta_{ig} = 1$, and sample size $n_i = \sum_g y_{ig}$. Given the θ_{ij}, the vectors $y_i = (y_{i1}, \ldots, y_{is})^T$ are taken to be independent. For $i = 1, \ldots, m$, consider m sets of multivariate logits $\{\gamma_{i1}, \ldots, \gamma_{is}\}$, satisfying

$$\theta_{ij} = \exp(\gamma_{ij}) / \sum_{g=1}^{s} \exp(\gamma_{ig}) \quad (j = 1, \ldots, s). \tag{3.1}$$

Suppose that the parameters of different multinomial distributions are a priori *exchangeable*, and that given μ and C, the $\gamma_i = (\gamma_{ij}, \ldots, \gamma_{is})^T$ are independent and possess multivariate normal distributions with common mean vector μ and covariance matrix C. This is a much more general specification than available (e.g. Leonard 1977b, c) via Dirichlet distributions, for the cell probabilities, since very general prior dependencies are permitted between the parameters. The γ_i are a posteriori independent, give μ and C, and the posterior density of γ_i is

$$\pi(\gamma_i | y_i) \propto \exp\{y_i^T \gamma_i - n_i D(\gamma_i) - (\gamma_i - \mu)^T C^{-1} (\gamma_i - \mu)/2\} \tag{3.2}$$

where

$$D(\gamma_i) = \log \sum_{g=1}^{g} \exp(\gamma_{ig}) \tag{3.3}$$

The posterior density (3.2) is non-normal owing to the presence of the $D(\gamma_i)$ term (3.3). Therefore, in the 1970s an exact posterior analysis was not available. However, two approximations (A and B) are applicable. Method (A) is based upon the normal approximation (Leonard 1973b):

$$\mathscr{L}^*(\gamma_i \mid y_i) = B(y_i)\exp\{-(\gamma_i - \hat{\gamma}_i)R_i(\gamma_i - \hat{\gamma}_i)/2\}, \quad (3.4)$$

$$= B(y_i)\exp\{-\sum_j y_{ij}(\gamma_{ij} - \log y_{ij} - d_j)^2/2\}, \quad (3.5)$$

to the likelihood of γ_i, given y_i. Here,

$$\hat{\gamma}_i = (\log y_{i1}, \ldots, \log y_{is})^T, \quad (3.6)$$

$$R_i = \text{diag}(y_{i1}, \ldots, y_{is}) - n_i^{-1} y_i y_i^T, \quad (3.7)$$

$$B(y_i) = n_i! n_i^{-n_i} y_{ij}^{-y_{ij}}/y_{ij}!, \quad (3.8)$$

and

$$d_j = \sum_j y_{ij}(\gamma_{ij} - \log y_{ij}). \quad (3.9)$$

The likelihood approximation (3.4) is reasonable if $y_{ij} \geq 5$ for all i, j. Consequently, the posterior distribution of γ_i, given μ and C, is approximately normal with mean vector

$$\gamma_i^* = (R_i + C^{-1})^{-1}(R_i\hat{\gamma}_i + C^{-1}\mu), \quad (3.10)$$

and covariance matrix

$$D_i = (R_i + C^{-1})^{-1}. \quad (3.11)$$

The expression in (3.10) expresses γ_i^* in the form of a matrix weighted average of γ_i and μ. A more precise procedure, (method B) which is suitable even if some of the y_{ij} are zero, involves taking the posterior distribution of γ_i, given μ and C, to be approximately normal with mean vector $\tilde{\gamma}_i$ and covariance matrix $\tilde{D}_i$, where $\tilde{\gamma}_i$ and $\tilde{D}_i$ are respectively the exact posterior mode vector and dispersion matrix of γ_i. These satisfy

$$n_i\tilde{\theta}_i = y_i - C^{-1}(\tilde{\gamma}_i - \mu), \quad (3.12)$$

and

$$D_i = n_i\{\text{diag}(\tilde{\theta}_{i1}, \ldots, \tilde{\theta}_{is}) - \tilde{\theta}_i\tilde{\theta}_i^T\} + C^{-1}, \quad (3.13)$$

where $\tilde{\theta}_i = (\tilde{\theta}_{i1}, \ldots, \tilde{\theta}_{is})^T$, with

$$\tilde{\theta}_{ij} = \exp(\tilde{\gamma}_{ij})/\sum_{g=1}^{g} \exp(\tilde{\gamma}_{ig}). \quad (3.14)$$

Note that (3.12) and (3.14) should be solved, using standard Newton-Raphson techniques for the $\tilde{\gamma}_i$. Both γ_i^* and $\tilde{\gamma}_i$ are, of course, highly dependent upon μ and C. Following I. J. Good's philosophy Fienberg and Holland (1973) showed in a much simpler context that it is possible to obtain estimators for the θ_{ij} with excellent mean squared error properties, by estimating the prior parameters empirically, from the current data set. Under a hierarchical Bayes procedure (e.g. Lindley and Smith 1972) it would be possible to assign further distributions to μ and C, at the second stage of the prior model. Before 1977, it seemed impossible, in the categorical data context, to find an estimation procedure for μ and C which worked in practical terms, if some of the cell frequencies were zero. For example, joint modal procedures (Leonard 1972, 1973a, b, 1975) tend to over-collapse the estimates of the variance components, as noted by Leonard (1976, 1977a).

However, the EM algorithm, summarized by Dempster, et al. (1977), and Laird (1978), tells us that we may sensibly estimate μ and C by equating prior and posterior expectations of $\gamma.\gamma.^T$ and $\sum_i \gamma_i \gamma_i^T$, where $\gamma. = m^{-1}\sum_i \gamma_i$. If performed exactly, this would lead to estimates $\tilde{\mu}$ and $\tilde{C}$ for μ and C maximizing their 'integrated likelihood', obtained from the joint distribution of $y_1, \ldots, y_m$ conditional only upon μ and C. Under the approximations developed above, for our method B, the EM algorithm yields the succinct expressions

$$\tilde{\mu} = \tilde{\gamma}., \qquad (3.15)$$

and

$$\tilde{C} = (m-1)^{-1} \sum_{i=1}^{m} (\tilde{\gamma}_i - \tilde{\gamma}.)(\tilde{\gamma}_i - \tilde{\gamma}.)^T + m^{-1} \sum_{i=1}^{m} \tilde{D}_i, \qquad (3.16)$$

which may be solved by cyclic substitution, combined with Newton–Raphson, for (3.12) and (3.14). Note the vital importance of the second term in (3.16) which avoids the difficult over-collapsing of the estimates of the variance components, mentioned earlier.

A numerical example of this procedure is described in section 18.4. Most of the first co-author's contributions of the 1970s follow naturally from Lindley (1964), together with Lindley and Smith (1972). For example, Lindley (1964), suggested smoothing the probabilities in a histogram by employing a multivariate normal prior distribution for a set of linearly independent log-contrasts, thus motivating Leonard (1973a, 1978). Both Leonard (1975) and Laird (1978) assign normal prior distributions to the parameters in Goodman's full rank interaction model (1.17). Nazaret (1987) later applied similar techniques to the analysis of three-way tables, but did not include Laird's important adjustments, paralleling (3.16) to the estimates of the variance components.

During the 1970s the Dirichlet prior approach was further developed (e.g. Good 1975, 1976; Crook and Good 1980; Lochner 1975). Gunel and Dickey (1974) combined Dirichlet priors with positive prior probabilities on null hypotheses of independence. Hickman and Miller (1977) further developed histogram smoothing, on an actuarial procedure for graduating mortality tables, and this is perhaps one of the most sucessful real applications of Bayesian procedures.

18.4 SMOOTHING GRADE DISTRIBUTIONS FOR 40 LONDON HIGH SCHOOLS

The observed percentages in the first six columns of Table 18.2 describe the percentages of students obtaining grades 1–6 in a mathematics test at 40 London schools. The underlying data were regarded as numerical realizations of the frequencies of a product multinomial model, i.e. 40 multinomial distributions, each with six cells. The smoothed percentages in the last six columns of Table 18.2 were calculated from $100\tilde{\theta}_{ij}$, where the $\tilde{\theta}_{ij}$ are the smoothed cell probabilities discussed in section 18.3. The sample sizes n_i in the 13th column describe the numbers of students taking the test at the 40 schools.

Table 18.2. Observed and smoothed percentages

	Observed						Smoothed						
i\j	1	2	3	4	5	6	1	2	3	4	5	6	n_i
1	6.7	17.8	24.4	28.9	6.7	15.6	6.6	18.8	24.2	28.2	7.2	15.0	45
2	0.0	21.6	24.3	18.9	13.5	21.6	3.8	16.9	22.2	26.2	9.9	20.9	37
3	5.3	15.8	42.1	26.3	5.3	5.3	8.3	21.4	29.1	26.5	5.8	8.9	19
4	22.2	25.9	29.6	11.1	7.4	3.7	19.5	26.9	26.5	17.9	4.2	4.9	27
5	16.7	33.3	11.1	16.7	5.6	16.7	12.7	25.4	22.8	22.3	5.7	11.1	18
6	5.9	7.4	26.5	33.8	8.8	17.6	4.8	12.9	23.8	31.5	8.9	18.0	68
7	31.7	17.1	14.6	19.5	9.8	7.3	24.1	22.1	21.7	19.1	5.7	7.4	41
8	4.5	4.5	22.7	31.8	9.1	27.3	3.6	12.4	20.2	29.7	9.6	24.5	22
9	16.1	45.2	19.4	9.7	3.2	6.4	17.5	33.6	23.8	16.7	3.4	5.0	31
10	13.0	31.5	31.5	24.1	0.0	0.0	14.9	30.3	29.1	20.5	2.4	2.9	54
11	23.5	35.3	20.6	20.6	0.0	0.0	22.7	31.7	24.3	16.5	2.2	2.6	34
12	22.8	26.3	21.1	15.8	3.5	10.5	19.8	26.2	23.0	18.7	4.4	7.8	57
13	7.1	14.2	14.2	32.1	10.7	21.4	5.3	15.7	20.5	29.0	9.0	20.4	28
14	13.9	33.3	25.0	19.4	0.0	8.3	14.2	29.2	25.6	20.9	3.7	6.5	38
15	0.0	18.2	13.6	54.5	9.1	4.5	4.2	17.2	23.2	34.7	7.4	13.4	22
16	12.5	31.3	18.8	18.8	6.3	12.5	11.3	24.8	24.6	23.5	5.7	10.1	16
17	0.0	9.4	25.0	50.0	9.4	6.3	3.3	14.3	24.4	36.3	8.0	13.7	32
18	0.0	5.3	15.8	21.1	26.3	31.6	2.0	9.1	16.1	26.6	13.9	32.3	18
19	0.0	3.7	7.4	14.8	3.7	70.4	0.7	4.6	8.4	20.0	9.4	57.1	27
20	2.4	7.3	19.5	36.6	4.8	29.3	2.7	11.3	18.8	31.9	8.4	27.0	42
21	29.2	16.7	8.3	37.5	4.2	4.2	19.5	23.5	22.8	23.3	4.4	6.5	24
22	14.2	21.4	42.9	21.4	0.0	0.0	14.8	26.4	29.0	21.2	3.7	4.8	14
23	0.0	17.4	34.8	17.4	17.4	13.0	4.8	17.2	25.5	26.8	9.7	15.9	23
24	5.6	19.4	30.1	22.2	11.1	11.1	7.1	19.9	26.5	26.1	7.9	12.5	36

25	31.3	18.8	28.1	15.6	3.1	3.1	25.6	25.3	25.5	16.7	3.2	3.7	32
26	9.5	19.0	42.9	14.3	0.0	14.3	10.1	22.9	28.4	23.5	5.2	9.8	21
27	5.6	37.0	22.2	25.9	3.7	5.6	8.8	30.0	25.4	24.5	4.3	7.1	54
28	0.0	30.8	7.7	30.8	0.0	30.8	4.4	17.7	20.2	28.7	7.4	21.6	13
29	4.0	12.0	16.0	32.0	8.0	28.0	3.7	13.6	19.3	29.5	9.1	24.8	25
30	0.0	16.7	27.8	33.3	5.6	16.7	4.5	17.1	23.9	30.1	7.6	16.8	18
31	3.7	11.1	37.0	37.0	0.0	11.1	5.8	18.0	27.6	30.9	5.8	11.9	27
32	0.0	42.9	28.6	21.4	7.1	0.0	9.6	27.6	27.4	23.6	4.8	6.9	14
33	0.0	14.3	28.6	21.4	21.4	14.3	4.5	15.8	23.2	28.1	10.1	18.2	14
34	19.5	39.0	17.1	22.0	0.0	2.4	19.4	33.0	23.4	18.1	2.5	3.5	41
35	37.9	13.8	37.9	6.9	3.4	0.0	31.8	24.6	26.2	12.8	2.4	2.1	28
36	0.0	18.8	6.2	25.0	6.2	43.8	2.5	11.9	15.9	27.2	9.4	33.1	16
37	13.3	20.0	20.0	26.7	6.7	13.3	9.8	21.6	24.5	25.7	6.4	11.8	15
38	16.7	37.5	41.7	4.2	0.0	0.0	20.3	32.4	28.3	14.3	2.2	2.3	24
39	18.3	31.7	21.7	15.0	6.7	6.7	17.5	28.7	24.0	18.4	4.8	6.5	67
40	0.0	11.1	11.1	33.3	22.2	22.2	3.4	13.3	19.7	29.4	10.5	23.7	9

The EM algorithm procedures of section 18.3 provided an empirical estimate

$$\tilde{\mu} = (-0.51, 0.41, 0.56, 0.59, -0.82, -0.24)^T \quad (4.1)$$

for the prior mean vector μ. This corresponds to a common prior estimate

$$\tilde{\xi} = (\exp[\tilde{\mu}_1], \ldots, \exp[\tilde{\mu}_6])^T / \sum_{j=1}^{6} \exp(\tilde{\mu}_j)$$
$$= (0.087 \quad 0.217 \quad 0.255 \quad 0.262 \quad 0.064 \quad 0.115)^T, \quad (4.2)$$

for the θ_i. An empirical estimate $\tilde{C}$ was also obtained for the prior covariance matrix C. This possessed diagonal terms

$$\mathrm{diag}(\tilde{C}) = (1.08 \quad 0.39 \quad 0.25 \quad 0.25 \quad 0.26 \quad 0.46 \quad 0.92), \quad (4.3)$$

with corresponding correlation matrix.

$$\tilde{B} = \begin{pmatrix} 1 & 0.79 & 0.57 & -0.07 & -0.38 & -0.57 \\ 0.79 & 1 & 0.79 & 0.29 & -0.14 & 0.31 \\ 0.57 & 0.79 & 1 & 0.61 & 0.21 & -0.01 \\ -0.07 & 0.29 & 0.61 & 1 & 0.68 & 0.62 \\ -0.38 & -0.14 & 0.21 & 0.68 & 1 & 0.83 \\ -0.57 & -0.31 & -0.01 & 0.62 & 0.83 & 1 \end{pmatrix}. \quad (4.4)$$

The correlation matrix $\tilde{B}$ indicates moderately high correlations for adjacent cells and negative correlations for all cells, more than two cells apart. The consequent smoothed percentages in Table 18.2 are quite complex. They combine empirical Bayes–Stein shrinkages which take into account the common estimate (4.2), across all schools, with histogram-style smoothing of the percentages for each individual school, which refer to the correlation matrix (4.4).

18.5 COMPUTATIONAL TECHNIQUES OF THE 1980s

Zellner and Rossi (1984) made perhaps the key contribution of the decade by pioneering exact Bayesian inferences for the linear logistic model (under either a uniform or multivariate normal prior), using a generalization of Monte Carlo simulation, referred to as 'importance sampling'. Geweke (1988, 1989) describes some precise theorems regarding the convergence of importance sampling. Similar techniques may be used to compute exact posterior inferences for many of the models reviewed in the current chapter. Hsu $et\ al.$ (1991) apply the methodology to general families of discrete distributions. Furthermore, the 'Gibbs sampler' is particularly useful for multinomial-Dirichlet models (see Gelfand and Smith 1990). Duffy and Santner (1988) also provide a Bayesian analysis for the linear logistic model.

Consider the multinomial model, analysed in section 18.3, and the posterior density (3.2), when μ and C are specified. It is virtually impossible to simulate a γ_i vector with density equal to (3.2). However, paralleling Zellner and Rossi, we can simulate a sequence of γ_i vectors from a multivariate t-distribution with ν (say $\nu = 20$), degrees of freedom, mean vector γ_i, and precision matrix $\nu \tilde{D}_i^{-1}(\nu + s)$, where $\tilde{\gamma}_i$ and $\tilde{D}_i^{-1}$ are calculated by the techniques of section 18.3.

Then importance sampling permits the computation of the exact posterior expectation, or posterior probabilities, for any parameter of interest, by reference to a simple reweighting formula, involving the ratio of the exact posterior density of γ_i and the above multivariate t-approximation. The degrees of freedom ν may be chosen pragmatically, to speed the convergence. In particular, the posterior expectations of $\gamma.\gamma.^T$ and $(\gamma_i - \gamma.)(\gamma_i - \gamma.)^T$ can be exactly computed, permitting the exact calculation of the integrated likelihood estimates for μ and C.

Rather than simulate, it is also possible to approximate closely the marginal posterior density of any parameter of interest by a continuous curve, with saddlepoint accuracy, using Laplacian/conditional maximization techniques (e.g. Leonard 1982; Tierney and Kadane 1986; Leonard, $et\ al.$, 1989). Extensions to hierarchical Bayesian models are described by Kass and Steffey (1989), who in particular discuss the logistic/multivariate normal prior formulation of section 18.3. Leonard, $et\ al.$ use Laplacian techniques to approximate the posterior density for a general measure of association, proposed by Altham (1970), for an $r \times s$ contingency table. We refer to the general Laplacian/conditional maximization approach as BCM (Bayesian conditional maximization).

Curve (a) of Figure 18.1 is also the BCM approximation to the posterior density of the log measure of association, and this is identical, to within three decimal point accuracy, of the exact curve. The BCM curve (a) for Figure 18.2 is even closer to the histogram (c), representing the exact curve, when compared to the normal approximation (b).

A variety of extensions of Good's Dirichlet prior approach are discussed by Albert (1983, 1985a, b, 1987a, b, 1988, 1990) and Albert and Gupta (1980, 1981, 1982, 1983a, b, c, 1985).

18.6 THREE-WAY TABLES AND SIMPSON'S PARADOX

Consider the data reported by Radelet (1981), and described in Table 18.3. Overall 11.88% of white defendants receive the death penalty, compared with 10.24% of black defendants. However, in cases with white victims these percentages should be replaced by 12.58% and 17.46%, and in the remaining cases, with black victims, they should be replaced by 0% and 5.83%. This is Simpson's paradox (as for example discussed by Lindley and Novick 1981). The conclusion of discrimination against white defendants suggested by the overall table is invalidated by a 'lurking variable' i.e. colour of victim; opposite conclusions are obtained in each of the two subtables.

Table 18.3 Racial characteristics, and imposition of death penalty

	Overall		White victim		Black victim	
	Death	Not death	Death	Not death	Death	Not death
White defendant	19	141	19	132	0	9
Black defendant	17	149	11	52	6	97

Simpson's paradox can be used to refute many analyses of contingency tables based upon non-randomized data. There may always exist a lurking variable which would lead to consistently opposite conclusions, if the contingency table was split according to that variable. However, if the data are appropriately randomized (e.g. if the white defendants and black defendants could have been chosen at random from subpopulations of white defendants and black defendants) then any lurking variable is much less likely to affect the analysis. Lindley and Novick alternatively argue that the problem can be minimized if the individuals can be subjectively regarded as 'exchangeable' members of a population. In any case, any conclusion obtained from a contingency table, based on non-randomized data, can at best be regarded as subjective.

It is difficult to think and interpret data in three dimensions, and in 1971, D. V. Lindley suggested a 'three-directional approach' to the first co-author. The co-authors have implemented this approach while teaching Statistics 421 at the University of Wisconsin, and it provides an alternative to log-linear models for three-way tables (e.g. Nazaret 1987) which we find a bit more difficult to interpret. We argue that the three-directional approach can be used to extract most of the important real-life conclusions from the data. Note that, for the data in Table 18.3, few of the normal approximations of section 18.1 are valid, owing to the occurrence of a zero cell frequency.

Direction 1 (Split according to 'colour of victim' variable).

Step A1 Investigate inequality of the unconditional cell probabilities of the two subtables in Table 18.3. One way of doing this is to consider the 2 × 4 table, with first row 19 132 11 52, and second row 0 9 6 97. Then investigate whether the interaction effects are zero, under Goodman's full rank interaction model (1.16). Curves (a), (b), (c) and (d) of Figure 18.3, describe our BCM approximation to the posterior densities of the interaction effects for the (1, 1), (1, 2), (1, 3), and (1, 4) cells of this 2 × 4 table, under a Jeffreys prior. The BCM approximations are virtually exact. Histograms representing exact, simulated results, are available from the authors. As zero lies in the extreme tail of three of the densities, we

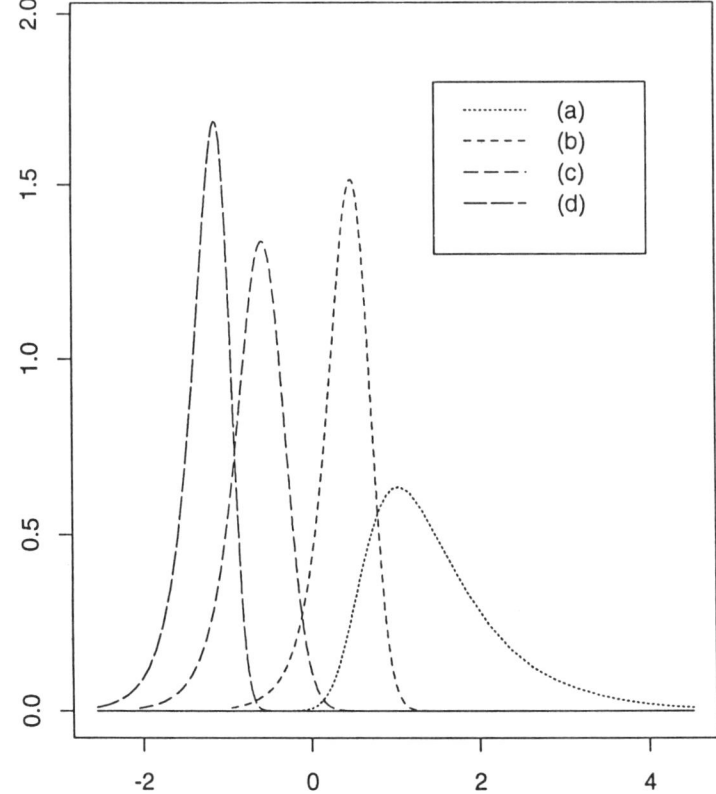

Figure 18.3 Posterior densities of interaction effects (split on colour of victim): (a) (1, 1) cell; (b) (1, 2) cell; (c) (1, 3) cell; (d) (1, 4) cell.

conclude that three of the interaction effects may be non-zero, and therefore that the unconditional cell probabilities in the two subtables are not identical.

Step B1 If step A1 concludes that each of the two subtables is unequal, analyse the two subtables, separately. Otherwise, just analyse the overall table in this direction. Curves (a) and (b) of Figure 18.4 describe the exact posterior densities of the log-measure of associations for our two subtables (white victim and black victim). We conclude that while there is some evidence of a negative association between colour of defendant, and death penalty, in each subtable, this evidence is not particularly significant, as zero does not lie in the tail of either density. Curve (c) describes the posterior density of the log-measure of association for the overall table, and it is not really relevant to consider this curve as the two subtables

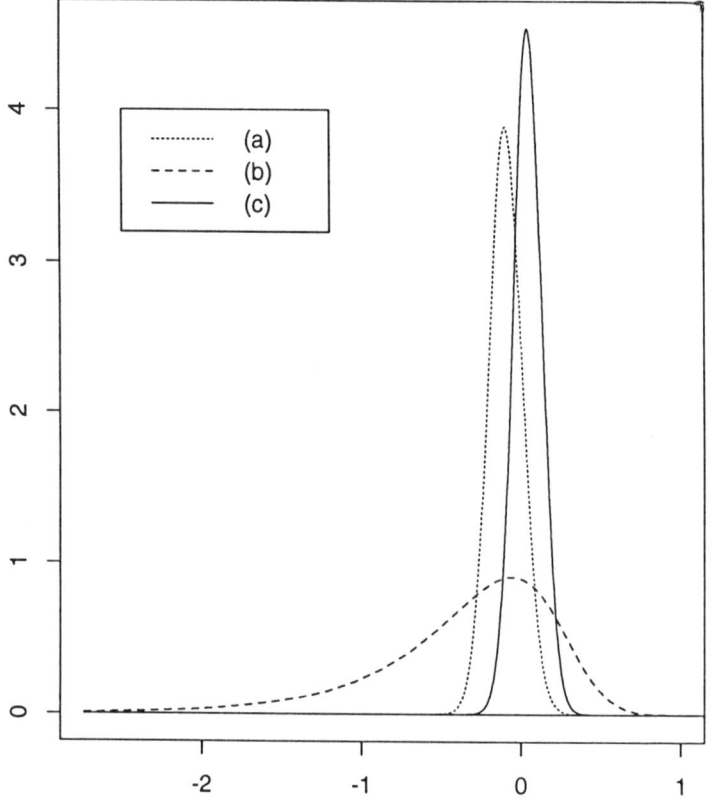

Figure 18.4 Posterior densities of log measures of association (split on colour of victim); (a) White victims; (b) Black victims; (c) overall.

have already been taken to be unequal. However, it does indicate a positive association, thus again highlighting Simpson's paradox.

Direction 2 (Repeat steps of direction 1, but with split according to death penalty variable. Each table or subtable cross-classifies colour of defendant against colour of victim).

Step A2 The locations of the posterior densities of the interaction effects in Figure 18.5 suggest than the data cannot strongly refute equality of the unconditional cell probabilities of the two subtables.

Step B2 Posterior density (c) of the log-measure of association for the overall table suggests that there is a noticeable negative association between colour of victim and colour of defendant. Curves (a) and (b) suggest that the death penalty variable does not affect this conclusion. Note that,

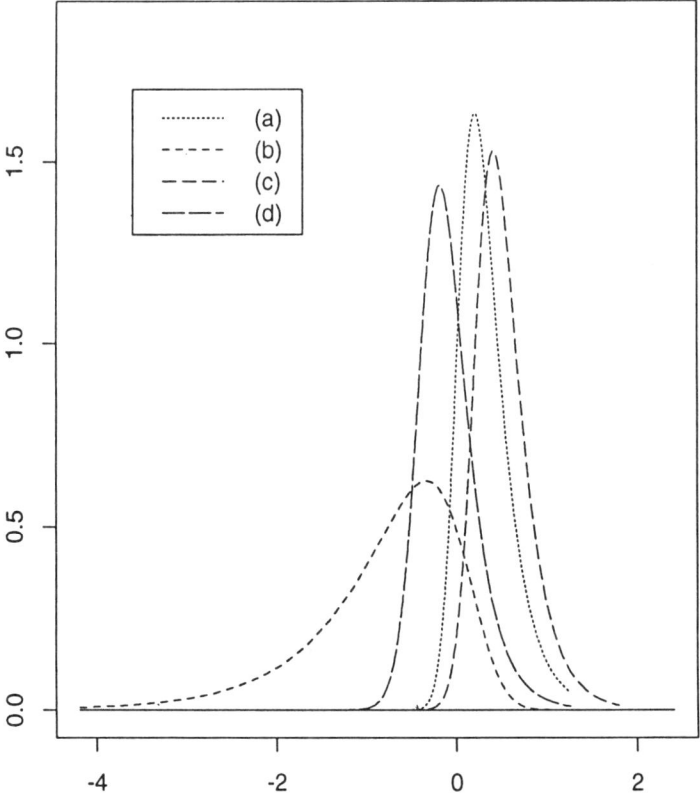

Figure 18.5 Posterior densities of interation effects (split on death penalty): (a) (1, 1) cell; (b) (1, 2) cell; (c) (1, 3) cell; (d) (1, 4) cell.

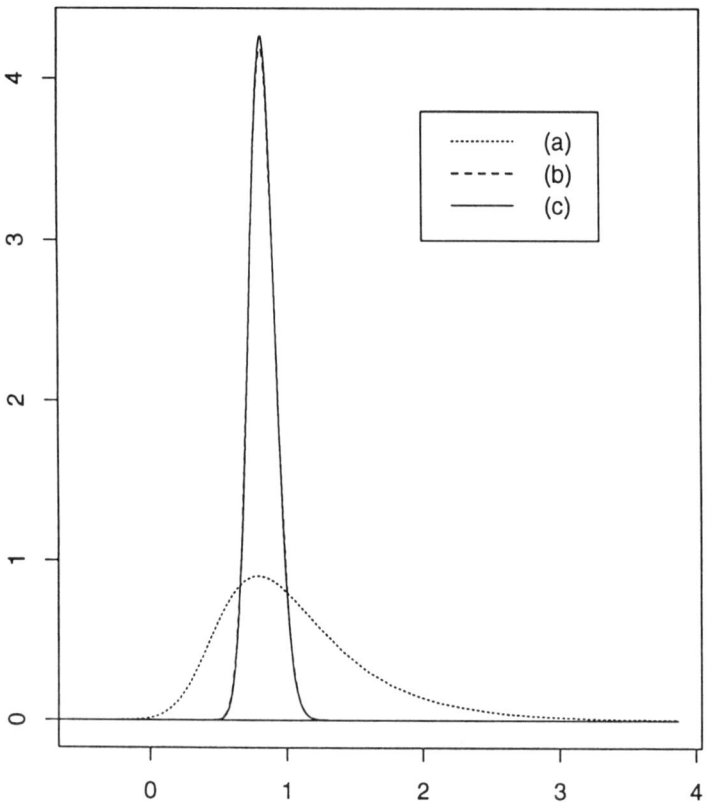

Figure 18.6 Posterior densities of log measures of association (split on death penalty): (a) death penalty; (b) no death penalty; (c) overall.

overall, 94.38% of white defendants have white victims, while 59.43% of black defendants have white victims i.e. black defendants have a stronger propensity to find victims of their own colour.

Direction 3 (Repeat steps of previous directions, but with split according to 'colour of defendant' variable. Each table as subtable cross-classifies colour of victim against death penalty variable).

Step A3 The posterior densities of the four indication effects, in Figure 18.7, indicates substantial differences between the two subtables.

Step B3 The posterior densities of the log-measures of association, in Figure 18.8, indicate that there is a strong positive association between colour of victim, and the death penalty, which is not refuted in either

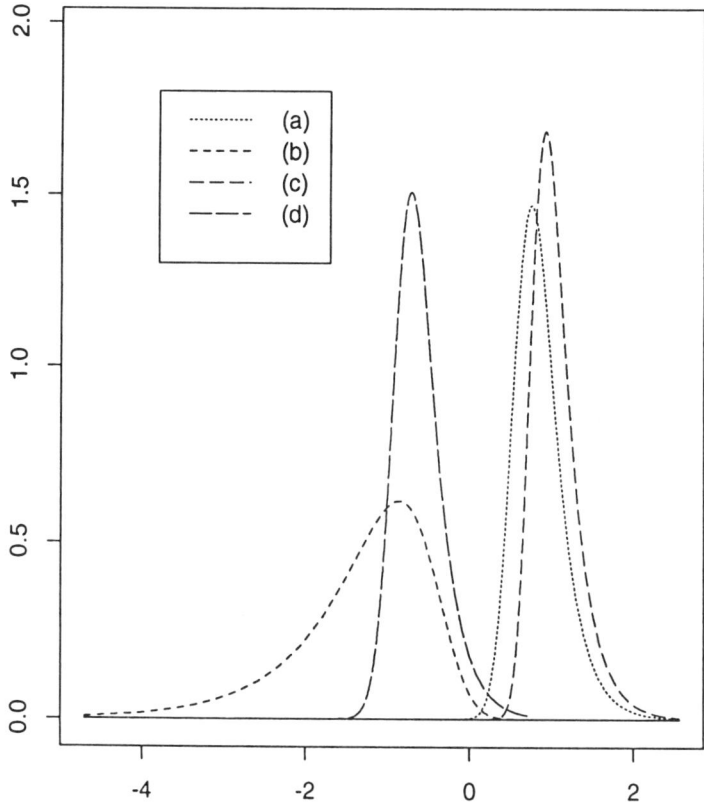

Figure 18.7 Posterior densities if interaction effects (split on colour of defendant): (a) (1, 1) cell; (b) (1, 2) cell; (c) (1, 3) cell; (d) (1, 4) cell.

subtable. Note that, overall, 94.38% of defendants with white victims, receive the death penalty, while only 61.17% of defendants with black victims receive the death penalty.

The key conclusions have been extracted from the data at steps B2 and B3 of our three-directional approach, with the help of virtually exact inferential techniques which are unavailable to non-Bayesians. Note that a fourth variable, e.g. socio-economic status of victim, might change the conclusions based upon the first three variables.

Next consider the Berkeley admissions data, reported by Freedman *et al.* (1978, p. 17). Out of 2691 men applying to the six largest graduate programme at Berkeley in 1973 44.5% were admitted, and out of 1835 women 30.4% were admitted. This apparent sex bias is spurious, as demonstrated by applying our three-directional approach to the $6 \times 2 \times 2$ table (graduate major versus gender of applicants versus acceptance/

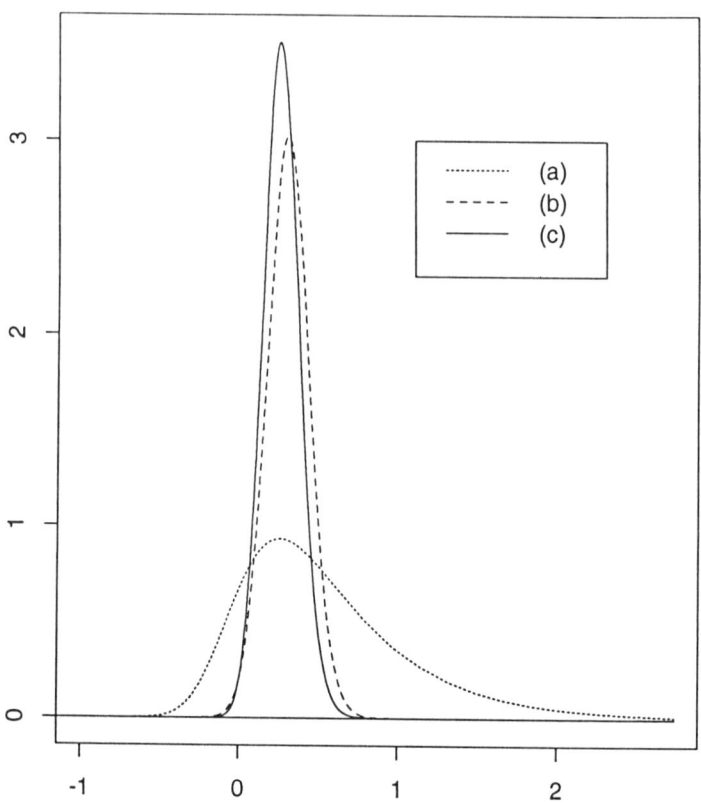

Figure 18.8 Posterior densities of log measures of association (split on colour of defendant); (a) Black defendant; (b) White defendant; (c) overall.

non-acceptance) approach. As all the cell frequencies are greater than 5, we do not report exact posterior densities, but instead refer to B^2, in (1.9) and (1.20), together with associated approximate normal statistics.

Direction 1 (Split according to major, each 2 × 2 subtable cross-classifies gender of applicant against acceptance/non-acceptance).

Step A1 Equality of the unconditional cell probabilities of the 2 × 2 tables for the six graduate majors A, B, C, D, E and F was investigated by calculating B^2, in (1.20) and under a Jeffreys prior for the 6 × 4 table, whose six rows each comprised the four entries for a 2 × 2 table for one of the graduate majors. This gave $B^2 = 1285.72$, with 15 degrees of freedom, clearly suggesting inequality of the unconditional cell probabilities of the six 2 × 2 tables, and hence indicating that any conclusion from the overall 2 × 2 table might be misleading.

Step A2 We analysed each of the 2 × 2 tables by calculating the approximate normal test statistic $B = \lambda^*/(v^*)^{1/2}$, for each table, where λ^* and v^* are defined in section 18.1. The B statistic for majors A, B, C, D, E and F, was evaluated, under a Jeffreys prior, as -3.97, -0.44, -1.77, -0.55, 1.01 and -0.62. Hence, there is only a significant association for major A, and this is negative (62% males admitted versus 82% females). All other majors give non-significant negative associations, apart from major E which is non-significant and positive. Hence, a version of Simpson's paradox occurs.

Direction 2 (Split according to gender, each 6 × 2 subtable cross-classifies graduate major against acceptance/non-acceptance).

Step A2 Equality of the unconditional cell probabilities of the 6 × 2 subtables for men and for women was investigated by calculating B^2 in (1.20) for an approximate 2 × 12 table, giving $B^2 = 861.35$ with 11 degrees of freedom, and hence refuting equality.

Step B2 The 6 × 2 table for men applicants, was analysed by calculating the approximate normal test statistics $b_{ij} = \tilde{\lambda}_{ij}/(v_{ij}^{AB})^{1/2}$, where $\tilde{\lambda}_{ij}$ and v_{ij}^{AB} are the approximate mean and variance of the interaction effects, introduced in section 18.1. For the first column of this 6 × 2 table, the b_{ij} were respectively 15.02, 13.84, -1.83, 0.11, -1.67 and -10.95. This suggests that, among the men, significantly more applicants were accepted for majors A and B, and significantly less were accepted for major F. For the first column of the corresponding 6 × 2 table for women the b_{ij} were respectively 8.88, 3.37, -1.62, -1.18, -5.56 and -10.77. This suggests that, among the women, significantly more applicants were accepted for majors A and B, and significantly less were accepted for majors E and F.

Direction 3 (Split according to acceptance non-acceptance each 6 × 2 subtable cross-classifies graduate major against gender).

Step A3 Equality of the unconditional cell probabilities, of the 6 × 2 subtables, for accepted applicants, and non-accepted applicants, was investigated by calculating B^2 in (1.20), for an appropriate 2 × 12 table, giving $B^2 = 581.33$, with 11 degrees of freedom, and hence refuting equality.

Step B3 The 6 × 2 table for accepted applicants, was analysed by calculating the b_{ij}. For the first column of this 6 × 2 table, the b_{ij} were respectively 9.39, 11.20, -9.23, -4.39, -7.35, -2.75, suggesting that significantly more men were accepted for majors A and B, and significantly more women were accepted for majors C, D, E and F. For the first column of the 6 × 2 table for non-accepted applicants, the b_{ij} were respectfully 9.35,

7.90, −11.53, −6.97, −11.43 and −7.68, suggesting that significantly more men were not accepted for majors A and B, and significantly more women were not accepted for majors C, D, E and F.

When performing the three-dimensional approach, it is important to also carefully investigate the raw data in each direction, to facilitate the extraction of all real-life conclusions from the data, and to compare conclusions from each direction. Anyway, it is clear that more men apply to the majors (A and B) with higher admission rates, and that the college perhaps tries to compensate for this by admitting a higher proportion of women to these majors. Majors C, D, E and F have much lower admission rates, but most of the women apply to these four majors.

18.7 FURTHER PROBLEMS WITH NON-RANDOMIZED DATA

Cell frequencies $y_1, \ldots, y_s$ may be taken to possess a multinomial distribution, with cell probabilities $\theta_1, \ldots, \theta_s$, and sample size n, if the n individuals were chosen at random, e.g. without replacement from a much larger population, and if $\theta_1, \ldots, \theta_s$ denote appropriate population proportions. Conversely, without randomization at the experimental design stage, a multinomial assumption may be both incorrect and misleading. However, Bayesian hierarchical models indicate possible ways of generalizing the multinomial sampling distribution which are given under the headings below.

18.7.1 The multinomial–Dirichlet distribution

Following Leonard (1977b, c) and Paul and Plackett (1978) suppose that, conditional on $\theta_1, \ldots, \theta_s$, the frequencies $y_1, \ldots, y_s$ possess a multinomial distribution, with cell probabilities $\theta_1, \ldots, \theta_s$, and sample size n, but where $\theta_1, \ldots, \theta_s$ possess a Dirichlet distribution, with parameters $\alpha \xi_1, \ldots, \alpha \xi_s$. Here $\xi_1 + \xi_2 + \ldots + \xi_s = 1$, and ξ_j is the expectation of the corresponding θ_j. Then the first and second moments of the y_j satisfy

$$E(y_j) = n\xi_j \quad (j = 1, \ldots, s), \tag{7.1}$$

$$\text{cov}(y_j, y_k) = n\tau^{-1}(\xi_j \delta_{jk} - \xi_j \xi_k) \quad (j = 1, \ldots, s; k = 1, \ldots, s), \tag{7.2}$$

where δ_{jk} denotes the Kronecker delta function, and $\tau = (1 + \alpha)/(n + \alpha)$. The covariance structure (7.2) is similar to the covariance structure for a multinomial distribution, but an over-dispersion factor $\tau = (n + \alpha)/(1 + \alpha)$ is also included. Hence the multinomial–Dirichlet model may provide a better fit to non-randomized data, when this possesses larger dispersion. Leonard and Novick (1986) show that the over-dispersion factor can be combined with a log-linear model for the ξ_j.

Note that, as n gets large, with τ fixed, the sampling distribution of τX^2 approaches a chi-squared distribution, with $s - 1$ degrees of freedom, where

$$X^2 = \sum_j (y_j - n\xi_j)^2/(n\xi)$$

denotes the usual chi-squared statistic. Since $\tau < 1$, goodness-of-fit tests based upon the multinomial–Dirichlet distribution tend to be less likely to yield significant results, when compared with tests based upon the multinomial distribution.

18.7.2 The multinomial logit–multivariate normal distribution

Under the multinomial assumptions in (A), conditional on $\theta_1, \ldots, \theta_s$, consider instead multivariate logits $\gamma_1, \ldots, \gamma_s$ satisfying $\theta_j = \exp(\gamma_j)/\sum \exp(\gamma_s)$, for $j = 1, \ldots, s$; and let $\gamma = (\gamma_1, \ldots, \gamma_s)^T$ possess a multivariate normal distribution, with mean vector μ, and covariance matrix C.

This sampling distribution also allows for over-dispersion, and possesses the advantage of permitting complex interdependencies between the frequencies, much more general than permitted by the restrictive covariance structure (7.2). This is in the spirit of other suggestions, permitting serial correlation, recommended by Tavaré and Altham (1983), which lead to alternative adjustments to the distribution of the chi-squared statistic. The methodology, in section 18.3, for the product–multinomial distribution, under a hierarchical prior, effectively provides a method for estimating μ and C, when there are several replications from the above distribution. For example, our analysis in section 18.4, of the London high school data could be regarded as a non-Bayesian analysis under a choice of sampling distribution which generalizes the multinomial, to compensate for the non-randomized nature of the data.

18.7.3 Lurking variables

In the State of Wisconsin the estimation of population gene frequencies, for purposes of HLA blood group testing (e.g. in parcentage and criminal cases) is based upon a non-random sample of 5500 white males attending blood-testing clinics in Milwaukee, plus a non-random sample of about 2000 patients at University of Wisconsin hospitals and clinics. This means that (a) binomial or multinomial assumptions are inappropriate, so that no valid statistical properties or standard errors are available for the estimation procedures, and (b) the conclusions are very much subject to the problem of lurking variables. Similar world-wide problems for the subject of genetics are summarized by Leonard (1991). For DNA testing, a non-random sample of about $n = 7500$ is used to estimate the USA population distribution of the allele lengths, for any particular DNA probe, with similar problems. For both HLA blood group testing, and DNA testing, a serious misapplication of Bayes's theorem is then used to calculate an alleged probability of guilt. Genetic testing, however, provides excellent real-life examples to illustrate the general point that, when considering categorical data, aspects of experimental design and choice of sampling distribution are often much more important than the precise details of a Bayesian analysis. For the gene frequency example of section 18.2 it is difficult to find a valid Bayesian or non-Bayesian analysis, unless the $n = 106$ individuals in the sample are assumed to be drawn at random from a population. In 1985 the first co-author was headlined in the Madison press as 'Statistics Professor asks "What do we really know?" ' in connection with a similar analysis of the local AIDS–HIV data.

In 1992 the Department of Sociology at the University of Wisconsin-Madison used logistic regression to investigate the gender equity data for men and women at the university. Data for the entire population were analysed, but some important variables, e.g. measuring merit, were not included in the analysis. Nevertheless, the conclusion was reached that 'women faculty are underpaid by 2.8% when compared with men'. Using the above concepts, we advised the Senate that 'any conclusion drawn from these data is at best subjective'. Clearly, the problems of non-randomization (e.g. reducing the apparent significance of any conclusion drawn from classical significance tests) and lurking variables, are overwhelming. This type of refutation of non-randomized data is not new; see, for example Fisher's 1936 treatise regarding Mendel's pea-breeding data.

ACKNOWLEDGEMENTS

The first co-author's contributions to this area would not have been possible without generous help and advice from Patricia Altham, Jack Good, Jim Hickman, Adrian Smith and Dennis Lindley. Both authors are indebted to Kam-Wah Tsui for his advice during the 1980s, and to two John Woods, the first for providing the London High School data, and the second for his advice on contingency tables and genetic testing.

REFERENCES

Albert, J. H. (1983) Bayesian estimation methods for incomplete two-way contingency tables using prior beliefs of association, *ASA Proc. of Survey Rsch. Methods Sect.*, pp. 738–42.

Albert, J. H. (1985a) Bayesian estimation methods for incomplete two-way contingency tables using prior beliefs of association, *Bayes Stat.* **2**, 589–602.

Albert, J. H. (1985b) Simultaneous estimation of Poisson means under exchangeable and independence models, *J. Statist. Computation and Simulation*, **23**, 1–14.

Albert, J. H. (1987a) Empirical Bayes estimation in contingency tables, *Communications in Stat., Part A, Th. and Meth.*, **16**, 2459–85.

Albert, J. H. (1987b) Bayesian estimation of odds ratios under prior hypotheses of independence and exchangeability, *J. Statist. Computation and Simulation*, **27**, 251–68.

Albert, J. H. (1988) Bayesian estimation of Poisson means using a hierarchical log-linear model, *Bayes Stat.*, **3**, 519–31.

Albert, J. H. (1990) A Bayesian test for two-way contingency table using independence priors, *Canadian J. Stat.*, **18**, 347–63.

Albert, J. H. and Gupta, A. K. (1980) Bayesian estimation in 2×2 contingency tables, *ASA Proc. of Social Stat. Sect.*, pp. 461–6.

Albert, J. H. and Gupta, A. K. (1981) Mixtures of Dirichlet distributions and estimation in contingency tables, *ASA Proc. of Social Stat. Sect.*, pp. 189–93.

Albert, J. H. and Gupta, A. K. (1982) Mixtures of Dirichlet distributions and estimation in contingency tables, *Ann. of Statistics*, **10**, 1261–8.

Albert, J. H. and Gupta, A. K. (1983a) Models for reflecting prior beliefs of association in two-way contingency tables, *Communications in Stat., Part A, Th. and Meth.*, **12**, 1241–59.

Albert, J. H. and Gupta, A. K. (1983b) Bayesian estimation methods for 2×2 contingency tables using mixtures of Dirichlet distributions, *J. Amer. Statist. Assn.*, **78**, 708–17.

Albert, J. H. and Gupta, A. K. (1983c) Estimation in contingency tables using prior information, *JRSS-B*, **45**, 60–9.

Albert, J. H. and Gupta, A. K. (1985) Bayesian methods for binomial data with applications to a nonresponse problem, *J. Amer. Statist. Assn.*, **80**, 167–74.

Altham, P. M. E. (1969) Exact Bayesian analysis of a 2×2 contingency table, and Fisher's exact test, *J. Roy. Statist. Soc.*, **B31**, 261–9.

Altham, P. M. E. (1970) The measurement of association of rows and columns of an $r \times s$ contingency table, *J. Roy. Statist. Soc.*, **B32**, 63–73.

Bloch, D. A. and Watson, G. (1967) A Bayesian study of the multinomial distribution, *Ann. Math. Stat.*, **38**, 1423–35.

Crook, J. F. and Good, I. J. (1980) On the application of symmetric Dirichlet distributions and their mixtures to contingency tables: Part II, *Annals of Statistics*, **8**, 1198–1218.

Dempster, A. P., Laird, N. M. and Rubin, D. B. (1977) Maximum likelihood estimation from incomplete data via the EM algorithm (with discussion), *J. Roy. Statist. Soc.*, **B39**, 1–38.

Duffy, D. E. and Santner, T. J. (1988) Estimating logistic regression probabilities, *V*, **1**, 177–94.

Fienberg, S. E. and Holland, P. W. (1973) Simultaneous estimation of multinomial cell probabilites, *J. Amer. Statist. Assoc.*, **68**, 683–9.

Fisher, R. A. (1936) Has Mendel's work been rediscovered? *Ann. Sci.*, **1**, 115–37.

Freedman, D., Pisani, R., Purves, R. and Adhikari, A. (1978) *Statistics*, 2nd edn, Norton.

Gelfand, A. E. and Smith, A. F. M. (1990) Sampling based approaches to calculating marginal densities, *J. Amer. Statist. Assn.*, **85**, 398–409.

Geweke, J. (1988) Antithetic acceleration of Monte Carlo integration in Bayesian inference. *J. Econometrics*, **38**, 73–90.

Geweke, J. (1989) Bayesian inference in econometric models using Monte Carlo integration, *Econometrica*, **57**, 1311–70.

Good, I. J. (1965) *The Estimation of Probabilities*, MIT Press.

Good, I. J. (1967) A Bayesian significance test for multinomial distributions (with discussion), *J. Roy. Statist. Soc.*, **B29**, 399–431.

Good, I. J. (1975) The Bayes factor against equiprobability of a multinomial distribution using a symmetric Dirichlet prior, *Annals of Statistics*, **3**, 246–50.

Good, I. J. (1976) On the application of symmetric Dirichlet distributions and their mixtures to contingency tables, *Annals of Statistics*, **4**, 1159–89.

Goodman, L. A. (1964) Interactions in multidimensional contingency tables, *Ann. of Math. Stat.*, **35**, 632–46.

Goodman, L. A. (1968) The analysis of cross-classified data: independence, quasi-independence, and interactions in contingency tables with or without missing entries, *J. Amer. Statist. Assn.*, **63**, 1091–1131.

Gunel, E. and Dickey, J. M. (1974) Bayes factors for independence in contingency tables, *Biometrika*, **61**, 545–57.

Hickman, J.C. and Miller, R. B. (1977) Notes on Bayesian graduation (with discussion), *Trans. Soc. Actuaries*, **29**, 7–49.

Hsu, J. S. J. (1990) Bayesian inference and marginalization, PhD thesis, University of Wisconsin-Madison.

Hsu, J. S. J., Leonard, T. and Tsui, K. W. (1991) Statistical inference for multiple choice tests, *Psychometrika*, **56**, 327–48.

Johnson, W. E. and Braithwaite, R. B. (1932) Appendix to 'Probability, deductive, and inductive Problems', *Mind*, **41**, 421–3.

Kass, R. E. and Steffey, D. (1989) Approximate Bayesian inference in conditionally independent hierarchical models (parametric empirical Bayes models), *J. Amer. Statist. Assn.*, **84**, 717–26.

Laird, N. M. (1978) Empirical Bayes methods for two-way contingency tables, *Biometrika*, **65**, 581–90.

Leonard, T. (1972) Bayesian methods for binomial data, *Biometrika*, **59**, 581–9.

Leonard, T. (1973a) A Bayesian method for histograms, *Biometrika*, **60**, 197–308.

Leonard, T. (1973b) Bayesian methods for the simultaneous estimation of several parameters, PhD thesis, University of London.

Leonard, T. (1975) Bayesian estimation methods for two-way contingency tables, *J. Roy. Statist. Soc.*, **B37**, 23–37.

Leonard, T. (1976) Some alternative approaches to multi-parameter estimation, *Biometrika*, **63**, 69–75.

Leonard, T. (1977a) An alternative Bayesian approach to the Bradley–Terry model for paired comparisons, *Biometrics*, 121–30.
Leonard, T. (1977b) Bayesian simultaneous estimation for several multinomial distributions, *Comm. Statist.*, **A6**, 610–30.
Leonard, T. (1977c) A Bayesian approach to some multinomial estimation and pre-testing problems, *J. Amer. Statist. Assn.*, **72**, 865–8.
Leonard, T. (1978) Density estimation, stochastic processes, and prior information (with discussion), *J. Roy. Statist. Soc.*, **B40**, 113–46.
Leonard, T. (1982) Comment on the paper by Lejeune and Faulkenberry, *J. Amer. Statist. Assn.*, **77**, 657–8.
Leonard, T. (1985) Comment on the paper by Diaconis and Efron, *Annals of Statistics*, **13**, 893–8.
Leonard, T. (1991) Commentary on 'Paternity probability; an unnecessary artifact', *J. Undergraduate Mathematics and Its Application*, **12**, 69–72.
Leonard, T., Hsu, J. S. J. and Tsui, K. W. (1989) Bayesian marginal inference, *J. Amer. Statist. Assn.*, **84**, 1051–8.
Leonard, T. and Novick, M. R. (1986) Bayesian full rank marginalization for two-way contingency tables, *J. of Educ. Stat.*, **11**, 33–56.
Lindley, D. V. (1964) The Bayesian analysis of contingency tables, *Am. Math. Statist.*, **35**, 1622–43.
Lindley, D. V. (1965) *Introduction to Probability and Statistics from a Bayesian Viewpoint, Part II: Inference*, CUP.
Lindley, D. V. (1971) The estimation of many parameters, in *Foundations of Statistical Inference*, V. Godamke and D. A. Sprott, eds, Holt, Rinehart and Winston, pp. 435–55.
Lindley, D. V. and Novick, M. R. (1981) The role of exchangeability in inference, *Ann. of Statistics*, **9**, 45–58.
Lindley, D. V., and Smith, A. F. M. (1972) Bayes estimates for the linear model (with discussion), *J. Roy. Statist. Soc.*, **B34**, 1–41.
Lochner, R. H. (1975) A generalized Dirichlet distribution in life testing, *JRSSB*, **37**, 103–13.
Nazaret, W. A. (1987) Bayesian log linear estimates for three-way contingency tables, *Biometrika*, **74**, 401–10.
Paul, S. R. and Plackett, R. L. (1978) Inference sensitivity for Poisson mixtures, *Biometrika*, **65**, 591–602.
Radelet, M. (1981) Racial characteristics and imposition of the death penalty, *American Sociological Review*, **46**, 918–27.
Ritter, C. (1992). Modern inference in non-linear least squares regression, PhD thesis, University of Wisconsin-Madison.
Tavaré, S. and Altham, P. M. E. (1983) Serial dependence of observations leading to contingency tables, and corrections to chi-squared statistics, *Biometrika*, **70**, 139–44.
Tierney, L. and Kadane, J. B. (1986) Accurate approximations for posterior moments and marginal densities, *J. Amer. Statist. Assn.*, **81**, 82–6.

Zellner, A. and Rossi, P. E. (1984) Bayesian analysis of dichotomous quantal response models, *J. of Econometrics*, **25**, 365–93.

Department of Statistics
University of Wisconsin-Madison
1210 West Dayton Street
Madison
WI 53706-1693
USA

Department of Statistics and Applied Probability
University of California at Santa Barbara
Santa Barbara
CA 93106-3110
USA

CHAPTER 19

Conflicting Information and a Class of Bivariate Heavy-tailed Distributions

Anthony O'Hagan and Huiling Le
University of Nottingham

19.1 HEAVY-TAILED BAYESIAN MODELLING

If x is normally distributed with mean μ and variance 1, and if the prior distribution of μ is $N(0, 1)$, then the posterior distribution is $N(x/2, 1/2)$. Elementary introductions to Bayesian statistics often present this example as typical of how the two sources of information, prior and data, are synthesized by Bayes's theorem. In particular, the posterior mean is a compromise between the prior mean and the data estimate. The compromise seems natural enough, but is questionable when the two sources conflict. If $x = 10$, for instance, the prior which asserts that μ is very unlikely to lie outside $[-3, 3]$ conflicts with the observation which is very unlikely for μ outside $[7, 13]$. The posterior distribution $N(5, 1/2)$ then claims that μ is almost certain to lie in $[3, 7]$, a region which is not supported by either prior or likelihood.

Normal-theory models, and all exponential family models with conjugate priors, have this property of compromising between the various sources of information, even when they conflict. Replacing normal distributions by distributions with heavier tails, such as t-distributions, will cause the posterior distribution to respond very differently to conflict. In the above example, if we replace both the prior distribution and the likelihood by

Aspects of Uncertainty edited by P. R. Freeman and A. F. M. Smith. © 1994 John Wiley & Sons Ltd.

t-distributions then three different kinds of resolution to the conflict are possible as $x \to \infty$. If the likelihood has more degrees of freedom than the prior, then the prior information will be rejected in favour of the observation, the posterior distribution tends to the likelihood (normalized to integrate to one), with a median of x. If the prior has more degrees of freedom then the observation is rejected and the posterior distribution tends to the prior distribution. Between these two cases, if both sources of information have exactly the same degrees of freedom, the posterior distribution has two modes, one around $\mu = 0$ and the other around $\mu = x$, which become further and further apart as $x \to \infty$. In this case $E(\mu \mid x) = x/2$, which agrees with the normal theory, but $\text{Var}(\mu \mid x) \to \infty$ shows that the failure to resolve the conflict in favour of one or other source of information causes increasing uncertainty over the value of μ.

This simplest form of conflict, between a prior distribution and a single observation on a scalar parameter, was studied by David (1973) and Hill (1974). O'Hagan (1979) generalized to many observations, so dealing with conflicts between one or more outlying observations and the prior, or between the data themselves. O'Hagan (1988) considered more general Bayesian modelling based on t-distributions and introduced the term 'credibility', which was renamed *credence* in O'Hagan (1990). The credence of a t-distribution is one plus its degrees of freedom, but the term is applied more generally so that a density whose tails decay like $|x|^{-c}$ is said to have credence c. O'Hagan (1990) shows generally that credences of non-conflicting information sources add, and that conflict between groups of information sources is resolved in favour of the group with greatest total credence.

The use of heavy-tailed distributions is therefore a valuable tool in developing robust Bayesian procedures, limiting the influence that extreme information sources can have on posterior inferences. Routine use of heavy-tailed distributions like the t or Cauchy (which is a t with 1 degree of freedom, or credence 2) can achieve a kind of automatic robustness; see for instance Angers and Berger (1991), Carlin and Polson (1991), Fan and Berger (1992), Geweke (1992), Meinhold and Singpurwalla (1989), Zen (1991). However, we have seen that there are always several possible ways to resolve any conflict, and in any application careful consideration of natural kinds of resolution can be a powerful aid to modelling. O'Hagan (1988) shows how very different posterior behaviour can be achieved from two different heavy-tailed distributions for the same problem, and that more interesting results are obtained by thinking beyond simple substitution of t-distributions for normals.

In the present chapter we consider multivariate heavy-tailed distributions. Section 19.2 demonstrates the diversity of tail forms that are possible in multivariate distributions, and defines an equivalence relation between densities whose tail shapes are the same in all directions. A particular class of bivariate heavy-tailed distributions is introduced in section 19.3, and it is shown that such distributions can arise through realistic Bayesian modelling. Through examples in sections 19.4 and 19.5, these distributions are shown to offer interesting and complex resolution of conflicts. Section 19.6 gives some concluding remarks.

19.2 MULTIVARIATE HEAVY-TAILED DISTRIBUTIONS

Multivariate distributions admit a much greater diversity of tail behaviour than univariate distributions, even if we continue to restrict attention to tails which decay polynomially. The multivariate t family is well known, with density generally of the form

$$f(\mathbf{x}) \propto \{1 + (\mathbf{x} - \boldsymbol{\mu})' \Sigma^{-1} (\mathbf{x} - \boldsymbol{\mu})\}^{-c/2}. \qquad (2.1)$$

As a function of any x_i, with other elements of $\mathbf{x}$ fixed, $f(\mathbf{x})$ is asymptotically proportional to $|x_i|^{-c}$. We contrast (2.1) with a product of independent t-distributions

$$f(\mathbf{x}) \propto \prod_{i=1}^{k} \{1 + (x_i - \mu_i)^2/\sigma^2\}^{-c_i/2}. \qquad (2.2)$$

If we set $c_i = c$ for all i, $f(\mathbf{x})$ is again asymptotically proportional to $|x|^{-c}$ as a function of each x_i, yet even then (2.2) is very different from (2.1) if we look at how its tails decay in a direction that is not parallel to one of the axes. For instance, let $x_i = y + z_i$ with $\sum z_i = 0$ and look at $f(\mathbf{x})$ as a function of y for fixed z_i's. Then (2.2) is asymptotically proportional to $|y|^{-kc}$, while for (2.1) we obtain instead $|y|^{-c}$. The multivariate t-distribution's tails decay polynomially according to the same power c in every direction, while those of a product of independent t-distributions decay at different polynomial rates in different directions.

We can now see the multivariate t-distribution as a very special case. In general, a multivariate heavy-tailed distribution might have tails decaying at different rates in every direction, and even in two dimensions there are an infinite number of directions to consider. We can expect distributions with different 'patterns' of tail thickness to have qualitatively different behaviour in conflict situations, and this is confirmed by O'Hagan (1988) by considering the following two distributions for a parameter vector $\boldsymbol{\theta} = (\theta_1, \theta_2, \ldots, \theta_k)$ in a one-way analysis of variance model. The first distribution is specified hierarchically by letting

$$f(\boldsymbol{\theta} \mid \xi) \propto \prod_{i=1}^{k} \{c + s(\theta_i - \xi)^2\}^{-c/2}, \qquad (2.3)$$

a product of independent t-distributions, and then giving ξ a uniform prior distribution at the second stage of the hierarchy. The second distribution is

$$f(\boldsymbol{\theta}) \propto \prod_{i<j} \{c + s(\theta_i - \theta_j)^2\}^{-c/(2k)}, \tag{2.4}$$

a product of $k(k-1)/2$ terms involving all pairs (θ_i, θ_j). O'Hagan (1988) shows by a numerical example that (2.4) is capable of producing 'multiple shrinkage' in cases of conflict in the data, which does not happen with the simpler prior distribution (2.3). The argument there is heuristic, and such differences between the two distributions have not been fully characterized theoretically. The present chapter is part of an effort to build a theory capable of dealing with heavy-tailed distributions as complex as (2.3) or (2.4).

In one dimension, a t-distribution with credence c has heavier tails than one with credence c' if and only if $c < c'$. We introduce the following general definition.

Definition f dominates g, written $f \geq g$, if there exists a constant $k > 0$ such that $f(\mathbf{x}) \geq kg(\mathbf{x})$ for all $\mathbf{x}$; f is equivalent to g, written $f \approx g$, if both $f \geq g$ and $g \geq f$.

If f dominates g the tails of f are at least as heavy in every direction as those of g. If f is equivalent to g their tails decay at the same rates in all directions. Dominance and equivalence are transitive, so that all those distributions equivalent to a given f form its equivalence class. If f dominates g then every distribution in the equivalence class of f dominates every distribution in the equivalence class of g. For instance, univariate t-distributions with different credences form different equivalence classes. O'Hagan (1990) presents some theory for those classes of distributions, with particular reference to conflicts. As an illustration of the power of looking at distributions in terms of their equivalence classes, it is easily shown that the sum of two independent t random variables does not have a t-distribution, but O'Hagan (1990) proves that if X and Y are independent with densities equivalent to t-distributions with credences c and c', then the density of $X + Y$ is equivalent to a t-distribution with credence $\min(c, c')$.

Our dominance relation does not provide a complete ordering of distributions. Even in one dimension it is possible to find two distributions such that neither dominates the other. Such cases are easy to construct using distributions whose right and left tails decay at different rates, but we can also find symmetric distributions that are not ordered by dominance. In two or more dimensions these problems arise even with the obvious classes of distributions like (2.1) and (2.2). We easily see that (2.1) dominates (2.2) if $c_i = c$ for all i, but consider the bivariate distributions

$$f(x, y) \propto (1 + x^2 + y^2)^{-3},$$
$$g(x, y) \propto (1 + x^2)^{-2}(1 + y^2)^{-2},$$
$$h(x, y) \propto (1 + x^2)^{-3}(1 + y^2)^{-1}.$$

None of these three distributions dominates either of the other two. The sheer diversity of tail forms in multivariate distributions makes it unrealistic to look for simple, general theory.

19.3 A CLASS OF BIVARIATE DISTRIBUTIONS

Consider a bivariate density function f of the form

$$f(x, y) \approx (1 + x^2 + y^2)^{-c/2}(1 + x^2)^{-c_1/2}(1 + y^2)^{-c_2/2} \quad (3.1)$$

for some c, c_1, c_2. If f satisfies (3.1) we will write $f \approx T(c, c_1, c_2)$. This class of distributions is studied by Le and O'Hagan (1993), which we refer to hereafter as LOH. The $T(c, c_1, c_2)$ distributions generalize the multivariate t-distributions (2.1) and the products of independent t-distributions (2.2) in a simple way, but are already sufficiently complex to yield some unexpected results (and some awkward mathematics). For instance, consider the marginal distribution

$$f_1(x) = \int_{-\infty}^{\infty} f(x, y) dy,$$

which exists if $c + c_2 > 1$ (otherwise the integral diverges). LOH prove that:

(a) if $c_2 > 1$, $f_1(x) \approx (1 + x^2)^{-(c+c_1)/2}$;
(b) if $c_2 = 1$, $f_1(x) \approx (1 + x^2)^{-(c+c_1)/2}\log(2 + x^2)$;
(c) if $c_2 < 1$, $f_1(x) \approx (1 + x^2)^{-(c+c_1+c_2-1)/2}$.

We thereby find that (3.1) is a proper distribution if $c + c_1 > 1$, $c + c_2 > 1$ and $c + c_1 + c_2 > 2$. Case (b) is interesting, because in this case $f_1(x)$ is not equivalent to a t-distribution. Its tails are heavier than those of the t-distributions with credence $c + c_1$, but lighter than those of a t-distribution with credence $c + c_1 - \varepsilon$ for any $\varepsilon > 0$. The distinction between

$$f_1(x) \approx (1 + x^2)^{-(c+c_1)/2} \quad \text{and} \quad f_1(x) \approx (1 + x^2)^{-(c+c_1)/2}\log(2 + x^2)$$

is obviously fine, but is potentially significant in practice. We are not aware of distributions like this, which lie between 'adjacent' t-distributions, having previously been considered in Bayesian statistics. There is, however, a connection with the theory of regular variation; see Bingham et al. (1987). The function $\log(2 + x^2)$ is slowly varying, and hence $f_1(x)$ is regularly varying with index $c + c_1$ both when $c_2 > 1$ and when $c_2 = 1$.

Our interest in $T(c, c_1, c_2)$-distributions, as in O'Hagan (1990), is to see how conflicts between two such distributions are resolved. Consider a location parameter (θ, ϕ) with prior density $f \approx T(c, c_1, c_2)$ and an observation (x, y) distributed with density

$$p(x, y \mid \theta, \phi) = g(x - \theta, y - \phi)$$

and suppose that $g \approx T(c', c_1', c_2')$. The prior $f(\theta, \phi)$ asserts that (θ, ϕ) is not expected to lie far from its prior mean, and the likelihood $g(x - \theta, y - \phi)$ asserts that (θ, ϕ) is unlikely to be far from (x, y). If x or y, or both, are large these two sources of information conflict. We examine the limiting form of the posterior distribution as x or y or both tend to infinity, but first we consider how such problems might arise in practice.

One way in which (3.1) might arise is in a normal model with unknown but correlated variances. Suppose that x and y are independent normal random variables with zero means and unknown precisions w_1 and w_2. That is, $x \sim N(0, w_1^{-1})$ and $y \sim N(0, w_2^{-1})$. If we assume that $w_1 = w_2 = w$ and give w the χ^2_{c+2} distribution we obtain the marginal distribution

$$f(x, y) \propto (1 + x^2 + y^2)^{-c/2}$$

so that $f \approx T(c, 0, 0)$. If we do not assume equal precision, but instead give the w_is independent $\chi^2_{c_i+1}$-distributions we find $f(x, y) \approx T(0, c_1, c_2)$. These are the bivariate cases of (2.1) and (2.2). In practice, neither assuming equal precisions or treating them as independent may truly represent one's beliefs about how x and y vary. We may not believe $w_1 = w_2$, but may expect them to be similar. One way to express this is to write $w_i = \alpha + \beta_i$, and to give α, β_1 and β_2 independent distributions. Correlation between the w_i's is then induced by the common component α. Now

$$p(x, y \mid \alpha, \beta_1, \beta_2) \propto \{(\alpha + \beta_1)(\alpha + \beta_2)\}^{1/2} \exp[-\{x^2(\alpha + \beta_1) + y^2(\alpha + \beta_2)\}/2],$$

and let

$$p(\alpha, \beta_1, \beta_2) \propto \{(\alpha + \beta_1)(\alpha + \beta_2)\}^{-1/2} \alpha^{c/2-1} \beta_1^{c_1/2-1} \beta_2^{c_2/2-1}$$
$$\exp\{-(\alpha + \beta_1 + \beta_2)/2\}.$$

Multiplying these and integrating out α, β_1 and β_2 yields the marginal distribution

$$f(x, y) \propto (1 + x^2 + y^2)^{-c/2}(1 + x^2)^{-c_1/2}(1 + y^2)^{-c_2/2}.$$

Apart from the first term (which has only a small effect unless c, c_1 or c_2 are small), $p(\alpha, \beta_1, \beta_2)$ represents a product of independent χ^2 distributions. Thus the $T(c, c_1, c_2)$ distributions can arise in simple and natural statistical modelling.

An alternative is to consider (3.1) as arising from the synthesis of two other sources of information. Suppose that we begin with independent t prior distributions for (θ, ϕ),

$$f_0(\theta, \phi) \propto (1 + \theta^2)^{-c_1/2}(1 + \phi^2)^{-c_2/2}. \qquad (3.2)$$

Then, prior to (x, y) we obtain an observation (u, v) with

$$p(u, v \mid \theta, \phi) \propto \{1 + (u - \theta)^2 + (v - \phi)^2\}^{-c/2}$$

(which might arise from giving u and v independent normal distributions with equal but unknown variances). Then the posterior distribution

$$f(\theta, \phi) \propto \{1 + (u - \theta)^2 + (v - \phi)^2\}^{-c/2}(1 + \theta^2)^{-c_1/2}(1 + \phi^2)^{-c_2/2}$$

becomes the prior distribution before observing (x, y). It is easy to prove now that for any fixed (u, v), $f \approx T(c, c_1, c_2)$. Now when we consider conflict between (x, y) and the prior mean, we are actually looking at three sources of information, the mean $(0, 0)$ of the original prior (3.2), the first observation (u, v) and the new observation (x, y). We are treating (u, v) as fixed and not conflicting with the prior mean $(0, 0)$, but as either x or y become large we have a conflict between (x, y) and both the other sources of information. In particular, there is conflict between the two parts (u, v) and (x, y) of the data. (Conflicts within the data are usually considered as outlier problems.) As in O'Hagan (1990), the non-conflicting sources combine to yield a single information source $f \approx T(c, c_1, c_2)$.

19.4 A SIMPLE EXAMPLE

Suppose that the prior distribution for (θ, ϕ) is a bivariate t-distribution

$$f(\theta, \phi) \propto (1 + \theta^2 + \phi^2)^{-3}, \tag{4.1}$$

and that x and y are independent Cauchy random variables

$$g(x - \theta, y - \phi) \propto \{(1 + (x - \theta)^2\}^{-1}\{1 + (y - \phi)^2\}^{-1}. \tag{4.2}$$

Then $f \approx T(6, 0, 0)$ and $g \approx T(0, 2, 2)$. Consider the marginal distribution of (x, y),

$$h(x, y) = \int_{-\infty}^{\infty} \int_{-\infty}^{\infty} f(\theta, \phi) g(x - \theta, y - \phi) \, d\theta \, d\phi.$$

LOH show that $T(0, 2, 2) \geqslant T(6, 0, 0)$ and thereby that in this case $h \approx T(0, 2, 2)$, i.e.

$$h(x, y) \approx (1 + x^2)^{-1}(1 + y^2)^{-1}. \tag{4.3}$$

Now the posterior distribution is

$$p(\theta, \phi \mid x, y) = f(\theta, \phi)g(x - \theta, y - \phi)/h(x, y),$$

and (4.3) allows us to examine this as a function of x and y as well as θ and ϕ.

First let $x \to \infty$. Now the prior (4.1) suggests that θ should be near zero (or at least finite), while the likelihood (4.2) suggests that θ is near x. We can now see which source of information is rejected as $x \to \infty$. First consider finite θ. For any fixed θ,

$$\{1 + (x - \theta)^2\}^{-1}/(1 + x^2)^{-1} \to 1$$

Table 19.1 Posterior moments for the first example, $x \to \infty$.

	$E(\theta \mid x, y)$	$E(\phi \mid x, y)$	$\text{Var}(\theta \mid x, y)$	$\text{Var}(\phi \mid x, y)$	$\text{Cov}(\theta, \phi \mid x, y)$
$x = 0, y = 1$	0.000	0.237	0.246	0.304	0.000
$x = 1, y = 1$	0.250	0.250	0.325	0.325	0.014
$x = 3, y = 1$	0.339	0.269	0.648	0.367	0.045
$x = 4, y = 1$	0.286	0.266	0.706	0.364	0.046
$x = 6, y = 1$	0.193	0.260	0.685	0.352	0.036
$x = 10, y = 1$	0.107	0.255	0.589	0.340	0.019
$x = 20, y = 1$	0.049	0.253	0.504	0.333	0.008
$x = \infty, y = 1$	0.000	0.252	0.464	0.332	0.000

as $x \to \infty$. Therefore for finite θ, ϕ and y we have

$$p(\theta, \phi \mid x, y) \propto (1 + \theta^2 + \phi^2)^{-3}\{1 + (y - \phi)^2\}^{-1} \qquad (4.4)$$

as the limiting form when $x \to \infty$. Now instead of finite θ consider θ near x. If $x \to \infty$ and $\theta \to \infty$ such that $x - \theta$ is fixed,

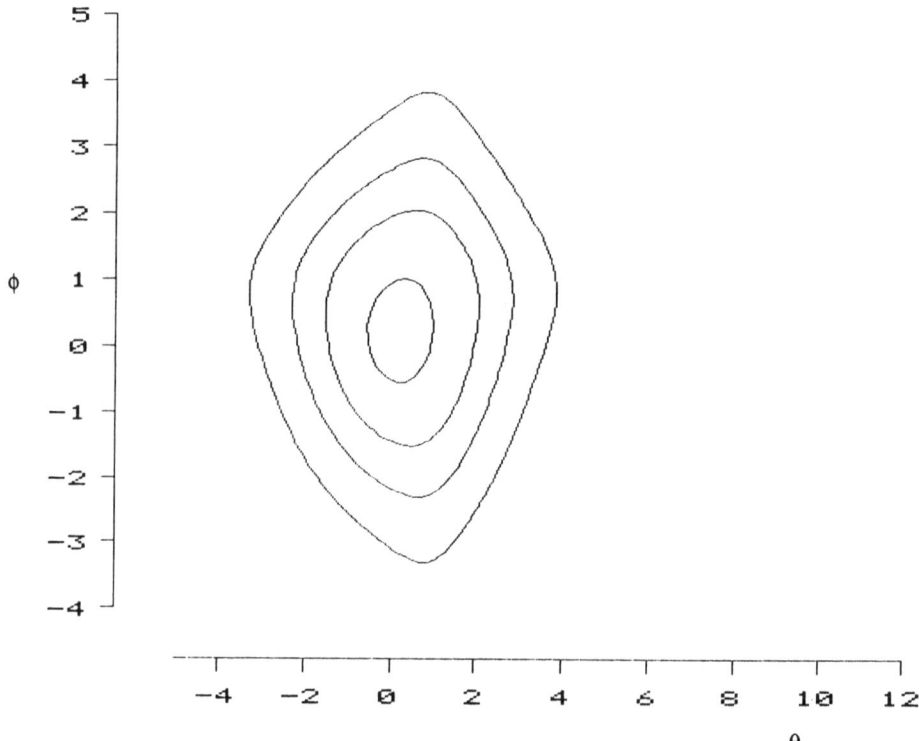

Figure 19.1 Density for first example: $(x, y) = (1, 1)$.

$$(1 + \theta^2 + \phi^2)^{-3}/(1 + x^2)^{-1} \to 0,$$

and this allows us to show that the posterior probability that θ is in any neighbourhood of x tends to zero as $x \to \infty$. Therefore the limiting distribution is indeed (4.4), and the observation x is rejected if it conflicts with the prior information. Table 19.1 shows values of the posterior means, variances and covariances of θ and ϕ for all $y = 1$ and various values of x. For $x = \infty$, the moments are those of (4.4). Figures 19.1–19.4 show contour maps of some of these posterior distributions.

By symmetry, as $y \to \infty$ the posterior is asymptotically proportional to

$$(1 + \theta^2 + \phi^2)^{-3}\{1 + (x - \theta)^2\}^{-1}.$$

If both x and y go to infinity we can prove in the same way that in the limit the whole of the likelihood (4.2) is rejected and the posterior distribution tends to the prior distribution. Posterior means of both θ and ϕ tend to zero, but in this case we find that the limiting posterior variances and covariances are not those of the limiting posterior distribution. To see

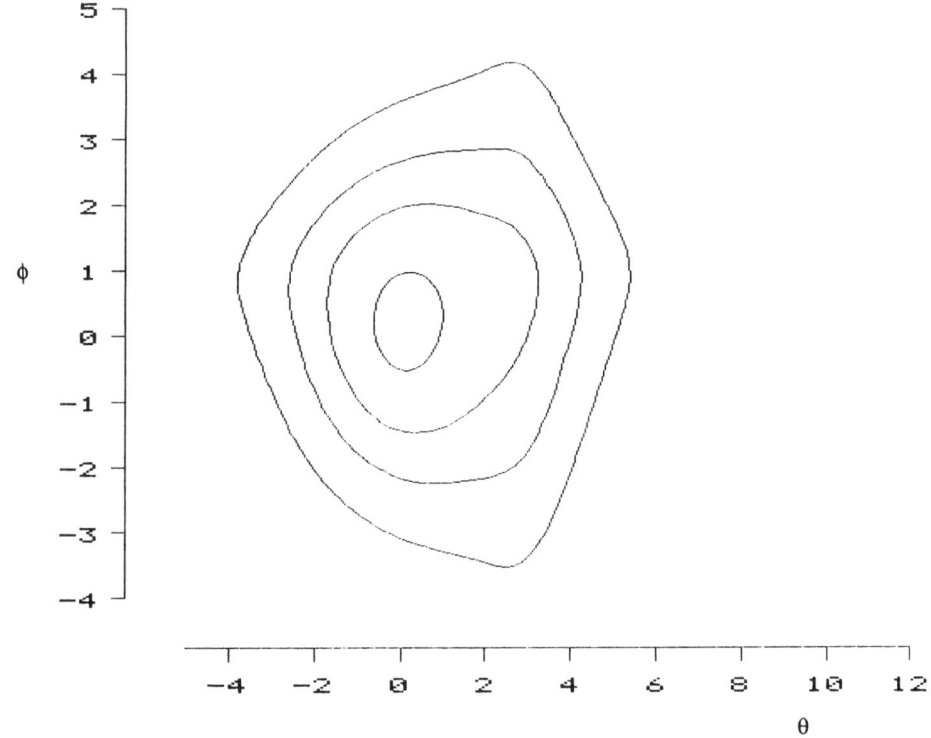

Figure 19.2 Density for first example: $(x, y) = (3, 1)$.

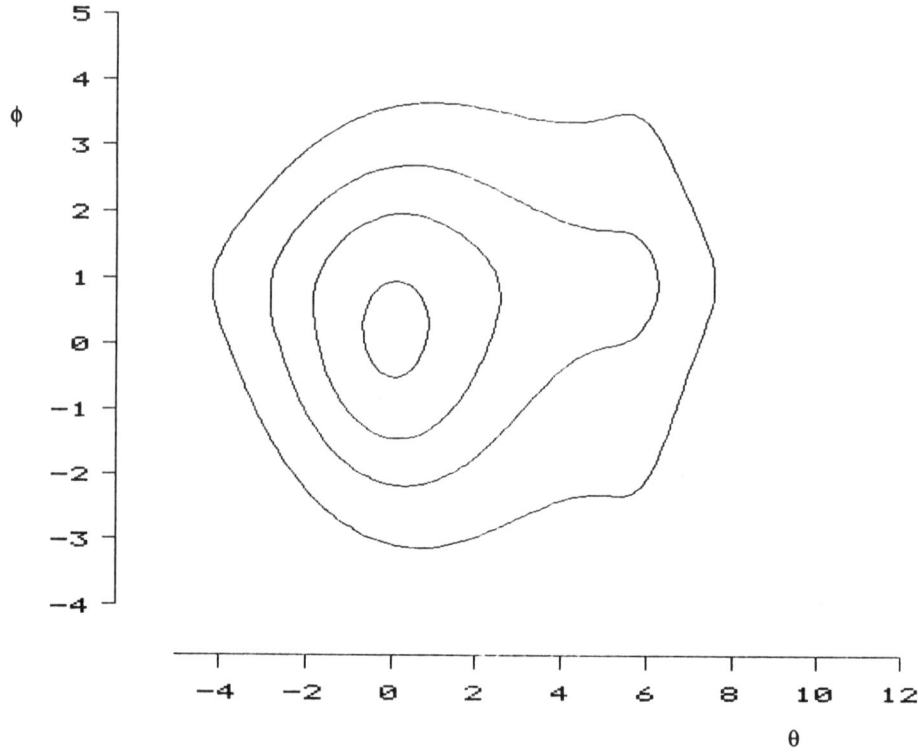

Figure 19.3 Density for first example: $(x, y) = (6, 1)$.

why this happens, consider the posterior density at $\theta = \phi = z$ when $x = y = z$. This is proportional to

$$(1 + 2z^2)^{-3}/(1 + z^2)^{-2}$$

Table 19.2 Posterior moments for the first example, $x = y \to \infty$.

	$E(\theta \mid x, y)$	$E(\phi \mid x, y)$	$\mathrm{Var}(\theta \mid x, y)$	$\mathrm{Var}(\phi \mid x, y)$	$\mathrm{Cov}(\theta, \phi \mid x, y)$
$x = y = 0$	0.000	0.000	0.230	0.230	0.000
$x = y = 1$	0.250	0.250	0.325	0.325	0.014
$x = y = 3$	0.431	0.431	0.855	0.855	0.212
$x = y = 4$	0.406	0.406	1.080	1.080	0.348
$x = y = 6$	0.318	0.318	1.311	1.311	0.537
$x = y = 10$	0.198	0.198	1.397	1.397	0.680
$x = y = 20$	0.095	0.095	1.352	1.352	0.747
$x = y = 40$	0.046	0.046	1.301	1.301	0.760
$x = y = 100$	0.018	0.018	1.266	1.266	0.756

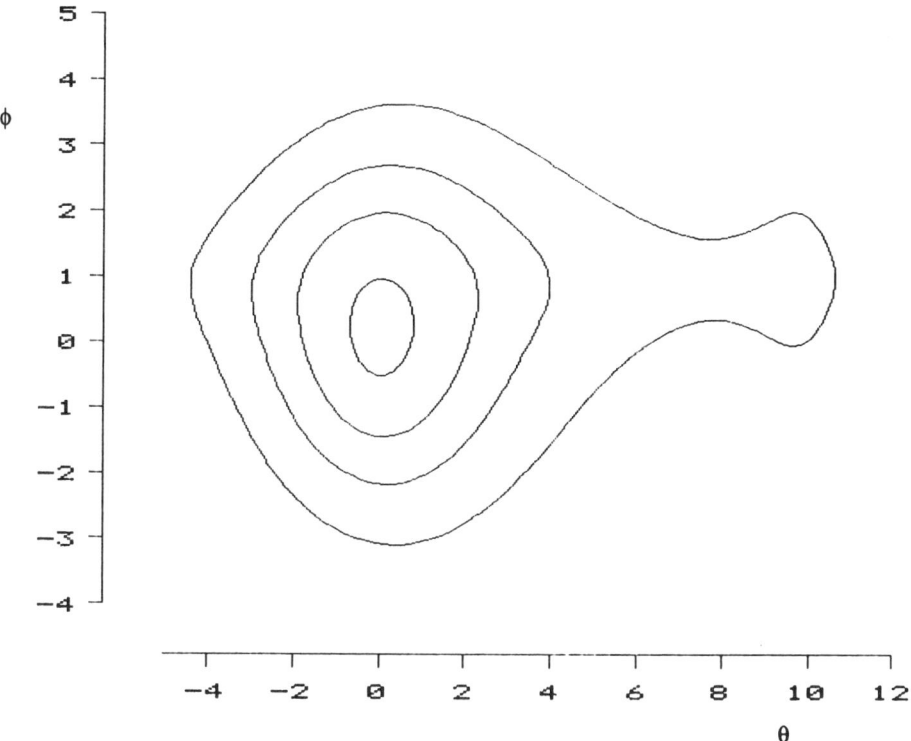

Figure 19.4 Density for first example: $(x, y) = (10, 1)$.

which is of order z^{-2}, and so tends to zero as $z \to \infty$. But although this means that the posterior probability that (θ, ϕ) is near (x, y) tends to zero (and the posterior distribution thereby tends to the prior distribution), it does so at a rate such that this region makes a non-zero contribution to the expectations of θ^2, ϕ and $\theta\phi$. Table 19.2 shows posterior moments for various $x = y$. In the limiting distribution we have variances of 0.495 and a zero covariance, but the limiting variances and covariances appear to be approximately 0.75 larger. Third-order moments of the limiting distribution exist, but posterior third-order moments go to infinity as both x and y go to infinity.

19.5 A MORE COMPLEX EXAMPLE

The resolution of conflict by rejecting the observations x and y in the first example is straightforward, and similar to the one-dimensional theory of O'Hagan (1990). We now consider an example showing more complicated behaviour. Let

$$f(\theta, \phi) \propto (1 + \theta^2 + \phi^2)^{-2}(1 + \theta^2)^{-4}, \qquad (5.1)$$
$$g(x - \theta, y - \phi) \propto \{1 + (x - \theta)^2 + (y - \phi)^2\}^{-1}\{1 + (x - \theta)^2\}^{-2}$$
$$\{1 + (y - \phi)^2\}^{-2}, \qquad (5.2)$$

so that $f \approx T(4, 8, 0)$ and $g \approx T(2, 4, 4)$. In this case, LOH prove that $h \approx T(0, 6, 4)$, or

$$h(x, y) \approx (1 + x^2)^{-3}(1 + y^2)^{-2}.$$

If we let x tend to infinity,

$$\{1 + (x - \theta)^2 + (y - \phi)^2\}^{-1}\{1 + (x - \theta)^2\}^{-2}/(1 + x^2)^{-3}$$

tends to one, so that $p(\theta, \phi | x, y)$ will be asymptotically proportional to

$$(1 + \theta^2 + \phi^2)^{-2}(1 + \theta^2)^{-4}\{1 + (y - \phi)^2\}^{-1}$$

for finite θ, ϕ and y. Conversely, if we let $x - \theta$ be fixed as $x \to \infty$ then

$$(1 + \theta^2 + \phi^2)^{-2}(1 + \theta^2)^{-4}/(1 + x^2)^{-3}$$

tends to zero, so the posterior probability that θ is near x also tends to zero. As $x \to \infty$ the two components of the likelihood (5.2) which involve θ are rejected.

Now consider y tending to infinity.

$$\{1 + (x - \theta)^2 + (y - \phi)^2\}^{-1}\{1 + (y - \phi)^2\}^{-2}/(1 + y^2)^{-2}$$

tends to zero, so that the posterior probability for finite ϕ tends to zero. However, if we let ϕ also tend to infinity, with $y - \phi$ finite, we note that $(1 + \theta^2 + \phi^2)^{-2}/(1 + y^2)^{-2}$ tends to one. We see that the asymptotic posterior distribution is now

$$p(\theta, \phi | x, y) \propto (1 + \theta^2)^{-4}\{1 + (x - \theta)^2 + (y - \phi)^2\}^{-1}\{1 + (x - \theta)^2\}^{-2}$$
$$\times \{1 + (y - \phi)^2\}^{-2}.$$

This conflict is therefore resolved in favour of the likelihood, with the term in the prior (5.1) which involves ϕ being rejected.

The situation changes again as we let both x and y tend to infinity. Now $g(x - \theta, y - \phi)/h(x, y)$ tends to one as x and y go to infinity, and so $p(\theta, \phi | x, y)$ is asymptotically proportional to $f(\theta, \phi)$ for finite θ and ϕ. However, suppose that θ is finite but ϕ tends to infinity with $y - \phi$ fixed. Then

$$(1 + \theta^2 + \phi^2)^{-2}\{1 + (x - \theta)^2 + (y - \phi)^2\}^{-1}\{1 + (x - \theta)^2\}^{-2}/h(x, y)$$

also tends to one. Therefore for (θ, ϕ) in a neighbourhood of $(0, \phi)$ asymptotically

$$p(\theta, \phi | x, y) \propto (1 + \theta^2)^{-4}\{1 + (y - \theta)^2\}^{-2}.$$

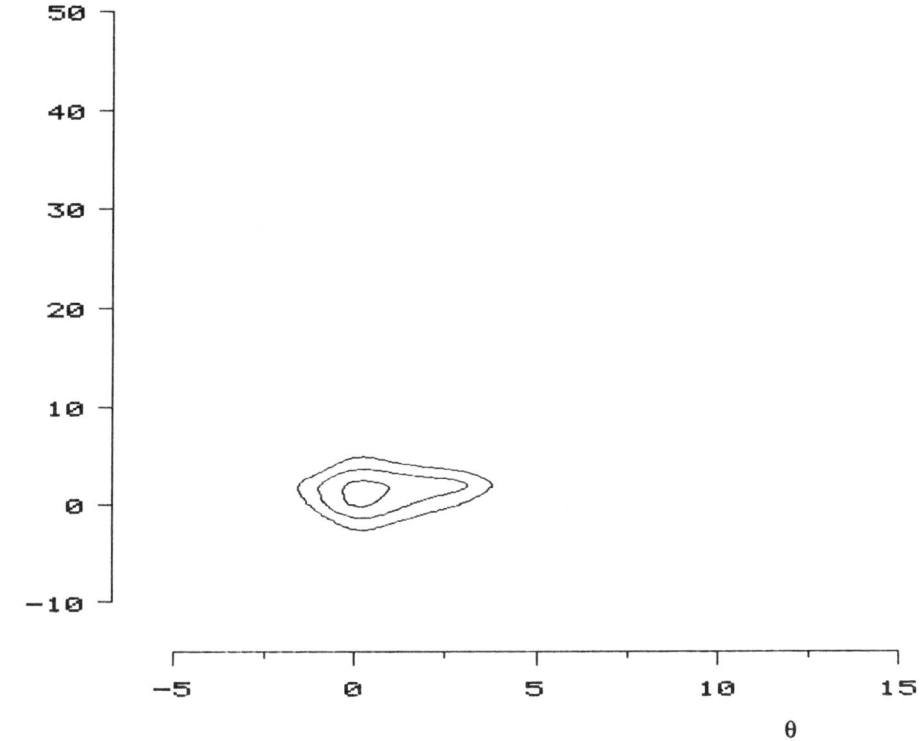

Figure 19.5 Density for second example: $(x, y) = (3, 2)$.

We find posterior probabilities tend to zero for θ in the neighbourhood of x for any φ (finite or infinite).

In this case there is no clear resolution of the conflict. The terms

$$\{1 + (x - \theta)^2 + (y - \phi)^2\}^{-1}\{1 + (x - \theta)^2\}^{-2}$$

involving θ in the likelihood are rejected, but the posterior distribution continues to give non-zero probability to finite φ and to φ near y as x and y tend to infinity. Figures 19.5 to 19.8 show a typical sequence of posterior contour maps, letting x and y tend to infinity with x/y fixed at 1.5. Figure 19.5 shows a beginning of conflict between a strong prior distribution in terms of θ, with prior mean $E(\theta) = 0$, and the likelihood, maximized at $\theta = x = 3$. In Figure 19.6 this is being resolved in favour of the prior distribution. The distribution is clearly trimodal, but the mode around $\theta = x$ is already very small. Figures 19.7 and 19.8 show two distinct posterior modes forming around the prior mode (0, 0) and around (0, y). The second mode is interesting because it is a region that is not strongly

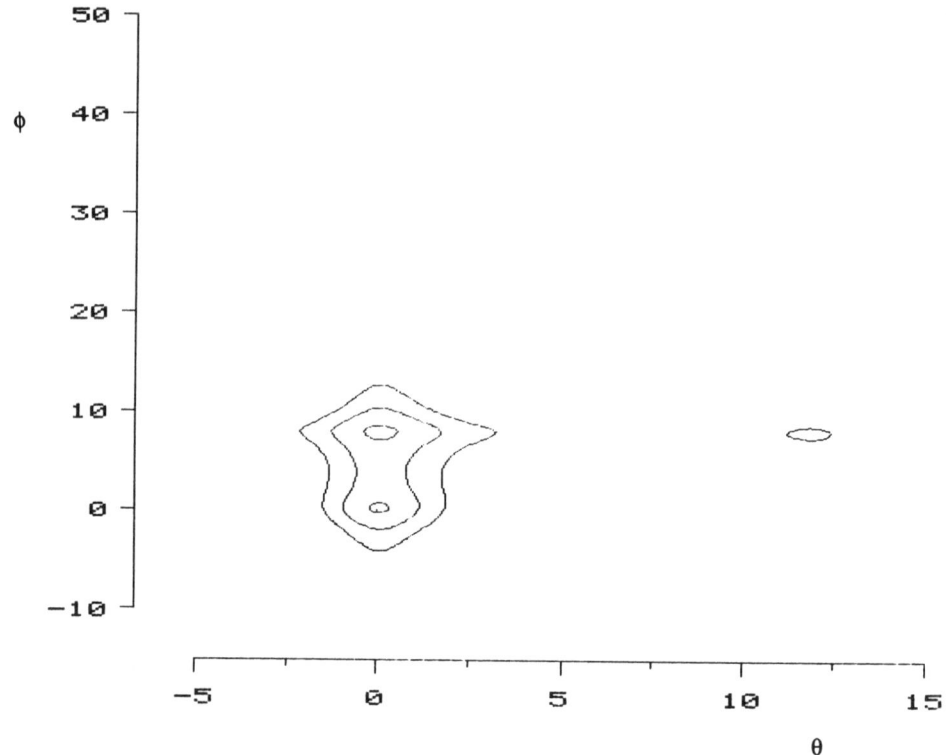

Figure 19.6 Density for second example: $(x, y) = (12, 8)$.

supported by either the prior or the likelihood. It is, however, supported by the likelihood if we delete the information about θ coming from x. A similar phenomenon occurs in the multiple shrinkage example of O'Hagan (1988), where the posterior distribution shows partial rejection of the prior information.

This example shows three different forms of behaviour as the observation tends to infinity. As $x \to \infty$ the θ part of the likelihood is rejected, and as $y \to \infty$ the ϕ part of the prior is rejected. As both x and y tend to infinity the θ part of the likelihood is again rejected, but now both the prior and likelihood information about ϕ are retained and the posterior has two diverging modes.

19.6 FINAL REMARKS

The class of distributions (3.1) far from exhausts the possibilities in two dimensions. We could, for instance, generalize further to

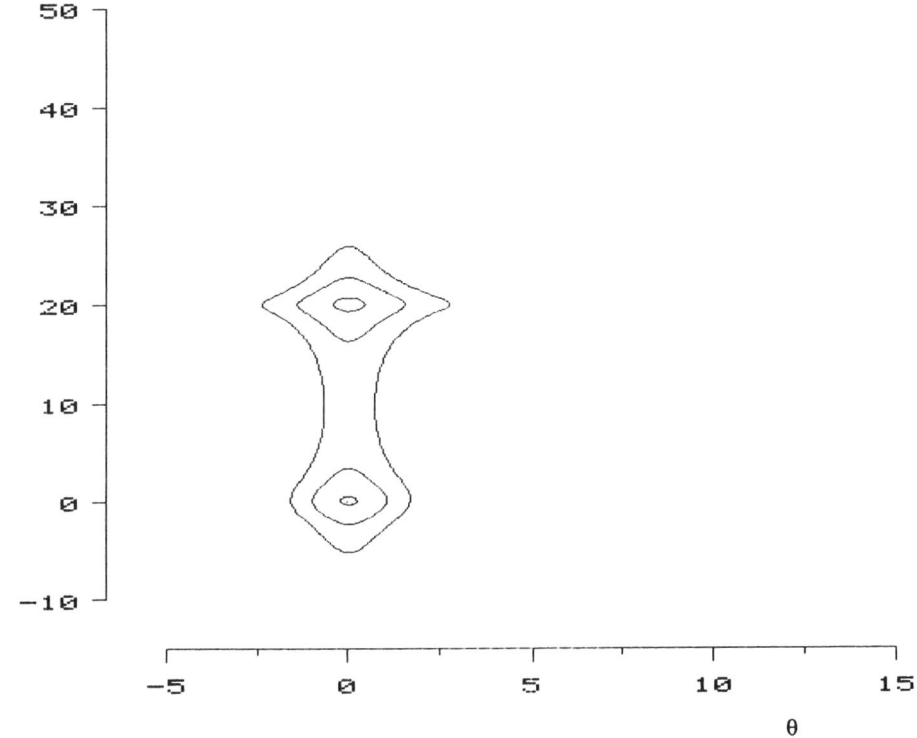

Figure 19.7 Density for second example: $(x, y) = (30, 20)$.

$$f(x, y) \propto (1 + x^2 + y^2)^{-c/2}(1 + x^2)^{-c_1/2}(1 + y^2)^{-c_2/2}\{1 + (x - y)^2\}^{-c_3/2},$$

which is not equivalent to $T(c', c'_1, c'_2)$ for any (c', c'_1, c'_2). Yet the posterior behaviour illustrated in the examples of sections 19.4 and 19.5 indicates the possible complexity available with multivariate distributions generally. There are two main reasons for studying these and more complex problems. First, we have shown that such distributions can arise in realistic Bayesian modelling. Second, by understanding these distributions we will be able to build models that have specific desired responses to conflicting information. We may thereby not only obtain inherently robust procedures, but also create quite novel Bayesian analyses.

ACKNOWLEDGEMENT

The authors are grateful to other members of the Bayesian robustness group at Nottingham University for several helpful comments.

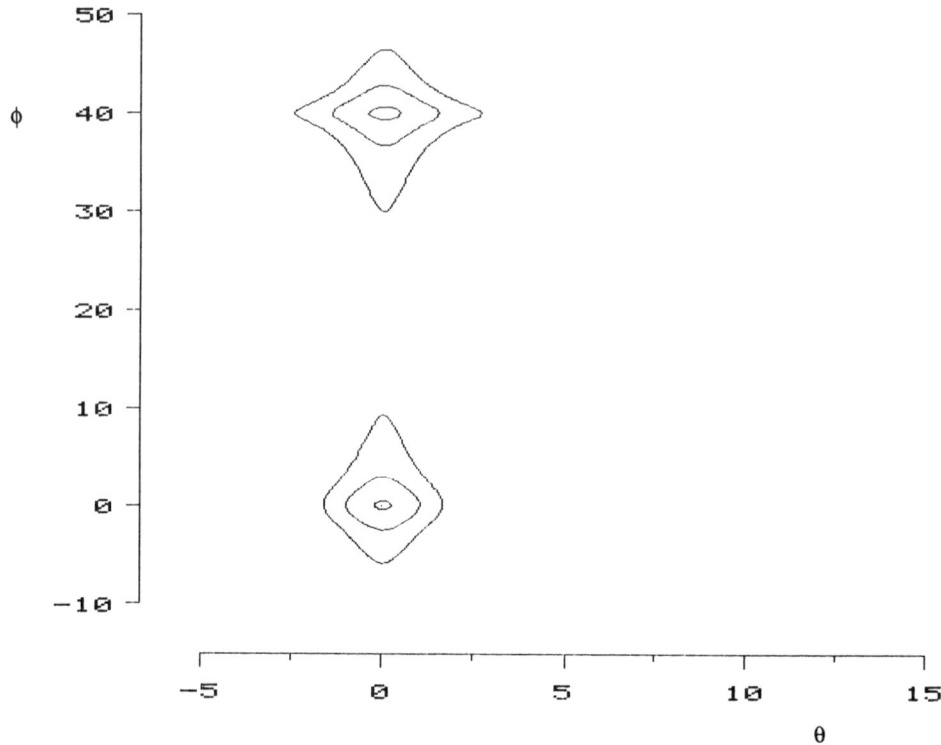

Figure 19.8 Density for second example: $(x, y) = (60, 40)$.

REFERENCES

Angers, J-F. and Berger, J. (1991) Robust hierarchical Bayes estimation of exchangeable means, *Canad. J. Statist.*, **19**, 39–56.

Bingham, N. L., Goldie, C. M. and Teugels, J. L. (1987) *Regular Variation*, Cambridge University Press.

Carlin, B. and Polson, N. G. (1991) Inference for non-conjugate Bayesian models using the Gibbs sampler, *Canad. J. Statist.*, **19**, 399–405.

Dawid, A. P. (1973). Posterior expectations for large observations, *Biometrika*, **60**, 664–7.

Fan, T-H. and Berger, J. O. (1992) Behaviour of the posterior distribution and inferences for a normal mean with t prior distributions, *Statistics and Decisions*, **10**, 99–120.

Geweke, J. (1992) Priors for macroeconomic time series and their applications, discussion paper 64, Institute for Empirical Macroeconomics, Federal Reserve Bank of Minneapolis.

Hill, B. M. (1974) On coherence, inadmissibility and inference about many parameters in the theory of least squares, in *Studies in Bayesian Econometrics and Statistics*, S. E. Feinberg and A. Zellner, eds, pp. 555–84, Amsterdam, North-Holland.

Le, H. and O'Hagan, A. (1993) A class of bivariate heavy-tailed distributions (in preparation).

Meinhold, R. J. and Singpurwalla, N. D. (1989) Robustification of Kalman filter models, *J. Amer. Statist. Assoc.*, **84**, 479–86.

O'Hagan, A. (1979) On outlier rejection phenomena in Bayes inference, *J. Roy. Statist. Soc.*, **B41**, 358–67.

O'Hagan, A. (1988) Modelling with heavy tails, in *Bayesian Statistics 3*, J. M. Bernardo et al. eds., pp. 345–359, Oxford University Press.

O'Hagan, A. (1990) On outliers and credence for location parameter inference, *J. Amer. Statist. Assoc.*, **85**, 172–6.

Zen, M-M. (1991) *Point Estimation of Multivariate Normal Mean Using t Priors*, Technical Report 91–43, Purdue University.

Department of Mathematics
University of Nottingham
University Park
NOTTINGHAM
NG7 2RD
UK

CHAPTER 20

Applications of Lindley Information Measure to the Design of Clinical Experiments

Giovanni Parmigiani and Donald A. Berry[†]
Institute of Statistics and Decision Sciences, Duke University

20.1 INTRODUCTION

From the point of view of decision-making, information is anything that enables us to make a better decision, that is, a decision with a higher expected utility. For example, an experiment that, irrespective of the outcome, will lead to the same decision that we would make prior to observing it, has no information content. Conversely, experiments able to lead to different decision are potentially of benefit. The expected change in utility can actually be used as a quantitative measure of the worth of an experiment in any given situation. This idea is about as old as Bayesian statistics (see Ramsey 1990) and is discussed by Raiffa and Schlaifer (1961) and DeGroot (1984).

The well-known measure of information proposed by Lindley (1956) is the object of investigation in this chapter. It can be seen as a very important special case of this general approach. Consider the decision problem of reporting a distribution regarding an unknown quantity Θ with values in Ω. This is a Bayesian way of modelling a situation in which data are not

---

[†] Research of D. A. Berry supported in part by the US Public Health Service under grant HS 06475-01. We are grateful to Chengchang Li, Peter Müller, Saurabh Mukhopadhyay and Dalene Stangl for helpful discussions.

Aspects of Uncertainty edited by P. R. Freeman and A. F. M. Smith. © 1994 John Wiley & Sons Ltd.

gathered to solve a specific decision problem, but rather to learn about the world or to provide several decision-makers with useful information in different decision problems.

Let the reported distribution be ϕ, and let the utility function be $U(\theta, \phi) = \log \phi(\theta)$. The Bayes decision is $\phi = \pi$, where π is the current distribution on θ (Good 1969; Bernardo 1979). Because of the desirable property of inducing the decision-maker to declare his or her own current distribution, this utility function is called proper. The value associated with the decision problem is

$$\mathscr{I}_0 \equiv E\{\log \pi(\Theta)\}. \quad (1)$$

This quantity is known as negative entropy of the distribution π, and has a fundamental role in information theory (see Cover and Thomas 1991).

Assume now that the outcome of the experiment $\mathscr{E}$, consisting of observing X from the distribution $p(x \mid \theta)$, becomes known. By the same argument as above, the value associated with the decision problem is

$$\mathscr{I}_1(x) \equiv E\{\log \pi(\Theta \mid X = x) \mid X = x\}.$$

The expected worth of the experiment $\mathscr{E}$ before performing it, is

$$\mathscr{I}(\mathscr{E}) \equiv E\{\log \pi(\Theta \mid X)\} - E\{\log \pi(\Theta)\} = E\left\{\log\left(\frac{\pi(\Theta \mid X)}{\pi(\Theta)}\right)\right\}, \quad (2)$$

where expectations are taken with respect to the joint distribution of Θ and X; $\mathscr{I}(\mathscr{E})$ was proposed as a measure of the information in $\mathscr{E}$ by Lindley (1956), and we shall refer to it as Lindley information measure. Lindley's original motivation differs from that given here and is based directly on more classic information theoretic arguments (see also Kullback and Leibler 1951; Kullback 1959).

The decision-theoretic and the information-theoretic interpretations are both forceful arguments in favour of $\mathscr{I}$. It is the method of choice for the measurement of the information contained in an experiment when the purpose of the investigation is not tied to a specific decision. See Verdinelli (1992) for a review of recent developments in information-theoretic Bayesian design.

The object of this chapter is to explore some of the uses of Lindley information in the design of clinical experiments. This field is ripe with applications. And pre-experimental information is frequently available for formulating prior distributions. Moreover, many biostatisticians are now using or at least willing to tolerate others using Bayesian methods.

There are two main practical difficulties in using information-theoretic measures in design. The first is computing. In realistic problems, closed-form expressions for $\mathscr{I}$ are typically hard to derive, and one has to resort to numerical methods. Recent progress in Bayesian numerical integration and optimization can be exploited to this purpose. Müller and Parmigiani (1994) discuss using Markov-chain Monte Carlo methods to evaluate information-theoretic measures such as entropy and Lindley information. Müller and Parmigiani (1992) introduce fast algorithms for optimization problems that arise in Bayesian design. We apply some of these methods in this chapter.

The second problem is combining information with other important aspects of clinical problems, such as the costs to the patients, the right of patients to receive the best available care and the obligation of clinicians to deliver it, the need for a short trial, etc. Information is expressed in units that arise very naturally in information theory (see Cover and Thomas 1991). These units are not very familiar to biostatisticians or clinicians. Conversion factors to combine information with, say, monetary costs, may be difficult to elicit. However, the problem of conversion is common to combining other measures of the outcome of a clinical trial, such as comparing a life saved with monetary costs of treatment, and solutions are available.

One approach to the combination problem is to use constrained optimizations. For example one can design a trial to maximize information for a fixed cost. A slightly more general version of this approach is to produce a family of optimal information–cost pairs for varying values of the Lagrange multiplier, as discussed, for example, in Parmigiani and Polson (1992) or Verdinelli and Kadane (1992). A simpler, and in some cases very effective approach consists of graphing the expected information of alternative experiments, and using the graph as a decision support tool.

An important aspect of the modern design of clinical trials is interim analysis. Interim analyses affect frequentist error rates (Berry 1988). So the number and timing of these analyses must be completely specified in the protocol in advance of the trial. Deviations from protocol are not allowed. Deviations that do occur compromise frequentist inferences from the results of the trial.

An enormous advantage of the Bayesian approach over the frequentist approach is that conclusions do not require conditioning on particular values of parameters. Posterior distributions depend on the accumulating data only through the likelihood function. So interim analyses need not be specified in advance of the trial (Berry 1987). This flexibility frees investigators to examine accumulating evidence at will and to modify the trial's design accordingly—this evidence may come from without as well as from within the trial.

The Bayesian approach is consistent with an attitude that is typical in scientific investigation: *experiment until you have enough information.* As we have indicated, Lindley's is an ideal measure of amount of information. It can be monitored as the data accrue and the trial can be stopped or otherwise modified accordingly.

The plan of the chapter is as follows. In section 20.2 we derive and interpret general first-order conditions for the optimality of a design using Lindley information. In section 20.3 we consider optimal fixed-sample-size problems and we derive expressions for Lindley information in the exponential and binomial models. In section 20.4 we consider the optimal design of duration and followup-times when the end-point is survival. Finally, in section 20.5 we discuss the trade-off between number of patients and number of centres in the design of multicentre clinical trials.

20.2 FIRST-ORDER CONDITIONS

Consider a family of experiments $\mathcal{E}_D$ indexed by the design variable D. The likelihood function for the experiment $\mathcal{E}_D$ is $p_D(x|\theta)$, $x \in \mathcal{X}$. Let $\pi_D(\theta|x)$ be the posterior distribution resulting from the observation of $X = x$ and let $m_D(x)$ the marginal distribution of X. The expected information is then given by

$$\mathcal{I}(D) = \int_\Theta \int_\mathcal{X} \log\left[\frac{\pi_D(\theta|x)}{\pi(\theta)}\right] p_D(x|\theta)\pi(\theta)\,dx\,d\theta. \tag{3}$$

In this section we study first-order conditions for the case of D one-dimensional and continuous or approximable by a continuous variable. We also assume that $p_D(\theta|x)$ is differentiable with respect to D. Define:

$$p'_D(x|\theta) \equiv \frac{\partial p_D(x|\theta)}{\partial D} \qquad \pi'_D(\theta|x) \equiv \frac{\partial \pi_D(\theta|x)}{\partial D}.$$

Then

$$\frac{d\mathcal{I}(D)}{dD} = \int_\Theta \int_\mathcal{X} \log\left[\frac{\pi_D(\theta|x)}{\pi(\theta)}\right] p'_D(x|\theta)\pi(\theta)\,dx\,d\theta. \tag{4}$$

To see that (4) holds, differentiate the integrand of (3) to obtain

$$\left[\log\left[\frac{\pi_D(\theta|x)}{\pi(\theta)}\right] p'_D(x|\theta) + p_D(x|\theta)\frac{\pi'_D(\theta|x)}{\pi_D(\theta|x)}\right]\pi(\theta).$$

From Bayes' rule

$$\pi'_D = \pi \frac{p'_D m_D - p_D m'_D}{m_D^2}.$$

Thus

$$\frac{p_D \pi'_D}{\pi_D} = \frac{m_D \pi'_D}{\pi} = \frac{p'_D m_D - p_D m'_D}{m_D} = p'_D - \pi p_D \frac{m'_D}{m_D}.$$

Now from the area condition,

$$\int_\Theta \int_\mathscr{X} p_D(x\,|\,\theta)\pi(\theta)\,\frac{m'_D(x)}{m_D(x)}\,\mathrm{d}x\,\mathrm{d}\theta = \int_\mathscr{X} m'_D(x)\mathrm{d}x = 0,$$

$$\int_\Theta \int_\mathscr{X} p'_D(x\,|\,\theta)\pi(\theta)\mathrm{d}x\,\mathrm{d}\theta = 0,$$

and (4) follows. □

With simple manipulations, (4) can be written:

$$\frac{\mathrm{d}\mathscr{I}(D)}{\mathrm{d}D} = \int_\Theta \int_\mathscr{X} \log[p_D(x\,|\,\theta)]p'_D(x\,|\,\theta)\mathrm{d}x\,\mathrm{d}\theta - \int_\mathscr{X} \log[m_D(x)]m'_D(x)\mathrm{d}x. \quad (5)$$

This expression parallels the alternative expression for (3), given by Lindley (1956):

$$\mathscr{I}(D) = \int_\Theta \int_\mathscr{X} \log[p_D(x\,|\,\theta)]p_D(x\,|\,\theta)\mathrm{d}x\,\mathrm{d}\theta - \int_\mathscr{X} \log[m_D(x)]m_D(x)\mathrm{d}x. \quad (6)$$

Formulas (4) and (5) can be used in calculating the optimal D whenever Lindley information enters as a part of the design criterion. For example, if D is sample size, and the cost of observation c is fixed, then the real-valued approximation to the optimal sample size satisfies $\mathrm{d}\mathscr{I}(D)/\mathrm{d}D = c$.

An interesting special case arises when $\mathscr{I}(D)$ has a maximum. Consider, for example, finding the most informative binary partition for an observation Y. This problem arises in applications when a continuous variable has to be turned into a categorical (in this case binary) variable. So let the random variable Y have conditional density $f(y\,|\,\theta)$. We are interested in learning about Θ. Assume that y cannot be observed exactly: we may only observe whether or not it is greater than some point D. What is the best D? Define X to be a Bernoulli random variable with success probability $F(D\,|\,\theta)$. Using (5), a necessary condition for D to be optimal is

$$\int_\Theta \log\left[\frac{F(D\,|\,\theta)}{1 - F(D\,|\,\theta)}\right]\pi(\theta\,|\,y = D)\mathrm{d}\theta = \log\frac{F(D)}{1 - F(D)}; \quad (7)$$

in words, we must choose the design so that the expected conditional logit is equal to the marginal logit.

As an example, take Y to be exponential with unknown failure rate Θ. Let the prior of Θ be a conjugate gamma (a, b). Then, using (7) the optimal D is the unique solution of

$$b^{a+1}\zeta(a + 2)\frac{\exp(1 + b/D)}{D^{a+1}} = \log\left[\left(\frac{b + D}{b}\right)^a - 1\right],$$

where ζ is Riemann's zeta function. Because the left-hand side is always positive, the optimal D is greater that the marginal median of Y.

An alternative approach to the derivation of the same equations, based on a Kullback–Leibler divergence between the actual posterior distribution and the posterior distribution that would be available if Y could be observed exactly, is discussed in Parmigiani (1992).

20.3 INFORMATION AND SAMPLE SIZE

A common design problem in clinical and other applications is determining the best fixed sample size for the experiment. To make the notation more standard, let $D = n$. If the data to be collected are exchangeable, Theorem 4 in Lindley applies to show that $\mathscr{I}(n)$ is increasing and convex in n. Therefore, if the cost of observation increases linearly or faster in n and if information and cost are additive, there are at most two (adjacent) solutions. In this section we study $\mathscr{I}(n)$ for exchangeable exponential and Bernoulli data.

20.3.1 Exponential data

We consider first exponential survival data with unknown failure rate Λ and conjugate prior gamma (a, b). Let $t = \sum_i^n x_i$. The expected information in a sample of size n is

$$\mathscr{I}(n) = \int_\infty^0 \int_\infty^0 \log\left[\frac{\Gamma(a)(b+t)^{a+n}}{\Gamma(a+n)b^a} \lambda^n \exp(-\lambda t)\right]$$

$$\times \frac{b^a}{\Gamma(a)\Gamma(n)} \lambda^{a+n-1} t^{n+1} \exp[-\lambda(b+t)] dt\, d\lambda$$

$$= \log\frac{\Gamma(a)}{\Gamma(a+n)} + (a+n)\psi(a+n) - a\psi(a) - n \qquad (8)$$

where ψ is the digamma function. Related calculations are discussed in Aitchison (1975) and Polson (1988) in the context of model choice.

The value of $\mathscr{I}(n)$ does not depend on b. The amount one expects to learn depends solely on the number of prior failures and not on the prior total survival. A useful corollary of (8) is the expected information from the next observation, which is

$$\mathscr{I}(1) = \frac{1}{a} + \psi(a) - \log a.$$

If $a = 1$, then $\mathscr{I}(1) = 1 - C = 0.4288$, where C is Euler's constant.

The following large n approximation brings out further interesting features of $\mathscr{I}(n)$. Using

$$\log \Gamma(n) \approx 1/2 \log 2\pi - n + (n - 1/2)\log n$$

and

$$\psi(n) \approx \log n - \frac{1}{2n},$$

we get
$$\mathcal{I}(n) \approx K(a) + 1/2 \log(a + n),$$
where
$$K(a) = \log \Gamma(a) - a(\psi(a) - 1) - 1/2 \log 2\pi - 1/2.$$

Using this approximation one can solve analytically for the optimal fixed sample size with sampling cost c (in information units) per observation. In particular, taking n to be continuous, the optimum is

$$n^* = \frac{1}{2c} - a.$$

Consider now evaluating the observed information $A(t, n)$ from outcome t when the sample size is n. Following Raiffa and Schlaifer (1961), we compute it by comparing the value of the decision problem under the current distribution for Λ with the current expected utility of the optimal prior decision. DeGroot (1984) shows that for the utility we are considering the observed information coincides with the Kullback–Leibler divergence between posterior and prior, obtained using the posterior as the baseline distribution. Therefore

$$A(t, n) = \int_\Theta \log\left[\frac{\Gamma(a)(b+t)^{a+n}}{\Gamma(a+n)b^a} \lambda^n \exp(-\lambda t)\right]$$

$$\times \frac{(b+t)^a}{\Gamma(a+n)} \lambda^{a+n-1} \exp[-\lambda(b+t)] d\theta$$

$$= \log \frac{\Gamma(a)}{\Gamma(a+n)} + a \log\left(\frac{b+t}{b}\right) + n\psi(a+n) - t\frac{a+n}{b+t}.$$

The function A is not monotone in t for a fixed sample size. By a simple minimization the least informative value occurs when $t/n = b/a$, that is, when the observed failure rate and the prior failure rate coincide. There is no upper bound.

20.3.2 Binomial data

Consider now the case of binary data with unknown success probability Θ, and conjugate beta(a, b) prior. The observed information is

$$A(x, n) = \int_0^1 \log\left(\frac{B(a, b)}{B(a+x, b+n-x)} \theta^x (1-\theta)^{n-x}\right) \frac{\theta^{a+x-1}(1-\theta)^{b+n-x-1}}{B(a+x, b+n-x)} d\theta$$

$$= \log \frac{B(a, b)}{B(a+x, b+n-x)} + x[\psi(a+x) - \psi(a+b+n)]$$

$$+ (n-x)[\psi(b+n-x) - \psi(a+b+n)].$$

If the prior is symmetric, that is, if $a = b$, then $A(x, n) = A(n - x, n)$: the observed information is the same from x successes and from $n - x$ successes. Also, in the uniform prior case,

$$A(x, n) = -\log B(x + 1, n - x + 1) - \sum_{k=1}^{n-x+1} \frac{x}{x + k} - \sum_{k=1}^{x+1} \frac{n - x}{n - x + k}.$$

Computing $\mathcal{I}(n)$ requires the expectation of A with respect to the marginal distribution of X, given by

$$m(x) = \frac{B(a + x, b + n - x)}{(n + 1)B(x + 1, n - x + 1)B(a, b)}.$$

Thus

$$\mathcal{I}(n) = \sum_{x=0}^{n} \frac{B(a + x, b + n - x)A(x, n)}{(n + 1)B(x + 1, n - x + 1)B(q, b)}.$$

Consider again the uniform prior case. Then

$$\mathcal{I}(n) = \frac{1}{n + 1} \sum_{x=0}^{n} A(x, n).$$

For $n = 1$

$$A(0, 1) = A(1, 1) = \log 2 - 1/2 = 0.1931.$$

So $\mathcal{I}(1) = 0.1931$. This provides a benchmark for calibration and elicitation of trade-off values for information versus costs.

Li et al. (1994) discuss computational methods and alternative utility functions for optimal sample size in comparing binomials. The paper includes an S implementation of the results of this section.

20.4 DURATION AND FOLLOW-UP TIME

An important design problem in studies with survival as end-point is choosing the duration of the study. Assume that patients enter the study at random times $a_1, a_2, \ldots$ governed by a Poisson process with unknown rate α (accrual rate). The numerical results are similar if the accrual rate is non-stochastic, with α patients arriving per unit of time. Patients arriving before time τ are admitted in the study. All patients are then followed until time $\tau + \delta$. Let S be the number of patients who enter the study. Patient survivals $x_1, \ldots, x_S$ are exchangeable exponentials with unknown failure rate λ.

We consider the problem of finding the information about Λ as a function of τ and δ. The result can be used to solve various decision problems. For illustration we will consider the case in which there is a cost c associated with treating each patient, and there is a cost w associated with each unit of time of the study. So the objective function to be maximized is

APPLICATIONS OF LINDLEY INFORMATION MEASURE 337

$$U(\tau, \delta) = \mathcal{T}(\tau, \delta) - c\tau E(\Lambda) - w(\tau + \delta). \quad (9)$$

Let $I = \{ : a_i + x_i < \tau + \delta\}$ be the set of subscripts of completely observed survivals, and let k be the number of elements in I. If α and λ are *a priori* independent and λ is gamma(a, b) then the marginal posterior distribution of λ is gamma(a', b'), with

$$a' = a + k$$
$$b' = b + \sum_{i \in I} x_i + \sum_{i \notin I} (\tau + \delta - a_i).$$

Based on this model we evaluated numerically $\mathcal{T}(\tau, \delta)$ using smoothing of Monte Carlo experiments. The method is based on simulating draws from the joint parameter/sample space and evaluating the observed value of $\log(\pi(\lambda \mid x)/\pi(\lambda))$. Fitting a smooth surface through these simulated points serves to estimate the expected information surface. Further details are discussed in Müller and Parmigiani (1992).

We first discuss an example based on a prior on α which is $\Gamma(20, 6)$, corresponding to an expected accrual rate of 3.3 patients per unit of time. The prior on λ is $\Gamma(2, 2)$. This prior corresponds to rather vague initial knowledge.

Figure 20.1 illustrates the estimation procedure used to evaluate the expected information about λ as a function of τ. Each of the points is the

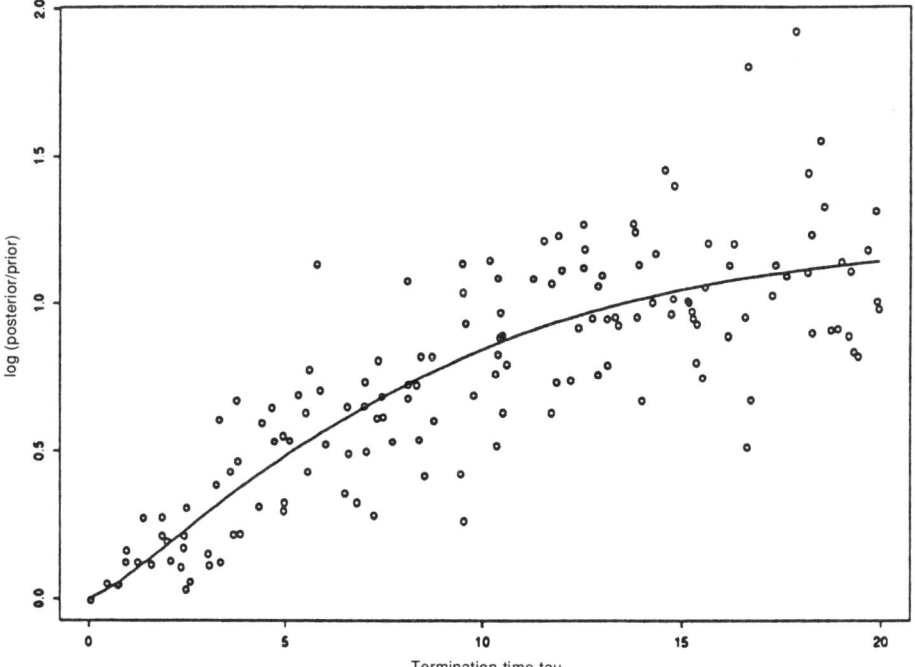

Figure 20.1 Smoothed Monte Carlo estimation of $\mathcal{T}(\pi, 0)$.

average of 15 draws. The smooth line is the estimate of $\mathscr{I}(\tau, 0)$ and is obtained by a parametric non-linear model. We added points in batches of 50 and stopped when adding would not make a distinguishable difference in the curve. We also compared the non-linear model fit with non-parametric curve estimates obtained with *loess* smoothing. The two methods gave very similar results.

The resulting curve brings out interesting qualitative features of $\mathscr{I}$. For very small values of τ information increases more than linearly in τ. Contiguous units of accrual time are not exchangeable: each additional unit contributes in expectation as many new patients as the previous one, plus some additional survival information from previously entered patients. As we learn more about the failure rate, though, this effect is outweighed by the typical concavity of the information curve. The initial convexity becomes more pronounced with smaller accrual rates and less pronounced with larger δ, as expected.

Figure 20.2 compares the estimated $\mathscr{I}(\tau, 0)$ and $\mathscr{I}(\tau, 4)$. The gain in expected information achieved by a larger follow-up time $\delta = 4$ is clearly null at $\tau = 0$, it increases in the middle part of the range, and then levels off when the duration is large and the loss of information due to censored observation is less important. This graph can also be used to compare the

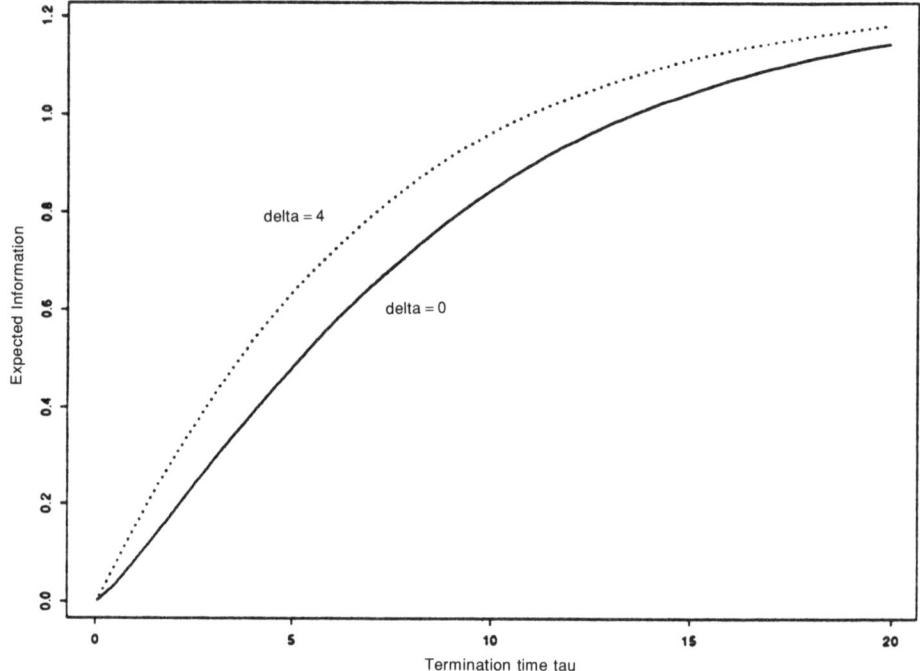

Figure 20.2 Comparison of estimated $\mathscr{I}(\pi, 0)$ and $\mathscr{I}(\pi, 4)$.

APPLICATIONS OF LINDLEY INFORMATION MEASURE

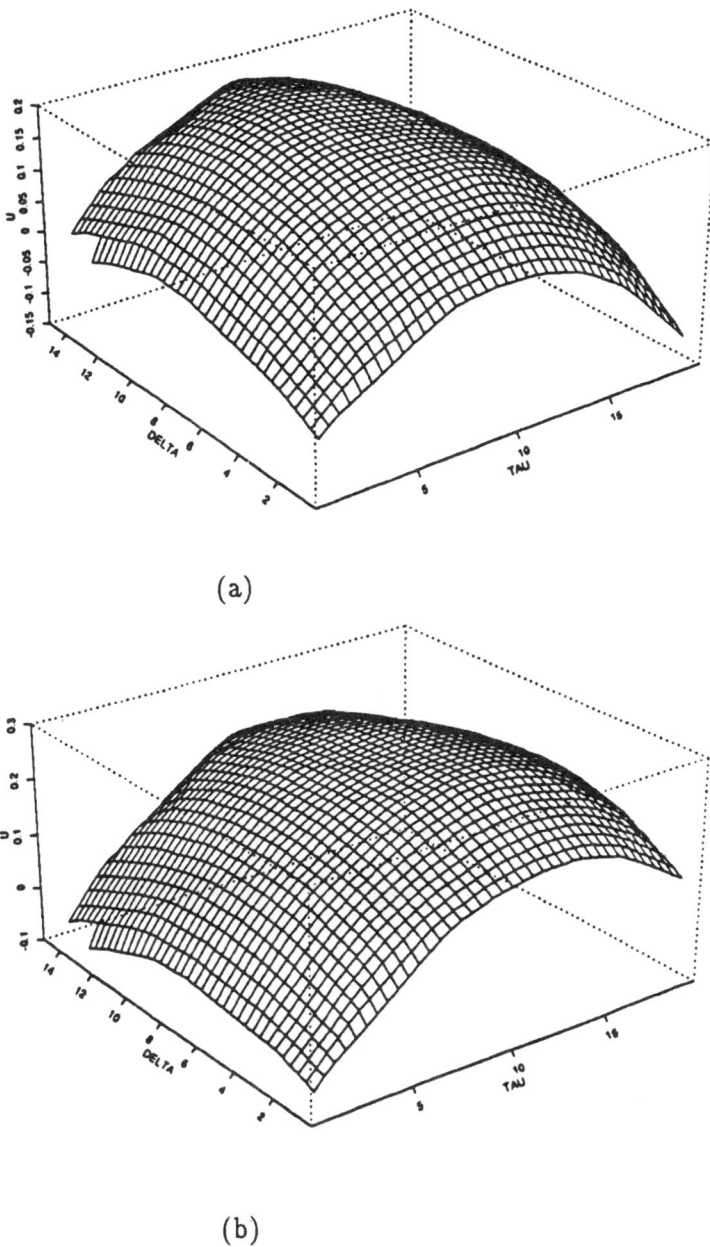

(a)

(b)

Figure 20.3 Estimated utility surface as a function of the duration π of the study and of the follow-up time δ. Values of the cost parameters are indicated. (a) $c = 0.255$, $w = 0.01$; (b) $c = 0.15$, $w = 0.015$.

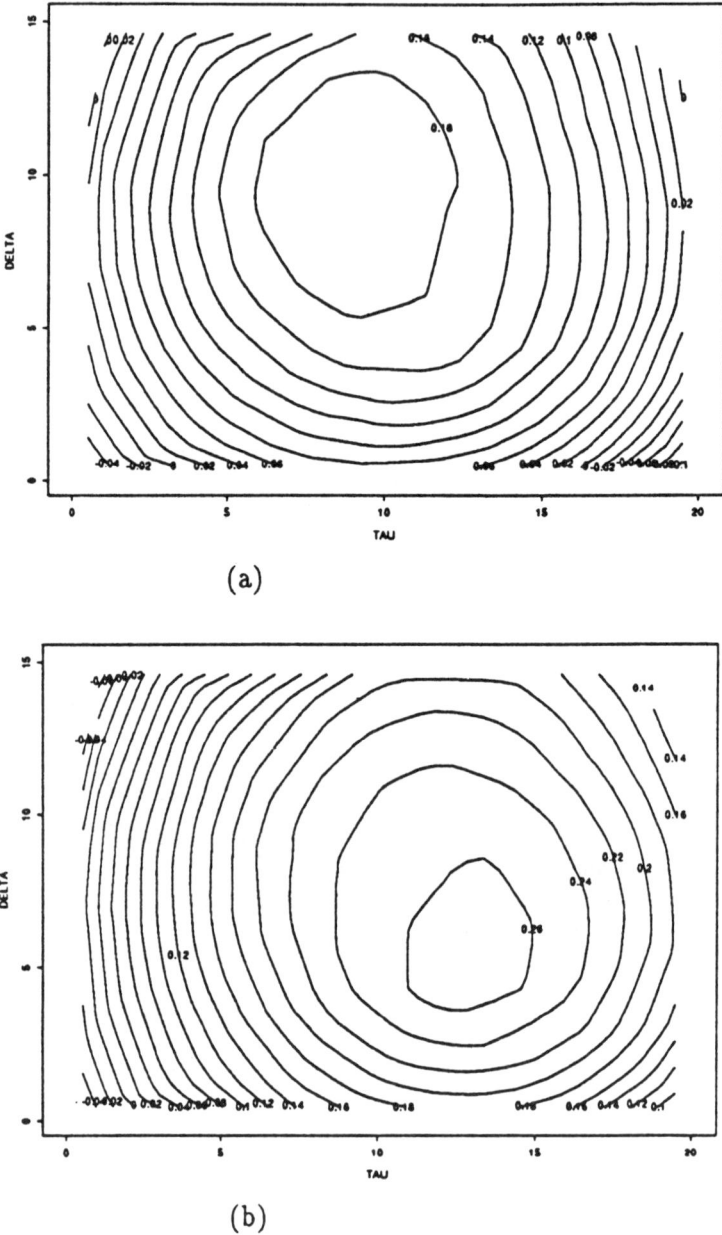

Figure 20.4 Estimated utility contours as a function of the duration of the study and of the follow-up time δ. Roughness in the contours is an artefact of the plotting procedure. (a) $c = 0.255$, $w = 0.1$; (b) $c = 0.15$, $w = 0.015$.

two combinations of τ and δ that lead to the same expected information. For example, if it is desired to achieve $\mathscr{I} = 0.6$, this can be done by δ = 0 and τ ≈ 7 or by δ = 4 and τ ≈ 5. Balancing this trade-off is an important design problem. We illustrate it in more detail in the following example.

Assume a prior on α which is Γ(6, 6), corresponding to an expected accrual rate of one patient per unit of time. The prior on λ is Γ(2, 10). Using smoothing of Monte Carlo points by two-dimensional loess, we estimated the surface U(τ, δ) for two different pairs of cost parameters: (c = 0.225, w = 0.01) and (c = 0.15, w = 0.015). The surfaces are shown in Figure 20.3. The optimal designs are respectively (τ = 8.43, δ = 9.65) and (τ = 13.42, δ = 4.99).

An important aspect of design problems is robustness of the solution to small perturbation. The two contour plots of the utility surfaces are shown in Figure 20.4. The figure illustrates the fact that the expected utility surface not very sensitive to changes of τ and δ around the optimum. This is a very useful property in practice, as it eases implementation of the design.

20.5 MULTICENTRE CLINICAL TRIALS

A controversial issue in multicentre trials is the appropriate number of centres. Should there be few centres with many patients in each, or many centres, each with few patients? The framework lends itself to using Bayesian hierarchical modelling (see Stangl 1992). We consider Bernoulli data. Within each centre, patients are exchangeable with unknown success probability p_i, where i is the centre. In turn, centres are drawn from an infinite population distributed according to a beta(A, B) where A and B are unknown. The prior distribution of A and B is given by independent gamma(u_A, t_A) and gamma(u_B, t_B) distributions.

The objective is to determine the information associated with a pair n, k where k is the number of centres in the study and n is the number of patients in each centre, the same in every centre. Of critical importance is the loss structure assumed. There are various versions of $\mathscr{I}$ that may be worth considering, depending on the parameters of interest. We carry out the calculation for one of the more natural choices. Namely, we consider the expected information about A and B only. The aim is to learn about the general population of centres rather than about any specific centre. Alternative formulations may include explicit consideration of one or more specific centres. For example, Mukhopadhyay and Stangl (1993) propose a loss function that combines squared error terms for estimating population parameters with squared error terms for the centre parameters, with normal data.

Let x be a k-dimensional vector representing the number of successes in each centre. Then, following Berry and Berry (1992) we get

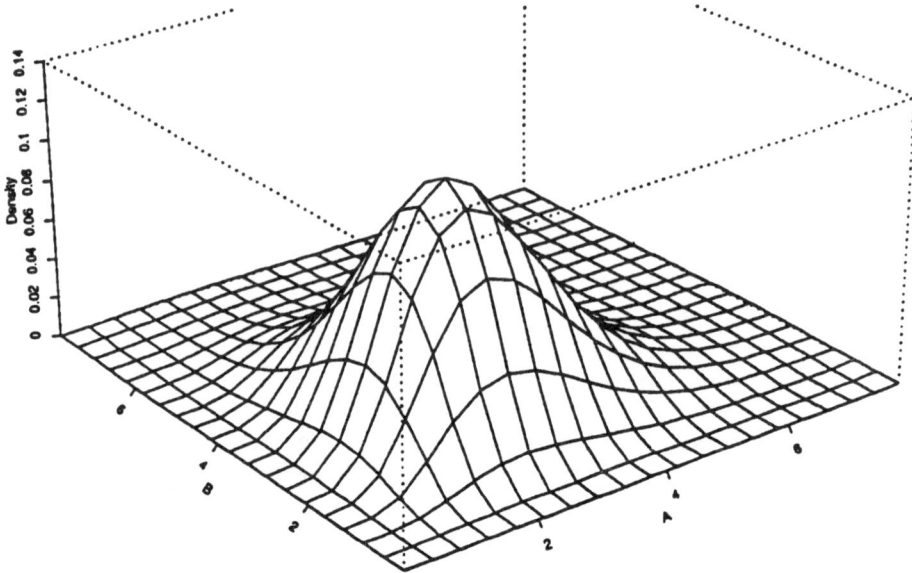

Figure 20.5 Prior distribution on A and B.

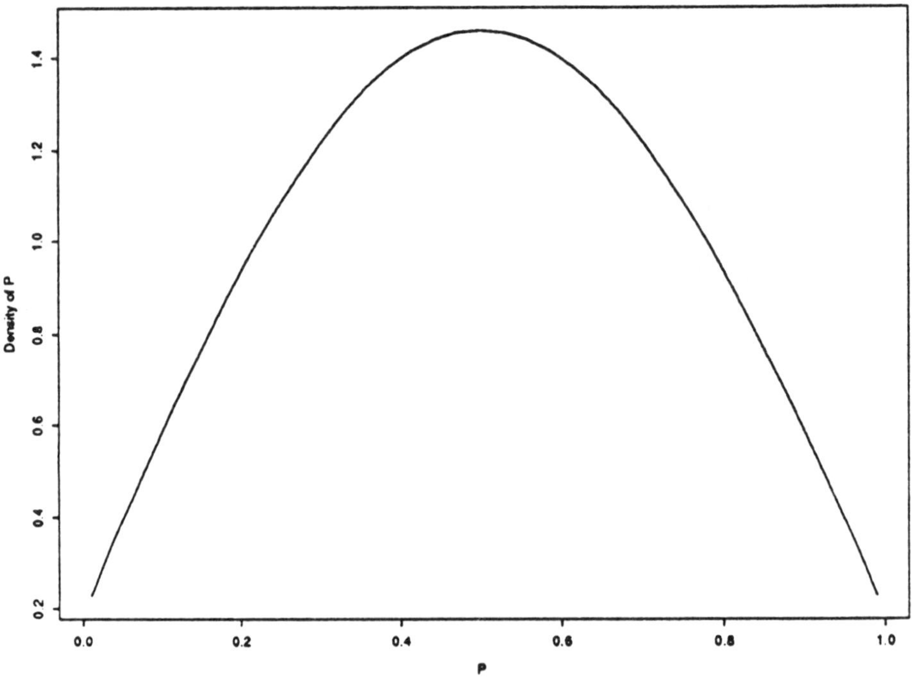

Figure 20.6 Prior predictive distribution of the success probability p.

$$m(x) = \int_{\infty}^{0} \int_{\infty}^{0} \left[B^{-k}(a, b) \prod_{i=1}^{k} C_{x_i}^{n} B(a + x_i, b + n - x_n) \right] \pi(a, b) \mathrm{d}a \mathrm{d}b$$

and

$$\pi(a, b \mid x) = \frac{\pi(a, b) \prod_{i=1}^{k} B(a + x_i, b + n - x_n)}{m(x) B^k(a, b)}.$$

We discuss in detail the case of a hyperprior with parameters $u_A = u_B = 6$ and $t_A = t_B = 2$, corresponding to relatively little knowledge about the population of centres. The distribution of A and B is graphed in Figure 20.5. The resulting prior predictive distribution on the probability of success p in a randomly selected centre is shown in Figure 20.6.

Figure 20.7 gives an example of the posterior distributions that can be obtained by allocating 16 patients in three different ways: all in one centre ($n = 16$, $k = 1$), four in each of four centres ($n = 4$, $k = 4$) and all in different centres ($n = 1$, $k = 16$). All three posteriors are based on 12 observed successes. As centres are exchangeable, the posterior distribution does not depend on the order of the elements of x. So for $k = 1$ and $k = 16$ the posterior graphed is the only one arising from 12 successes. For $k = 4$ the posterior shown is one of six possible. High values of k correspond, all else being equal, to a more concentrated posterior distribution, that is, to a better knowledge of the population.

An alternative way of illustrating the same fact is using the marginal predictive distribution of the success probability p_{k+1} under alternative designs. This is done in Figure 20.8.

In the remainder of this section we approximate k and n by real variables, and estimated the surface $\mathscr{T}(n, k)$. The computational technique is the same as in section 20.4. Figure 20.9 gives the contours of the surface. The dotted lines are hyperbolas representing designs with the same total number of patients. Consistent with the posterior plots of Figure 20.7, the expected information increases if one moves along a hyperbola, increasing k.

To illustrate the use of the surface $\mathscr{T}(n, k)$ in solving specific design problems, consider the case of a fixed cost c per patient and a fixed 'start-up' cost s per centre. Again, costs are expressed in information units. The objective function is

$$U(n, k) = \mathscr{T}(n, k) - cnk - sk.$$

For example if $s = 0.045$ and $c = 0.015$ the optimal design is ($n = 11$, $k = 6$). A contour plot of the utility surface is shown in Figure 20.10. Some choice of designs have negative expected utility and are therefore worse than no sampling. In particular, all designs with only one centre (the horizontal line at $k = 1$) are worse than no experimentation. This methodology can

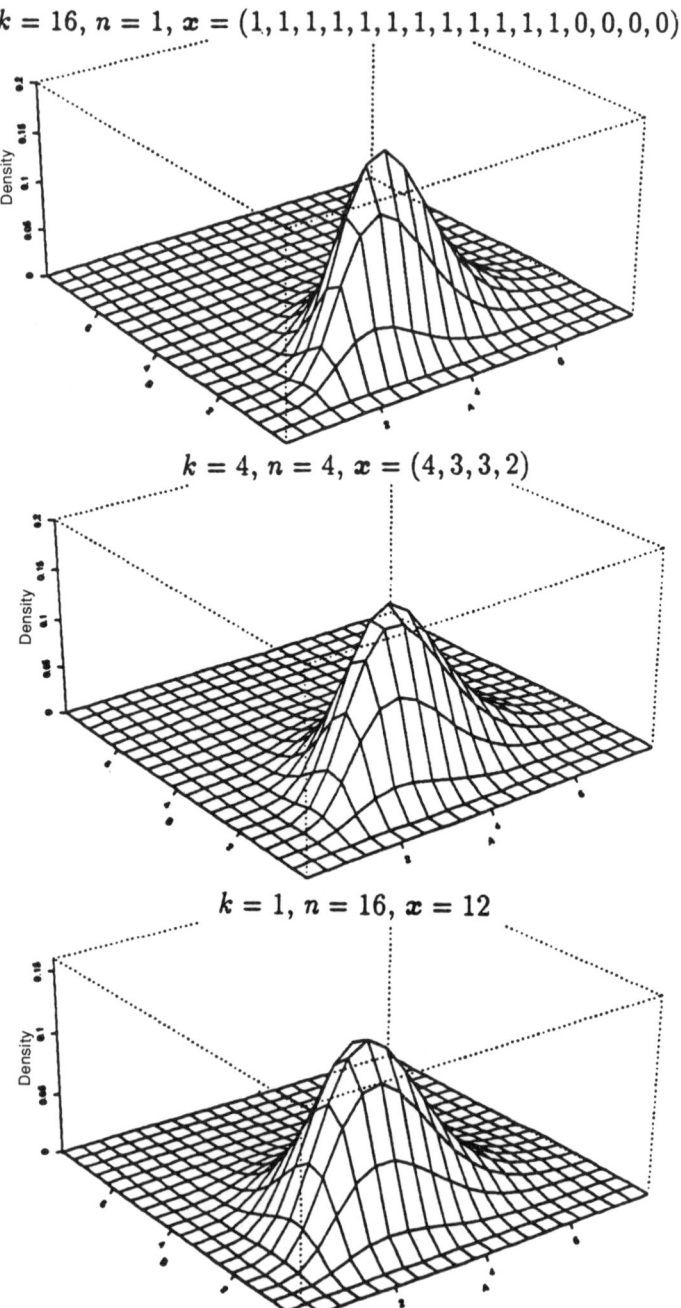

Figure 20.7 Posterior distributions on A and B under alternative designs; all are based on 12 successes in 16 patients. Higher values of the numbers of centres k correspond to more concentrated posterior distributions.

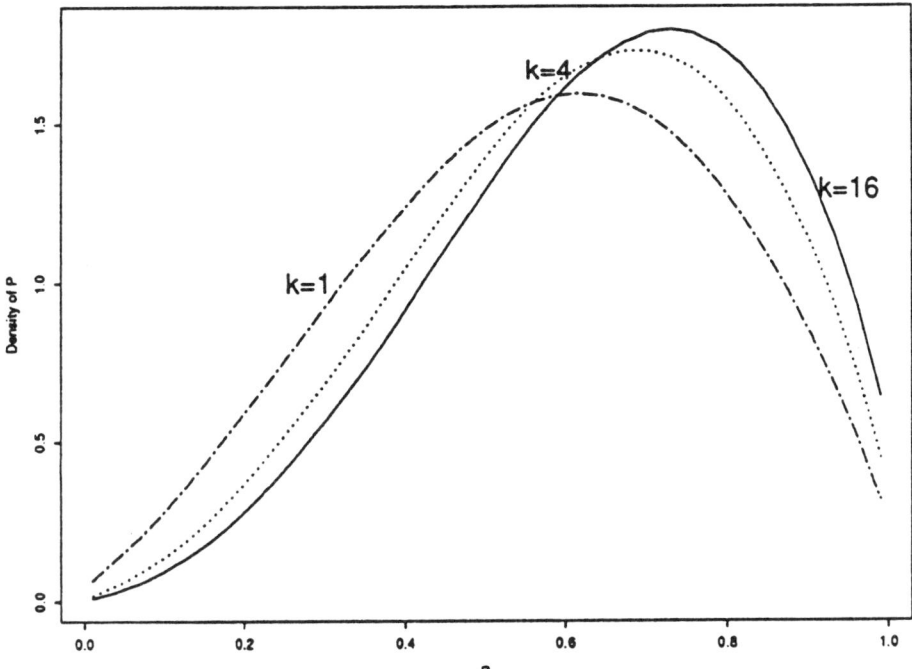

Figure 20.8 Marginal distributions of the success probability p_{k+1} under alternative designs; all are based on 12 successes in 16 patients. Outcomes are as in the previous figure.

provide the basis for interactive sensitivity analysis with respect to the prior hyperparameters and the cost parameters.

20.6 CONCLUSIONS

In this chapter we have illustrated the use of Lindley information in the design of clinical experiments. Lindley information is a natural objective function in the design of experiments aiming at learning about the world or providing several decision-makers with useful information in different decision problems.

We used it to explore properties of Bayesian designs in biostatistics applications. In particular, in the duration and follow-up time problem we illustrated how information may increase more than linearly for small durations. If costs are linear in duration, the point of inflection of the curve provides a lower bound to the optimal duration, independent of the cost. Also we illustrated how to construct pairs of duration and follow-up times that lead to the same expected information.

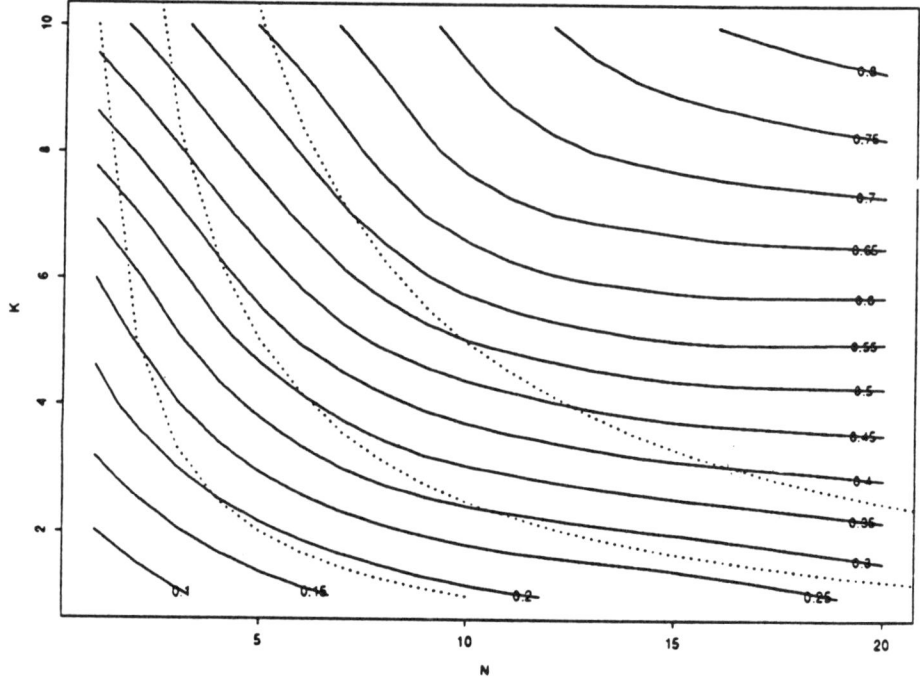

Figure 20.9 Contour plot of the estimated information surface as a function of the number of centres k and the number of patients per centre n; dotted lines identify designs with the same number of patients—respectively 10, 25 and 50.

In the multicentre clinical trial problem, we observed that if one wants to learn solely about the population parameters, and there are no 'set-up' costs for a centre, the most efficient way to allocate a given number of patients is to assign one per centre. This is illustrated for one specific prior but we believe it is true generally. A 'set-up' cost for a centre may lead to a different solution. The optimal balance is illustrated in one specific example.

This chapter can be developed in several directions. For example, a very widely applied criterion for making design decisions in clinical applications is power against a specific alternative to the null hypothesis of interest. Bayesian decision theory provides a number of alternatives to this approach, one of which is using Lindley information. We plan to investigate relations between designs maximizing information and design maximizing power and sequential problems, with particular attention to sample size determination.

REFERENCES

Aitchison, J. (1975) Goodness of prediction fit, *Biometrika* **62**, 547–54.

APPLICATIONS OF LINDLEY INFORMATION MEASURE

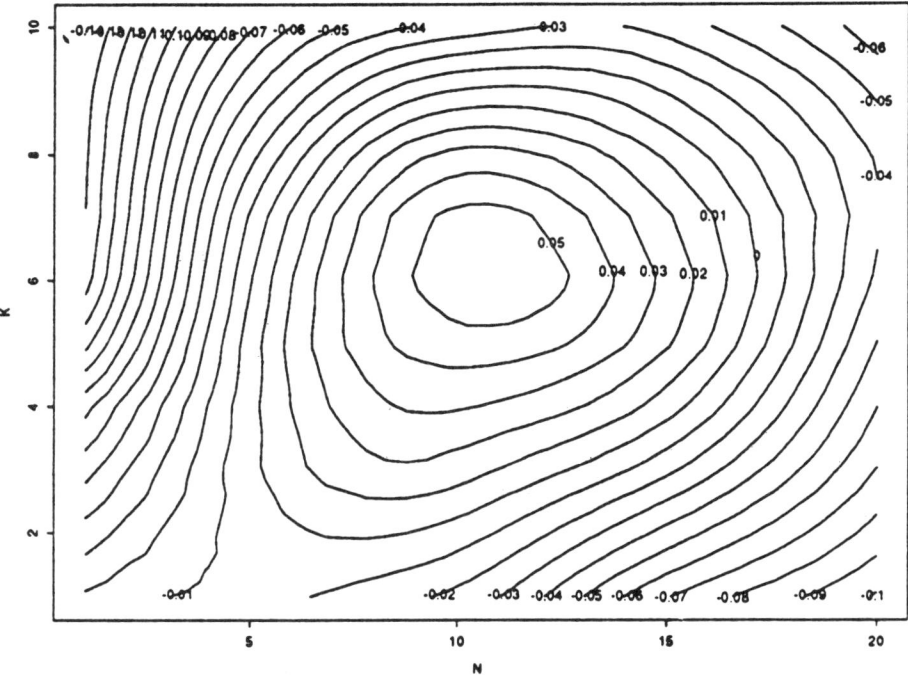

Figure 20.10 Contour plot of the estimated utility surface as a function of the number of centres k and the number of patients per centre n. Pairs inside the 0-level curve are better than no experimentation. Points outside are worse.

Bernardo, J. (1979) Expected information as expected Utility, *Annals of Statistics* **7**, 686–90.

Berry, D. A. (1987) Interim analysis in clinical trials: the role of the likelihood principle, *The American Statistician* **41**, 117–22.

Berry, D. A. (1988) Interim analysis in clinical research, *Cancer Investigation* **5**, 469–77.

Berry, D. A. and Berry, S. (1992) *Bayesian Metaanalisys for Treatment Comparisons: Dichotomous Responses.*

Cover, T. M. and Thomas, J. A. (1991) *Elements of Information Theory*, New York, John Wiley & Sons.

DeGroot, M. H. (1984) Changes in utility as information, *Theory and Decision* **17**, 287–303.

Good, I. J. (1969) What is the use of a distribution? *Multivariate Analysis II*, P. R. Krishnaiah, ed., New York, Academic Press, pp. 183–203.

Kullback, S. (1959) *Information Theory and Statistics*, New York, John Wiley & Sons.

Kullback, S. and Leibler, R. A. (1951) On information and sufficiency, *Annals of Mathematical Statistics* **22**, 79–86.

Li, C., Müller, P. and Parmigiani, G. (1994) Optimal sample size for the binomials, *ISDS Discussion Paper*, Duke University.

Lindley, D. V. (1956) On the measure of information provided by an experiment, *Annals of Statistics* **27**, 986–1005.

Lindley, D. V. (1957) Binomial sampling schemes and the concept of information, *Biometrika*, **44**, 179–186.

Mukhopadhyay, S. and Stangl, D. (1993) Balancing centers and observations in multicenter clinical trials, *ISDS, Discussion Paper*, Duke University.

Müller, P. and Parmigiani, G. (1992) Sequential design by curve fitting of Monte-Carlo experiments, *ISDS, Discussion Paper 92–A33*, Duke University.

Müller, P. and Parmigiani, G. (1994) Numerical evaluation of information theoretic measures, in *Bayesian Statistics and Econometrics: Essays in Honor of A. Zellner*, D. A. Berry, K. M. Chaloner and J. F. Geweke, eds.

Parmigiani, G. (1992) Design of partially observable survival data, *ISDS, Discussion Paper 92–A34*, Duke University.

Parmigiani, G. and N. G. Polson (1992) Bayesian design of random walk barriers, in *Bayesian Statistics IV*, J. O. Berger, J. M. Bernardo, A. P. Dawid and A. F. M. Smith, eds, Oxford University Press, pp. 715–21.

Polson, N. G. (1988) Some Bayesian perspectives on statistical modelling, PhD thesis, University of Nottingham.

Raiffa, H. and Schlaifer, R. (1961), *Applied Statistical Decision Theory*, Cambridge, Mass., Harvard University Press.

Ramsey, F. P. (1990) Weight or the value of knowledge, *British Journal for the Philosophy of Science*, **41**, 1–4.

Stangl, D. (1991) Modeling heterogeneity in multi-center clinical trials using Bayesian hierarchical survival models, PhD thesis, Department of Statistics, Carnegie Mellon University.

Verdinelli, I. (1992) Advances in Bayesian experimental designs, in *Bayesian Statistics IV*, J. O. Berger, J. M. Bernardo, A. P. Dawid and A. F. M. Smith, eds, Oxford University Press, pp. 467–82.

Verdinelli, I. and Kadane J. B., (1992) Bayesian designs for maximizing information and outcome, *Journal of the American Statistical Association*, 510–15.

Institute of Statistics and Decision Sciences
Duke University
Durham
NC 27706
USA

CHAPTER 21

On Two Classic Theorems Involving the Characteristic Function

Walter L. Smith
Chapel Hill, North Carolina

21.1 SOME GENERAL COMMENTS

In the Michaelmas term of 1950 I was beginning my stint at Cambridge as a graduate student and attended a course of lectures by Mr D. V. Lindley on probability theory (on Tuesday, Thursday and Saturday, in the Mill Lane Lecture Rooms, at noon). At a time when few people working in England on probabilistic problems had any knowledge of measure theory, he presented a fully measure-theoretic treatment of the main results of the subject, with many references to the work of Lévy, Kolmogorov, Feller, Cramér and other 'greats'. I have always believed that Lindley was pioneering, in England, by giving such a course. Here is not the place for extended reminiscences: I will merely state that I found the course most difficult to understand and the notes I took from it superlatively useful to me throughout the years since I wrote them down. Additionally, those lectures greatly intensified the enthusiasm I already felt for characteristic functions.

Two theorems I first learnt from Lindley hold special interest for me: the Lindeberg central limit theorem and Cramér's famous theorem about the normal distribution (if two independent random variables sum to a normal variable then each is normal). In the course of many years I have come across new ways to prove each of these theorems. My proof of the Lindeberg theorem has the elegant advantage that the necessity and sufficiency parts of the proof go hand in hand, and their relationship becomes

---
Aspects of Uncertainty edited by P. R. Freeman and A. F. M. Smith. © 1994 John Wiley & Sons Ltd.

almost transparently obvious. This 'new' proof of the Lindeberg theorem has been presented (admittedly, in a slightly less satisfactory form than presented here) in graduate courses at Chapel Hill for many years, but has never been published.

My unhappiness with the usual proof of the Cramér theorem is that it rests on an appeal to the Hadamard factorization theorem for entire functions. An understanding of this Hadamard theorem requires the expenditure of an amount of study of entire functions that is excessive if one's primary purpose is merely to understand, fully, the proof of Cramér's theorem. I have long felt a proof should be available which rests on more immediately accessible parts of function theory. Relatively recently, I discovered such a proof, that presented here.

I present these two proofs as a token of my deep appreciation for the encouragement and help Dennis Lindley has given me on a number of crucial occasions: two modest blooms from the seeds he sowed so long ago!

21.2 THE LINDBERG CENTRAL LIMIT THEOREM

The Lindeberg central limit theorem is one of a few theorems that should, and typically do, constitute the major landmarks, as it were, of a serious first graduate course in probability theory. Most of the published proofs of this important theorem involve Fourier analysis in that they employ characteristic functions. Two noteworthy exceptions are the proofs of Trotter (1959) and (for the somewhat restricted case of lattice random variables, identically distributed, and therefore hardly meriting to be referred to as the 'Lindeberg' theorem) one due to Petrovsky and Kolmogorov (given in Khinchin 1948 and presented very clearly, in English, in Rosenblatt 1974). However, the majority of the published proofs we have seen, including those in a number of much-used 'standard' texts, essentially involve two distinct proofs, each somewhat intricate; one deals with the sufficiency part of the proof, the other with the necessity part. It is many years since I introduced into the first graduate course on probability theory at Chapel Hill a proof of the Lindeberg theorem which makes use of so-called 'derived' distributions; this proof is strikingly straightforward, the necessity and the sufficiency parts of the theorem are established simultaneously.

21.2.1 Conditions for the theorem

Let $X_1, X_2, \ldots$ be an infinite sequence of mutually independent random variables with finite means and variances. With no loss of generality and some gain in ease of discussion we may suppose that $\mathscr{E} X_n = 0$ for all n; we shall write σ_n^2 for the variance of X_n and set

$$\tau_n^2 = \sigma_1^2 + \sigma_2^2 + \ldots + \sigma_n^2 \qquad (1)$$

for the variance of $S_n = X_1 + X_2 + \ldots + X_n$. Let us also write $F_n(x) = P\{X_n \leq x\}$ for the distribution function of X_n. Finally make the assumption of *asymptotic negligibility*:

$$\left\{ \max_{1 \leq j \leq n} \frac{\sigma_j^2}{\tau_n^2} \right\} \to 0, \quad \text{as } n \to \infty. \qquad (2)$$

Let us refer to the all the assumptions described in the preceding paragraph as the *global hypothesis*. We may now introduce the famous Lindeberg condition.

Condition (LC) For every fixed $\varepsilon > 0$, as $n \to \infty$,

$$\frac{1}{\tau_n^2} \sum_{j=1}^{n} \int_{|x| > \varepsilon \tau_n} |x|^2 F_j(\mathrm{d}x) \to 0. \qquad (3)$$

Lindeberg (1922) established the following:

Theorem I If the global hypothesis be taken to hold, then, in order for S_n/τ_n to be asymptotically $\mathcal{N}(0, 1)$ as $n \to \infty$ it is necessary and sufficient that condition (LC) hold.

21.2.2 Derived distributions

The author introduced the so-called 'derived' distributions in his PhD thesis (Smith 1953) and has used them since in various ways; see Smith (1959) and, for more general results, Smith (1966). For completeness of the present account we present briefly the necessary basic ideas, in spite of their availability in much greater generality elsewhere.

Let $F(\cdot)$ be the distribution function of a non-negative random variable whose first two moments μ_1 and μ_2 are both finite. It is well known that $\mu_1 = \int_0^\infty \{1 - F(x)\} \mathrm{d}x$; thus we can define a probability density function on $(0, \infty)$ by

$$f_{(1)}(x) = \frac{\{1 - F(x)\}}{\mu_1} \qquad (4)$$

and we shall call this the *first derived density function*. It is a trivial matter to show that this new density has its first moment finite and equal to $\mu_2/2\mu_1 = \mu_1^{(1)}$, say. Thus, if we write $F_{(1)}(\cdot)$ for the df associated with $f_{(1)}(\cdot)$ we can introduce a *second derived density* by

$$f_{(2)}(x) = \frac{\{1 - F_{(1)}(x)\}}{\mu_1^{(1)}}. \qquad (5)$$

It is a comparatively easy matter to verify the following equation which relates the tail probabilities of $f_{(2)}(\cdot)$ to the original distribution function $F(\cdot)$:

$$\int_x^\infty f_{(2)}(u)\,du = \frac{1}{\mu_2}\int_x^\infty (x-z)^2 F(dz). \tag{6}$$

Now let us introduce *characteristic functions*: write, for dummy real θ,

$$\phi(\theta) = \int_{0-}^\infty \exp(i\theta x)\,F(dx). \tag{7}$$

Then, in an obvious notation, one can obtain by an integration by parts the result

$$\phi_{(1)}(\theta) = \frac{1-\phi(\theta)}{-\mu_1 i\theta} \tag{8}$$

and, by the same token,

$$\phi_{(2)}(\theta) = \frac{1-\phi_{(1)}(\theta)}{-\mu_1^{(1)} i\theta}. \tag{9}$$

If we combine (8) and (9) we are led to the result

$$\phi(\theta) = 1 + \mu_1 i\theta + \frac{1}{2}\mu_2 (i\theta)^2 \phi_{(2)}(\theta). \tag{10}$$

This equation (10) is the basis of our proof; notice that it is valid for *all* θ and not, as is the case with the more familiar Taylor expansions of a characteristic function at the origin, for 'small' θ only. All that remains for us to do, before we turn to the actual proof of the central limit theorem proper, is to extract from (10) which refers to a distribution on the non-negative half-line, a similar result for a distribution on the whole real line. However, before we do that let us notice that $\phi_{(2)}(\theta)$ is the characteristic function of the pdf $f_{(2)}(\cdot)$. Moreover, $f_{(2)}(\cdot)$ is a pdf with special properties: it is (as a glance at (4) and (5) will reveal) non-increasing, with a non-increasing derivative which tends to zero as $x \to \infty$. Thus (10) is remarkable in that it connects the arbitrary characteristic function $\phi(\theta)$ to the cf of such a special convex pdf (whenever there is a finite variance, of course).

Now let X be an unrestricted (to the non-negative reals) random variable with zero mean and finite variance σ^2. Let us set

$$p = \mathbf{P}\{X \geq 0\} \quad \text{and} \quad q = \mathbf{P}\{X < 0\}.$$

Then we can regard X as arising from two non-negative random variables X^+ and X^-: with probability p, $X = X^+$; with probability q, $X = -X^-$. In an obvious notation we have

$$\mu_1 = p\mu_1^+ - q\mu_1^-$$

$$\mu_2 = p\mu_2^+ + q\mu_2^-$$

$$\phi(\theta) = p\phi^+(\theta) + q\phi^-(-\theta)$$

and from equations like (10) for both $\phi^+(\theta)$ and $\phi^-(\theta)$ and can then derive the result, remembering that $\mu_1 = 0$, $\mu_2 = \sigma^2$,

$$\phi(\theta) = 1 - \frac{1}{2}\sigma^2\theta^2\psi(\theta) \tag{11}$$

in which the function $\psi(\cdot)$ is a *characteristic function*:

$$\psi(\theta) = \frac{p\mu_1^+\phi_{(2)}^+(\theta) + q\mu_2^-\phi_{(2)}^-(-\theta)}{\sigma^2} \tag{12}$$

associated with a pdf $g(\cdot)$, say, which is convex and monotonically decreasing to zero as x increases to ∞ on the positive half-line, convex and monotonically decreasing to zero as x decreases to $-\infty$ on the negative half-line. This result (11) has been the object of the present section; it is a special case of a more general result (Smith 1966), but given here for completeness and ease. It is important to note that, like (10), (12) is true for *all* real θ.

From (12) and (6) we can see that, for any $\varepsilon > 0$,

$$\int_{|z| > \varepsilon} g(x)\mathrm{d}x = \frac{1}{\sigma^2}\int_{|x| > \varepsilon}(|x| - \varepsilon)^2 F(\mathrm{d}x). \tag{13}$$

21.2.3 Proof of Lindeberg's theorem

We must first prove a crucial lemma. Lemmas something like it occur throughout the literature of mathematics, but the special form of the lemma given here is important to us, in that it enables us to avoid ambiguities arising from the use of the logarithm.

Lemma 1 For $j = 1, 2, \ldots, n$; $n = 1, 2, \ldots$, ad inf., let $\{a_{jn}\}$ be complex numbers such that

$$\sum_{j=1}^{n}|a_{jn}| \leq A < \infty \tag{14}$$

and suppose further that

$$\alpha_n = \mathrm{def}\left\{\max_{1 \leq j \leq n}|a_{jn}|\right\} \to 0, \quad \text{as } n \to \infty. \tag{15}$$

Then, as $n \to \infty$,

$$\prod_{j=1}^{n}(1 - a_{jn}) \sim \exp\left(-\sum_{j=1}^{j=n}a_{jn}\right). \tag{16}$$

Proof Let

$$\frac{\exp(-a_{jn})}{(1-a_{jn})} = 1 + \xi_{jn}, \text{ say.}$$

Then

$$\xi_{jn} = \frac{\frac{1}{2!}a_{jn}^2 + \frac{1}{3!}a_{jn}^3 + \cdots}{1 - a_{jn}}$$

so

$$|\xi_{jn}| \leq \frac{\alpha_n |a_{jn}|}{2(1-\alpha_n)^2}.$$

We may thus conclude that

$$\sum_{j=1}^{n} |\xi_{jn}| \to 0, \text{ as } n \to \infty. \tag{17}$$

But

$$\left|\prod_{j=1}^{n}(1+\xi_{jn}) - 1\right| \leq \prod_{j=1}^{n}(1+|\xi_{jn}|) - 1$$

$$\leq \exp\left(\sum_{j=1}^{j=n} |\xi_{jn}|\right) - 1$$

which, in view of (17), tends to zero and proves the lemma. □

Let us now turn to the proof of Theorem I. In the notation already established, let us define the characteristic function of S_n/τ_n to be

$$\Phi_n(\theta) = \mathscr{E}\exp(i\theta S_n/\tau_n).$$

Then, because of the mutual independence of the $\{X_n\}$,

$$\Phi_n(\theta) = \prod_{j=1}^{n} \phi_j(\theta/\tau_j)$$

$$= \prod_{j=1}^{n} (1 - a_{jn}), \text{ say,}$$

where we have set

$$a_{jn} = \frac{1}{2}\sigma_{jn}^2\left(\frac{\theta^2}{\tau_n^2}\right)\psi_j(\theta/\tau_n).$$

The question is then whether the $\{a_{jn}\}$ so defined satisfy the conditions of Lemma 1. Because the ψ_j are characteristic functions it follows that for *all* θ we must have $|\psi_j(\theta)| \leq 1$ and so, for every fixed θ, we have

$$|a_{jn}| \leq \tfrac{1}{2} \sigma_{jn}^2 \left(\frac{\theta^2}{\tau_n^2}\right), \qquad (18)$$

from which it is plain that, because of (2), (15) is satisfied. But from (18) it is equally plain that

$$\sum_{j=1}^{n} |a_{jn}| \leq \tfrac{1}{2} \theta^2 < \infty, \qquad (19)$$

and so condition (14) of Lemma 1 is also satisfied.

Notice that from the global hypothesis alone we must have that, as $n \to \infty$, for every fixed θ,

$$\Phi_n(\theta) \sim \exp\{-\tfrac{1}{2} \theta^2 \Psi_n(\theta)\} \qquad (20)$$

where we have conveniently written

$$\Psi_n(\theta) = \sum_{j=1}^{n} \frac{\sigma_j^2}{\tau_n^2} \psi_j(\theta/\tau_n). \qquad (21)$$

From (21) it can be seen that $\Psi_n(\theta)$ is also a characteristic function! Thus the global hypothesis of Lindeberg's theorem implies the asymptotic relationship between two sequences of characteristic functions. Moreover, we see that $\Psi_n(\theta)$ is associated with a pdf with special convexity properties since it is a convex linear combination of densities like the generic $g(\cdot)$ of section 21.2.3 (especially after equation (12)).

Let us write G_n for the df associated with $\Psi_n(\theta)$ and U for the degenerate df (sometimes called the 'Heaviside unit function') which places all the unit probability on the origin. Then it is plain from (20) that

$$\Phi_n(\theta) \to \exp(-\tfrac{1}{2} \theta^2) \qquad (22)$$

if and only if

$$\Psi_n(\theta) \to 1. \qquad (23)$$

From the continuity theorem for characteristic functions it can be inferred that (22) holds if and only if the sequence of distribution functions $\{G_n\}$ converges weakly to U. This is essentially completes the proof of Lindeberg's theorem although, of course, something must be said to relate the condition '$G_n \underset{w}{\Rightarrow} U$', which the present argument shows to arise in a natural way as the crucial condition for asymptotic normality, and the classical condition (LC).

From (13) coupled with (21) we see that the condition $G_n \underset{w}{\Rightarrow} U$ is equivalent to the statement that, for all $\varepsilon > 0$,

$$\frac{1}{\tau_n^2} \sum_{j=1}^{n} \int_{|x| > \varepsilon \tau_n} (|x| - \varepsilon \tau_n)^2 F_j(\mathrm{d}x) \to 0, \quad \text{as } n \to \infty. \tag{24}$$

Obviously, condition (LC) implies (24). On the other hand, in the integrals in (24), for the range $|x| > 2\varepsilon\tau_n$ it is true that $|x| - \varepsilon\tau_n > \frac{1}{2}|x|$ and so it should be apparent that (24) implies (LC) (with the unimportant change of ε into 2ε in the latter condition). Thus the Lindeberg central limit theorem is proved.

21.2.4 Further remarks

Let us write K_n for the df of S_n/τ_n, with cf Φ_n, and let us continue to write G_n for the df associated with Ψ_n. The argument we have given in the preceding section proves the following theorem, more general that the Lindeberg theorem.

Theorem II Under the global hypothesis of section 21.1 $\{K_n\}$ converges weakly to a limit df K, say, if and only if $\{G_n\}$ converges weakly to a (possibly defective) df G, say. If $\Psi(\cdot)$ be the cf associated with G, then K will have a finite variance given by $\Psi(0)$.

Proof If K_n converges weakly to K then $\Phi_n(\theta)$ tends to $\exp\{-\frac{1}{2}\theta^2\Psi(\theta)\}$ where we must also have $\Psi_n(\theta) \to \Psi(\theta)$. The usual weak-compactness argument will then show that we must have G_n converging weakly to some, possibly defective, df G of which function Ψ is the Fourier–Stieljes transform.

Conversely, if the G_n converge weakly to some G then the $\{\Psi_n(\theta)\}$ tend to a limit $\Psi(\theta)$, the transform of G. Thus $\Psi(\theta)$ must be uniformly continuous and $0 \leq \Psi(0) \leq 1$. Thus $\Phi_n(\theta) \to \Phi(\theta)$ where the limit function must also be uniformly continuous and $\Phi(0) = 1$; the latter claim may easily be understood by letting $\theta \to 0$ in the equation $\Phi(\theta) = \exp\{-\frac{1}{2}\theta^2\Psi(\theta)\}$. From this result it follows that K must always be a proper distribution function,

It is well known, and easily proved, that if

$$\liminf_{\theta=0} \frac{1 - \Re\Phi(\theta)}{\theta^2} < \infty$$

then K, the df associated with Φ, must have a finite variance; and hence, from a representation like (11), but possibly including (for argument's sake) a term corresponding to a first moment, it follows that K must in fact have a zero first moment and a finite variance given by

$$\kappa^2 = \lim_{\theta=0} \frac{1 - \Phi(\theta)}{\frac{1}{2}\theta^2}.$$

Thus the claim $\kappa^2 = \Psi(0)$ also follows from equation $\Phi(\theta) = \exp\{-\frac{1}{2}\theta^2\Psi(\theta)\}$.

The approach to the Central Limit Theorem by means of the derived distributions makes it easy to construct examples of behaviour differing from the familiar convergence predicted by the Lindeberg theorem. We need two simple lemmas, however.

Lemma 2 Let $g(x)$ be an absolutely continuous function in $\mathscr{L}_1(0, \infty)$ such that both $g(x)$ and $|g'(x)|$ are non-increasing, and both tend to zero as $x \to \infty$. Also assume that $|g'(0)| \leq 1$. Then the distribution function $F(x; g) \equiv 1 - |g'(x)|$ on $[0, \infty)$ has its first two moments finite and given by

$$\mu_1(g) = g(0); \quad \mu_2(g) = 2\int_0^\infty g(x)\,\mathrm{d}x$$

and a second derived pdf $f_{(2)}(x)$ such that

$$\tfrac{1}{2}\mu_2(g)f_{(2)}(x) = g(x).$$

The proof of Lemma 2 is straightforward, in view of what has gone earlier in this note, and so we do not give details. Even simpler to prove is the following:

Lemma 3 If $g(x)$) satisfies the conditions of Lemma 2 and a is any constant $(0 < a \leq 1)$, then $g_a(x) \equiv ag(ax)$ also satisfies those conditions, and $\mu_2(g_a) = \mu_2(g)$. Furthermore, suppose g_1 and g_2 both satisfy the conditions of Lemma 2. Then, for any $0 \leq p, q \leq 1$ such that $p + q = 1$, $pg_1 + qg_2$ also satisfies Lemma 2, and

$$\mu_2(pg_1 + qg_2) = p\mu_2(g_1) + q\mu_2(g_2).$$

Plainly, if $g(x)$ on $(0, \infty)$ satisfies Lemma 2, we can use $g(|x|)$ to define $\sigma^2 f_{(2)}(x)$, where $f_{(2)}(x)$ is the second derived pdf of some symmetric pdf $f(x)$ on $(-\infty, +\infty)$. We have written σ^2 for the variance of this pdf $f(x)$. Let us also write $\sigma^2\psi(\theta)$ for the Fourier transform of $g(|x|)$; thus $\psi(\theta)$ is the characteristic function of the second derived pdf. We have, by Lemma 3, the equation

$$\sigma^2 = 2\int_0^\infty g(x)\,\mathrm{d}x,$$

and note that this variance is unchanged by scale changes made upon g.

Example 1 Let $g(|x|)$ and $\psi(\theta)$ be the functions introduced in the preceding paragraph, associated with a pdf $f(x)$ with zero mean and variance σ^2. Then, for $j = 1, 2, \ldots$ we can define the characteristic functions

$$\phi_j(\theta) = 1 - \tfrac{1}{2}\sigma^2\theta^2\psi(\theta\sqrt{j})$$

of random variables with zero means and variances all equal to σ^2. For this example we find

$$\Psi_n(\theta) = \frac{1}{n}\sum_{j=1}^{n}\psi(\theta\sqrt{j}/\sqrt{n}).$$

One can deduce from this that as $n \to \infty$

$$\Psi_n(\theta) \to \int_0^1 \psi(\theta\sqrt{u})du = \Psi(\theta), \text{ say}.$$

The limit cf Ψ is associated with the pdf

$$\frac{1}{\sigma^2}\int_0^1 \frac{g(|x|/\sqrt{u})}{\sqrt{u}}du.$$

Thus we have a situation in which the global hypothesis is satisfied but the condition (LC) is not: yet there is still convergence, though not to a normal limit. The characteristic function of the limiting distribution is

$$\exp\left\{-\tfrac{1}{2}\theta^2\int_1^0 \psi(\theta\sqrt{u})du\right\}.$$

It may be verified from this last formula that the non-normal limit K still has unit variance.

Example 2 We use Lemma 3. Pick $0 < \varpi < 1$ and set,

$$\psi_j(\theta) = \varpi\psi(\theta) + (1 - \varpi)\psi(\theta j),$$

where ψ is the characteristic function introduced in the previous example. In this case $\Psi_n(\theta) \to \varpi$. To see this let us begin by noting that, since $\psi(0) = 1$,

$$\frac{1}{n}\sum_{j=1}^{n}\psi(\theta/\sqrt{n}) \to 1, \quad \text{as } n \to \infty.$$

Further, by the Riemann–Lebesgue lemma, $\psi(\theta) \to 0$ as $|\theta| \to \infty$. Therefore, given a small $\varepsilon > 0$ we can find $\Delta(\varepsilon)$ such that $|\psi(\theta)| < \varepsilon$ for all $|\theta| > \Delta$. From this it is possible to show that, for any fixed θ,

$$\left|\frac{1}{n}\sum_{j=1}^{n}\psi(\theta j/\sqrt{n})\right| \leq \frac{\Delta\sqrt{n}}{n|\theta|} + \varepsilon.$$

Since ε is arbitrary the claim is established. It follows that for this example the limit distribution is normal, but the variance is strictly less than one, depending on our choice of ϖ. There is an example of this phenomenon in Feller (1971).

21.3 CRAMÉR'S THEOREM ON THE NORMAL DISTRIBUTION

21.3.1 General comments

A famous theorem of Harald Cramér states that if two independent random variables have a sum which is normally distributed then each of the two random variables is normally distributed. His original proof of this important result made use of the theory of entire functions of finite order and Hadamard's factorization theorem. Later accounts of the theorem in various books use much the same kind of proof. For this reason it is not usual to prove this result for students in any but advanced courses, and rarely even there, since the necessary knowledge of the theory of entire functions can seldom be assumed. The following proof of Cramér's theorem depends on relatively elementary results from function theory, results with which most graduate students will be familiar. The main tool is the following.

Theorem III Let $f(z)$ be an entire function such that $\overline{f(z)} \equiv f(\bar{z})$, or, equivalently, $f(z)$ is real when z is real; let $\{r_v\}$ be a sequence of reals such that $r_v \to \infty$ as $v \to \infty$; and suppose that for all $|z| = r_v$ and all v,

$$\Re f(z) \leq A |z|^2,$$

where A is some constant. Then $f(z)$ is a polynomial of degree at most two.

Proof For integer $n \geq 1$, by Cauchy's integral formula,

$$f^{(n+2)}(0) = \frac{(n+2)!}{2\pi i} \int_{|z|=r_v} \frac{f(z)}{z^{n+3}} dz. \tag{25}$$

Let us write $f(re^{i\theta}) = u(r, \theta) + iv(r, \theta)$. Because $\overline{f(z)} \equiv f(\bar{z})$, it follows that $u(r, \theta)$ is an even function of θ and $v(r, \theta)$ is an odd function of θ. We can rewrite (7.1) as follows:

$$f^{(n+2)}(0) = \frac{(n+2)!}{2\pi r_v^{n+2}} \int_0^{2\pi} \exp\{-(n+2)i\theta\}[u(r_v, \theta) + iv(r_v, \theta)]d\theta. \tag{26}$$

However, since $f(z)$ is entire, it must be true that

$$\int_{|z|=r_v} z^{(n+1)} f(z) dz = 0 \tag{27}$$

which is the same as

$$\int_0^{2\pi} \exp\{(n+2)i\theta\}[u(r_v, \theta) + iv(r_v, \theta)]d\theta = 0. \tag{28}$$

If we use the known parity properties of $u(r, \theta)$ and $v(r, \theta)$, it follows that

$$\int_{-2\pi}^0 \exp\{-(n+2)i\theta\}[u(r_v, \theta) - iv(r_v, \theta)]d\theta = 0. \tag{29}$$

Plainly the range of integration for the last integral can be changed to $(0, 2\pi)$ without changing the value of that integral. Thus (29) and (26) can be combined to yield the result:

$$f^{(n+2)}(0) = \frac{(n+2)!}{\pi r_v^{n+2}} \int_0^{2\pi} \exp\{-(n+2)i\theta\} u(r_v, \theta) d\theta. \tag{30}$$

Since we are given that $|u(r_v, \theta)| \leq A r_v^2$, it follows immediately from (30) that

$$|f^{(n+2)}(0)| \leq \frac{2(n+2)!}{r_v^n} A. \tag{31}$$

We can now let $r_v \to \infty$ and deduce that $f^{(n+2)}(0) = 0$ for all integer $n \geq 1$. This proves the theorem.

Now consider the main problem. Let X_1, X_2 be the independent random variables whose sum $X_1 + X_2$ is normally distributed. With no loss of generality we shall suppose $X_1 + X_2$ has zero mean and variance 2. We shall write v_1, v_2, for the medians of the random variables X_1, X_2, and $F_1(x)$, $F_2(x)$, respectively, for their distribution functions. We shall make use of the Laplace–Stieltjes integrals (which are nothing but expectations) denoted by

$$F_j^*(\zeta) = \int_{-\infty}^{\infty} \exp(-\zeta x) dF_j(x), \quad j = 1, 2, \tag{32}$$

where $\zeta = \xi + i\eta$. We shall also make use of the easily proved, and well-known, result that if the integral (31) converges for all *real* values of ζ then $F_j^*(\zeta)$ is an entire function.

Evidently, for $\xi \geq 0$,

$$F_2^*(\xi) \geq \int_{-\infty}^{v_2} \exp(-\xi x) dF_2(x) \geq 1/2 \exp(-v_2 \xi), \tag{33}$$

while, for $\xi < 0$, we have

$$F_2^*(\xi) \geq \int_{v_2}^{\infty} \exp(-\xi x) dF_2(x) \geq 1/2 \exp(-v_2 \xi). \tag{34}$$

Thus, whatever the signs of ξ and v_2,

$$F_2(\xi) \geq \frac{1}{2}\exp(-|v_2\xi|). \tag{35}$$

Since $X_1 + X_2$ is $\mathcal{N}(0, 2)$ we have that

$$F_1^*(\zeta)F_2^*(\zeta) = \mathscr{E}\exp\{-\zeta(X_1 + X_2)\} = \exp(\zeta^2), \tag{36}$$

so that, from (35), we deduce

$$F_1^*(\xi) \leq 2\exp(\xi^2 + |v_2\xi|). \tag{37}$$

From (37) we see that $F_1^*(\zeta)$ must be an entire function of ζ, since its defining integral converges for all real values of ζ. Moreover, from (36) it is plain that $F_1^*(\zeta)$ cannot have a zero in the finite part of the plane. We can also deduce from (37) as follows:

$$\left|\int_{-\infty}^{\infty}\exp(-\zeta x)\mathrm{d}F_1(x)\right| \leq \int_{-\infty}^{\infty}\exp(-\xi x)\mathrm{d}F_1(x) \leq 2\exp(|\zeta|^2 + |v_2\zeta|). \tag{38}$$

Because $F_1^*(\zeta)$ is entire and zero-free, we can introduce the logarithm $f_1(\zeta) = \log F_1^*(\zeta)$, selecting the principal value of the logarithm, which is real when $F_1^*(\zeta)$ is real and positive, and claim that $f_1(\zeta)$ is also an entire function. But from (38) it follows that, for all large ζ,

$$\Re f_1(\zeta) = O(|\zeta|^2). \tag{39}$$

Also, it is not hard to see that $f_1(\zeta)$ also has the property of being real when ζ is real. Thus we can apply Theorem III to $f_1(\zeta)$ to deduce that it must be a polynomial of degree at most two. The rest of the proof follows upon obvious lines; it is now easy to show that X_1 has a normal distribution.

REFERENCES

Feller, W. (1971) *An Introduction to Probability Theory and Its Applications*, 2nd edn, New York, John Wiley & Sons.

Khinchin, A. I. (1948) *Asymptotische Gesetze des Wahrscheinlichkeitsrechnung*, New York, Springer-Verlag.

Lindenberg, J. W. (1992) Eine reue Herleitung des Exponetialgesetzes in der Wahrscheinlichkeitsrechrung, *Math. Zeitschrift*, **15**, 211–25.

Rosenblatt, M. (1974) *Random Processes*, 2nd edn, New York: Springer-Verlag.

Smith, W. L. (1953) Stochastic sequences of events, PhD thesis, Cambridge University.

Smith, W. L. (1959) On the cumulants of renewal processes, *Biometrika*, **46**, 1–29.

Smith, W. L. (1966) A theorem on functions of characteristic functions and its application to some renewal theoretic random walk problems, *Proc. Fifth Berkeley Symposium*, vol. II, Pt. 2, Berkeley, California, University of California Press.

Trotter, H. F. (1959) An elementary proof of the Central Limit Theorem, *Archiv der Mathematik*, **10**, 226–34.

302 Meadow Lane
Chapel Hill
NC 27514
USA

CHAPTER 22

Hierarchical Priors and Mixture Models, With Application in Regression and Density Estimation

Mike West[†], Peter Müller[†] and Michael D. Escobar[‡]
[†]*Duke University, Durham, North Carolina*
[‡]*Carnegie Mellon University, Pittsburgh*

22.1 INTRODUCTION

In his 1972 review of Bayesian statistics, Dennis Lindley identified as a success story for Bayesian ideas the advances made in problems of many parameters and the growth of what is now referred to as Bayesian hierarchical modelling (Lindley 1972, section 8). In that same monograph, Lindley identified non-parametrics as an area notable for lack of Bayesian progress, bemoaning the fact that non-parametric statistics was a 'subject about which the Bayesian method is embarrassingly silent' (Lindley 1972, section 12.2). Our purpose here is to develop and review recent work with mixture priors that has substantially contributed to both hierarchical modelling and non-parametrics, the latter focused on problems usually referred to as non-parametric density estimation.

It is certainly the case that the wide application of hierarchical models is one of the major success stories of modern Bayesian statistics. In the early 1990s, we are experiencing tremendous growth in serious applied hierarchical modelling work, substantially driven by the invigorating

Aspects of Uncertainty edited by P. R. Freeman and A. F. M. Smith. © 1994 John Wiley & Sons Ltd.

breakthroughs in Bayesian computation via Markov chain simulation (Gelfand and Smith 1990; Gelfand, et al. 1990). Simultaneously, these computational methods allow development and application of data and prior models that significantly extend the scope of more traditional, mathematically tractable forms, providing scope for closer representation of real-world problems, routine sensitivity and robustness analysis. For example, long-suffered constraints to conjugate prior distributions in hierarchical models (Smith, in discussion of DuMouchel and Harris 1983; West 1984, 1985) can be relieved by the use of mixtures of conjugate priors with ease (Escobar and West 1992). Mixture priors, especially Dirichlet mixtures (Antoniak 1974; Escobar and West 1991), have also opened the way to serious Bayesian developments in (so-called) non-parametric modelling and density estimation and smooth regression estimation (Erkanli et al. 1992). It is our purpose in this chapter to exhibit a general framework for hierarchical linear modelling and density estimation, to show how posterior computations via Markov chain simulations can be routinely applied and to provide illustrations in each context.

Section 22.2 provides a rather general theoretical setting and summarizes key features of multivariate data models with hierarchical mixture priors. Section 22.3 discusses Markov chain simulation methods in these models, with special emphasis on models centred around traditional normal structures. Section 22.4 concerns an application in hierarchical regression, highlighting the use of mixture priors for robustness and sensitivity analysis, and section 22.5 develops an application to multivariate density estimation.

22.2 HIERARCHICAL MODELS WITH MIXTURE PRIORS

Consider a sample of n vector observations y_i in r dimensions, modelled as arising from a collection of distributions $y_i \sim F_i(\cdot \mid \phi_i, \sigma)$, where ϕ_i is a vector of parameters specific to case i, σ is a vector of parameters common to each case, F_i is of specified form and completely known when ϕ_i and σ are known and the y_i are conditionally independent given F_i, the ϕ_i and σ. Each F_i is assumed to have a density function $f_i(\cdot \mid \phi_i, \sigma)$. A regression example has $f_i(\cdot \mid \cdot)$ multivariate normal with mean $X_i \mu_i$ and variance matrix $\sigma^2 \Sigma_i$; here each X_i is a known regression matrix, $\phi_i = \{\mu_i, \Sigma_i\}$, and σ is a common scale parameter. Non-linear/normal regressions are obvious alternative examples. Generally the interpretation and dimensions of the ϕ_i may differ, though often they represent case-specific realizations of parameters with common interpretation across i; we make that assumption here so that the ϕ_i enter into the $f_i(\cdot \mid \cdot)$ through a common functional form, and they have common dimension p, say. A typical hierarchical model assumes the case-specific parameters ϕ_i are conditionally independently distributed according to some prior distribution $G(\cdot \mid \pi)$; here π represents a collection of hyperparameters and a three-stage hierarchical model is completed by specifying an appropriate hyperprior distribution for π.

Dirichlet process mixture models are based on Dirichlet process priors for the primary parameters ϕ_i. Such a model assumes that the prior distribution function G itself is uncertain, drawn from a Dirichlet process $G \sim D(\alpha G_0)$ (in standard notation, such as in Antoniak 1974). Here $E(G) = G_0$ is the *base prior*, and $\alpha > 0$ the scalar precision parameter. This is conditional on hyperparameters π which will include the precision α and determine the prior expectation G_0. In this section we suppress dependence on both σ and π in the notation, for clarity; it should be borne in mind that the following discussion is all conditional on both σ and π.

The following properties and reinterpretation of this model clearly elucidate the resulting structure (see Antoniak 1974, and West 1990).

1. Any realization of n case-specific parameters ϕ_i generated from G lie in a set of $k \leq n$ distinct values, denoted by $\theta = \{\theta_1, \ldots, \theta_k\}$.
2. The θ_i are a random sample from $G_0(\cdot)$.
3. $k \leq n$ is drawn from a prior distribution that is implicitly determined and depends only on n and the precision $\alpha > 0$; it is Poisson-like, though fatter tailed, with asymptotic mean $\alpha \log(1 + n/\alpha)$ for $n \to \infty$ (Antoniak 1974; West 1992).
4. Given k, the n values ϕ_i are selected from the set θ according to a uniform multinomial distribution.

Conditional on k, introduce indicators $\mathscr{S}_i = j$ if $\phi_i = \theta_j$ so that, given $\mathscr{S}_i = j$ and θ, $y_i \sim F_i(\cdot \mid \theta_j)$. The *configuration* $\mathscr{S} = \{\mathscr{S}_1, \ldots, \mathscr{S}_n\}$ (West 1990; MacEachern 1992) determines a one-way classification of the data $Y = \{y_1, \ldots, y_n\}$ into k distinct groups or clusters; the $n_j = \#\{\mathscr{S}_i = j\}$ observations in group j share the common parameter value θ_j. Write I_j for the set of indices of observations in group j, $I_j = \{i : \mathscr{S}_i = j\}$ and let $Y_j = \{y_i : \mathscr{S}_i = j\}$ be the corresponding group of observations. Then, as the θ_j are a random sample from the base prior G_0, posterior analysis devolves to a collection of k independent analyses; specifically, the θ_j are conditionally independent with posterior densities

$$p(\theta_j \mid Y, \mathscr{S}, k) \equiv p(\theta_j \mid Y_j, \mathscr{S}, k) \propto \left(\prod_{i \in I_j} f_i(y_i \mid \theta_j) \right) dG_0(\theta_j), \qquad (1)$$

for $j = 1, \ldots, k$.

A further feature of the Dirichlet structure of importance in posterior inference is the mechanism for allocating observations to groups, summarized through conditional distributions for individual ϕ_i given existing configurations of other parameters. For $i = 1, \ldots, n$, write

$$\phi^{(i)} = \{\phi_1, \ldots, \phi_{i-1}, \phi_{i+1}, \ldots, \phi_n\},$$

and denote by $\mathscr{S}^{(i)}$ the configuration of $\phi^{(i)}$ into some $k^{(i)}$ distinct values, with $n_j^{(i)}$ taking common value $\theta_j^{(i)}$. Then

$$(\phi_i | \phi^{(i)}, \mathscr{S}^{(i)}, k^{(i)}) \sim (\alpha + n - 1)^{-1}\alpha G_0 + (\alpha + n - 1)^{-1}\sum_{j \neq i} \delta(\phi_j)$$

$$\sim (\alpha + n - 1)^{-1}\alpha G_0 + (\alpha + n - 1)^{-1}\sum_{j=1}^{k^{(i)}} n_j^{(i)}\delta(\theta_j^{(i)}), \quad (2)$$

where $\delta(x)$ denotes the distribution degenerate at $\phi_i = x$. This shows that ϕ_i is distinct from the other parameters and drawn from G_0 with chance $\alpha/(\alpha + n - 1)$, otherwise it is chosen from existing realised values according to a multinomial allocation with probabilities proportional to existing groups sizes $n_j^{(i)}$.

Extension of n to $n + 1$ in (2) is directly relevant in predicting a further case $i = n + 1$; this requires the distribution of $(\phi_{n+1} | \phi, \mathscr{S}, k)$ which, incidentally, is just the posterior mean $E(G | \phi, \mathscr{S}, k)$, i.e.

$$(\phi_{n+1} | \phi, \mathscr{S}, k) \sim E(G | \phi, \mathscr{S}, k) \equiv (\alpha + n)^{-1}\alpha G_0 + (\alpha + n)^{-1}\sum_{i=1}^{k} n_i\delta(\theta_i). \quad (3)$$

where $\delta(\theta_i)$ denotes the distribution degenerate at $\phi_{n+1} = \theta_i$. As a result, predictive inference for future data y_{n+1} focuses on conditional predictive densities

$$(y_{n+1} | \phi, \mathscr{S}, k) \sim (\alpha + n)^{-1}\alpha F_{n+1}(\cdot | \theta_{k+1}) + (\alpha + n)^{-1}\sum_{i=1}^{k} n_i F_{n+1}(\cdot | \theta_i), \quad (4)$$

where θ_{k+1} is a further independent draw from G_0.

In some contexts where alternatives to the common 'shrinkage' effects induced through the use of a more traditional prior G_0 are sought, this mixture prior construction has appeal; it reflects automatic adaptive shrinkage induced by the implicit grouping of subsets of parameter, and provides mechanisms for assessing differences among the parameters. The problems of extreme or outlying parameters potentially corrupting inference (West 1984, 1985; O'Hagan 1988) are avoided.

22.3 POSTERIOR COMPUTATIONS

Escobar (1988) introduced Monte Carlo Markov chain methods in simplified versions of the above model, later extended to univariate normal mixtures for density estimation in Escobar and West (1991). Further extensions and applications in Escobar and West (1992), West and Cao (1992), Turner and West (1993), and elsewhere, are based on normal data distributions F_i and conditionally conjugate base priors $G_0(\cdot)$. Recent work of Erkanli, et al. (1992) uses a refined Monte Carlo Markov chain algorithm based on original work of MacEachern (1992), and also adopts non-conjugate base priors. The general Gibbs sampling strategy proposed by MacEachern is outlined here, followed by comments specific to normal linear models and assumed forms of G_0 and hyperparameter estimation.

22.3.1 Basic Gibbs sampling

As usual in Gibbs sampling, we identify collections of complete conditional posterior distributions that determine the full posterior for all parameters $\phi = \{\phi_1, \ldots, \phi_n\}$. (Note that we are continuing under the assumption that σ and π are known, which is relaxed below.) Note that knowledge of ϕ is equivalent to knowledge of k, $\mathscr{S}$ and $\theta = \{\theta_1, \ldots, \theta_k\}$. Original simulation methods in Escobar (1988) and Escobar and West (1991) simulate values in ϕ via posteriors based on the conditional priors (2). MacEachern (1992) suggested a more efficient algorithm that works in terms of the theoretically equivalent parameters k, $\mathscr{S}$ and $\theta = \{\theta_1, \ldots, \theta_k\}$. MacEachern's algorithm is generally preferable and recommended, for reasons given in his paper; it applies quite generally, as we illustrate here. From (2), we can immediately deduce sets of conditional posteriors

$$(\phi_i \mid Y, \phi^{(i)}, \mathscr{S}^{(i)}, k^{(i)}) \sim q_{i,0} G_{i,0} + \sum_{j=1}^{k^{(i)}} q_{i,j} \delta(\theta_j^{(i)}), \quad (5)$$

where the chances $q_{i,j}$ are given by

$$q_{i,j} = \begin{cases} c \cdot \alpha \cdot h_i(y_i) & \text{if } i = 0, \\ c \cdot n_j^{(i)} \cdot f_j(y_i \mid \theta_j) & \text{if } j > 0, \end{cases}$$

where $G_{i,0}$ denotes the posterior obtained by updating the prior G_0 via the likelihood function $f_i(y_i \mid \phi_i)$, namely

$$dG_{i,0}(\phi_i) \propto f_i(y_i \mid \phi_i) dG_0(\phi_i)$$

and subject to normalisation, $h_i(y_i)$ is a weight obtained via

$$h_i(y_i) = \int f_i(y_i \mid \phi_i) dG_0(\phi_i), \quad (6)$$

i.e. the marginal density of y_i, evaluated at the realized datum, under the base prior for ϕ_i and c is a constant of normalization.

Equation (5) immediately implies conditional posteriors for the configuration indicators, namely

$$P(\mathscr{S}_i = j | Y, \phi^{(i)}, \mathscr{S}^{(i)}, k^{(i)}) = q_{i,j}. \tag{7}$$

We may now successively sample sets of values k, $\mathscr{S}$ and $\theta = \{\theta_1, \ldots, \theta_k\}$ by iterating through the following sequence:

(a) Given old values θ and $\mathscr{S}$, generate a new configuration by sequentially sampling indicators from the posteriors (7), successively simulating and substituting $\mathscr{S}_1, \mathscr{S}_2, \ldots$; for any index i such that $\mathscr{S}_i = 0$, draw a new ϕ_i from $G_{i,0}$ in (5).

(b) Given k and $\mathscr{S}$, generate a new set of parameters θ by sampling each new θ_j from the relevant component posterior in (1), namely $(\prod_{r \in I_j} f_r(y_r | \theta_j)) dG_0(\theta_j)$.

Successive simulations eventually lead to sampled values drawn approximately from the joint posterior $p(k, \mathscr{S}, \theta | Y)$. General issues of convergence are covered in MacEachern and Müller (1994). Inference may be based on histograms of sampled values of individual parameters, or the more usual refinements based on Monte Carlo averages of analytically available conditional distributions and moments; the above reference gives further details. For example, averaging (4) with respect to simulated values of $\theta_1, \ldots, \theta_{k+1}$ provides an appropriate Monte Carlo estimate of the actual predictive density function $p(y_{n+1} | Y)$. Each such sampled mixture is based on appropriate values of $\theta_1, \ldots, \theta_k$ generated at the corresponding Gibbs iteration, and on an appropriate values of θ_{k+1} drawn directly from G_0.

The above discussion is all conditional on the common parameter σ. In some models, there will be no such parameter and nothing more need be said. If common parameters are included in the data model, then the simulation analysis extends as follows to include steps to generate σ values from appropriate conditional posteriors. Given a prior $p(\sigma)$, we have

$$p(\sigma | Y, \theta, \mathscr{S}, k) \propto p(\sigma) \prod_{i=1}^{n} f_i(y_i | \mu_i, \sigma) = \prod_{j=1}^{k} \prod_{i \in I_j} f_i(y_i | \theta_j, \sigma). \tag{8}$$

The successive substitution sampling scheme above then extends to include a simulated value of σ from this distribution, conditioning on the most recently sampled values of θ, $\mathscr{S}$ and k at each step.

22.3.2 Extension to hyperparameters

Usually the hyperparameters π will be uncertain, having a specified prior. In Escobar and West (1991) and some of the other references, for example, extensions to include π in analyses are detailed for normal models. More generally, the structure and consequent extensions of simulation analyses

HIERARCHICAL PRIORS AND MIXTURE MODELS 369

are detailed here. Discussion in section 22.3.1 is all conditional on π which includes the Dirichlet precision α and other parameters in G_0; call the latter parameters γ so that $\pi = \{\alpha, \gamma\}$ and now make explicit the dependence on hyperparameters in the notation $G_0(\cdot \mid \gamma)$. Distributions (1) and (7) simulated in steps (a) and (b) are conditional on π. Gibbs sampling is directly extensible to incorporate π simply by adding a further step to draw from an appropriate conditional posterior. Generally, the relevant distribution is $(\pi \mid Y, \phi, \sigma) \equiv (\pi \mid Y, \theta, \mathcal{S}, \sigma, k)$. This may be developed, as in the special cases in West (1992) and Escobar and West (1991), as follows.

First, we assume that α and γ are a priori independent with specified density

$$p(\alpha, \gamma) = p(\alpha)p(\gamma).$$

It follows, as in the above references, that α and γ are also conditionally independent given Y, θ, $\mathcal{S}$, σ and k; hence the hyperparameters α and γ can be considered separately.

For α, the Dirichlet process structure is such that only k is relevant in inference about α, so that

$$p(\alpha \mid Y, \theta, \mathcal{S}, \sigma, k) \equiv p(\alpha \mid k).$$

Developments in West (1992), applied in Escobar and West (1991) and other papers cited above, show how this posterior $p(\alpha \mid k)$ may be represented as a mixture of gamma distributions if the prior $p(\alpha)$ is a mixture of gammas. Applications to date have assumed a single gamma prior, in which case the posterior is easily simulated. Examples of applications appear in the above references.

For γ, note that this parameter enters only through G_0; then, using (1),

$$p(\gamma \mid Y, \theta, \mathcal{S}, \sigma, k) \equiv p(\gamma \mid \theta, \sigma, k) \propto p(\gamma) \prod_{j=1}^{k} \mathrm{d}G_0(\theta_j \mid \gamma). \qquad (9)$$

Base priors G_0 that enable direct sampling of posteriors (9) may be sought in any given application. Assume this is the case; then knowledge of Y, θ, $\mathcal{S}$, σ and k, reduced to the sufficient information set θ and k, leads immediately into simulation of a value for $\{\alpha, \gamma\}$, as follows:

(c) Given k, draw a new value of α from $p(\alpha \mid k)$ (West 1992; Escobar and West 1991); given k and θ draw a new value of γ from (9).

The resulting value for $\pi = \{\alpha, \gamma\}$ is then used in conditioning at (a) and (b) in the next iteration.

22.3.3 Normal linear models: conjugate base priors

The above development is quite general, and certain applications with non-normal models have been implemented. Most of the experience to date is, however, with normal linear models in regression and density estimation, and we now specialize to that context, assuming $f_i(\cdot \mid \phi_i)$ to be normal with mean $X_i\mu_i$ and variance matrix Σ_i here X_i is a known $r \times p$ design matrix, μ_i an uncertain p-vector, and Σ_i an uncertain $p \times p$ variance matrix. Also, $\phi_i = \{\mu_i, \Sigma_i\}$. Note particularly that there is no common, uncertain parameter σ in this case (we note an extension to include a common scale parameter at the end of this subsection). In the special case of random sampling, where multivariate density estimation is the focus, $r = p$ and X_i is the $p \times p$ identity matrix.

Immediately, equation (1) suggests a conjugate normal/inverse Wishart form for the base prior G_0. With such a form, posterior analysis for θ involves a set of k independent normal regression analyses with traditional conjugate priors. Resulting independent posteriors (1) for the individual parameters θ_i are then normal/inverse Wishart too. Full details appear in Escobar and West (1992). Thus (1) may be easily sampled. Furthermore, in the steps to simulate $\mathscr{S}$ and (as a result) k from (7), the required weights $q_{i,j}$ are easily computed. For each $j > 0$, this involves the evaluation of a multivariate normal density function; for $j = 0$ the required expression (6) reduces to a multivariate T density function, also easily evaluated.

In this context the hyperparameters γ determine the chosen normal/inverse Wishart base prior G_0 for each μ_i and Σ_i. Explicitly, such a prior has the form

$$\mu_i \sim N(m, b\Sigma_i) \quad \text{and} \quad \Sigma_i^{-1} \sim W(s, (sS)^{-1}) \tag{10}$$

with $\gamma = \{m, b, s, S\}$; here m is the p-vector prior mean for μ_i, b and s are positive scalars, S is a $p \times p$ variance matrix. Here $s \geq p$ is the Wishart degrees of freedom index, and the notation corresponds to density function

$$p(\Sigma_i^{-1} \mid s, S) = c \mid sS \mid^{s/2} \mid \Sigma_i \mid^{-(s-p-1)/2} \exp(-\operatorname{trace}(sS\Sigma_i^{-1})/2)$$

with $E(\Sigma_i^{-1}) = S^{-1}$; thus S is a prior estimate for Σ_i (the prior harmonic mean).

Early implementations by the authors use this structure (Escobar and West 1992), which is illustrated in section 22.4 below, often fixing γ at specified prior values. Realistically, uncertainty about these quantities is almost surely evident, so we turn to the issue of suitable priors for γ. In all cases, we assume s to be specified. For $\{m, b, S\}$, the structure of equation (9) is suggestive; we simply note that

1. A conjugate normal/inverse gamma prior for m and b implies a conjugate conditional posterior $p(m, b \mid \theta, s, S)$; and
2. A Wishart prior for S implies a Wishart conditional posterior $p(S \mid \theta, m, b, s)$.

Under such priors, with s fixed and $\{m, b\}$ a priori independent of S, the conditional posteriors may be sampled—first generating m, b given θ and the previously sampled value of S, then sampling S based on θ and the new values of m and b. This is easily implemented. Typically, a rather diffuse prior may be adopted for m and b—indeed, West and Cao (1993) use the traditional reference prior $p(m, b) \propto b^{-1}$—as long as the prior is relatively informative for S. All applications to date assume s specified; in applications of West and Cao (1993), and Turner and West (1993), substantial prior information is available in the assessment of s and S. In other applications, anticipated ranges of data values may be used to assess the prior for S, while the Wishart degrees of freedom parameter s will typically be taken in the low integers to reflect high uncertainty. Some further discussion of prior assessment for γ appears in sections 22.4 and 22.5 below.

In some applications, a common scale parameter (or variance matrix) may be desirable in the individual regression equations, so that extension of the above development is needed. Assuming a common scale factor $\sigma > 0$ implies $f_i(\cdot | \phi_i, \sigma)$ is normal with mean $X_i \mu_i$ and variance matrix $\sigma^2 \Sigma_i$. Now the above development of the simulation sequences applies all conditional on σ, and the analysis simply needs to be supplemented with simulation of σ values from the appropriate conditional posteriors in (8). We simply note that, if $p(\sigma^2)$ is inverse gamma, then (8) is immediately an inverse gamma, and so easily simulated too, neatly completing the Monte Carlo iteration.

22.3.4 Normal linear models: non-conjugate base priors

Most recent applications have adopted a non-conjugate base prior specification, described in this subsection, and used in illustrations in sections 22.4 and 22.5 below. The conjugate form has implications that are sometimes undesirable. Under the above prior specification, suppose the hyperprior for m and b is diffuse relative to the likelihood function in (9) when θ is known; then, for example, the conditional posterior mean for m is essentially a matrix weighted average of the μ_i with matrix weights proportional to Σ_i^{-1}. Consider a context in which one of the μ_i is truly quite separate from the rest, and that the number n_i of associated observations in that group is very small compared to n. The posterior for Σ_i will support small variation, hence high precision Σ_i^{-1}; as a result, the term $\Sigma_i^{-1} \mu_i$ in the conditional mean of m will be heavily influential. Indeed, the entire posterior for m and b given θ will be overly influenced by such an extreme and small group. This is a feature induced by the conjugate form of G_0, and is alleviated by assuming an alternative, non-conjugate and independent form

$$\mu_i \sim N(m, B) \quad \text{independently of} \quad \Sigma_i^{-1} \sim W(s, (sS)^{-1}), \tag{11}$$

where now B is a variance matrix, and so $\gamma = \{m, B, s, S\}$. The development of the previous sections can be followed through with certain obvious modifications to the iterative sampling of conditional posteriors, described below.

1. The component posteriors for the θ_i in (1) are no longer conjugate forms, and so each $\theta_i = \{\mu_i^*, \Sigma_i^*\}$ is updated by first sampling μ_i^* conditional on a value of Σ_i^*, and then Σ_i^* conditional on μ_i^*. The conditional posterior for ($\mu_i^* | \Sigma_i^*, Y_i, \mathscr{S}, \sigma, k, \gamma$) is multivariate normal, and the distribution ($\Sigma_i^* | \mu_i^*, Y_i, \mathscr{S}, \sigma, k, \gamma$) is inverse Wishart. Note that iterating these two draws sufficiently often would simulate an approximate draw from ($\mu_i^*, \Sigma_i^* | Y_i, \mathscr{S}, \sigma, k, \gamma$). While not generally necessary, this might be desirable if the previous draw for Σ_i^* was based on a very different configuration.

2. In sampling configurations using (7) at step (b) of the iterations, the new component probability $q_{i,0}$ requires evaluation of the integral in (6). This can be approximated numerically using quadrature methods or Monte Carlo integration. The latter may be implemented by replacing the integral in (6) with an average over draws from the base prior G_0. Alternatively, the integration in (6) can be partially performed analytically over μ_i.

For the components m, B and S of the hyperparameter γ, the above development is modified usefully in non-conjugate form as follows: m has a normal prior, hence a normal posterior when B and S are fixed; B has an inverse Wishart prior, hence an inverse Wishart posterior when m and S are fixed; the prior and conditional posteriors for S are Wishart, as in the previous section.

As noted at the end of section 22.3.3, some applications involve a common scale parameter multiplying the individual variance matrices Σ_i. This is the case in an example in the following section. The immediate extension of the simulation scheme to incorporate a common scale factor and extend the simulation analysis applies here as discussed in section 22.3.3: if the model is rephrased in terms of $(y_i | \phi_i, \sigma) \sim N(X_i\mu_i, \sigma^2\Sigma_i)$, the above development applies conditional on σ, and values of σ are simulated from the conditional posteriors in (8); if $p(\sigma^2)$ is inverse gamma, then (8) is also inverse gamma, and so trivially simulated. Other priors are possible, of course, as (8) is a univariate density.

22.4 A REGRESSION EXAMPLE

In the development of mixture priors in hierarchical models of Escobar and West (1992), one particular theme was the use of Dirichlet process priors as a way of modelling uncertainty about the functional form of the usually assumed prior G. In particular, traditional normal hierarchical models assume G to be either conjugate normal/inverse Wishart or independent normal (for regression vectors) and inverse Wishart (for variance matrices) (e.g. Gelfand *et al.* 1990). The Dirichlet mixture structure elaborates this traditional assumption by using one such form for the baseline prior G_0 so that G is initially expected to be in some appropriate neighbourhood of the usual form. Posterior inference about G then permits sensitivity analysis to assess the traditional assumption, and resulting posterior and predictive inferences will not be so dependent on the assumed forms if they are a posteriori revealed to be inappropriate. We illustrate these concepts here through a direct application of the hierarchical regression analysis in section 22.3.4. In addition to providing an interesting example and sensitivity analysis, we elaborate the structure of the conditional posteriors used in Gibbs sampling in some detail.

Gelfand *et al.* (1990) considered analysis of a growth curve problem where individual growth curves of several subjects are modelled using a standard linear regression with normal error. The assumed priors for the regression coefficients are there taken to be conditionally normal—the traditional baseline analysis. Our analysis makes this assumption for the actual baseline distribution G_0, supposing that the prior G is 'near' a bivariate normal, and hence admitting uncertainty about the between-group population distribution in these models. The data set in Gelfand *et al.* contains the weights of rats ($n = 30$ taken from the control group in that paper), measured at the same five time points. Let $y_{i,j}$ denote the jth measurement on the ith rat, collected in the vector $y_i = (y_{i,1}, \ldots, y_{i,5})'$, for $i = 1, \ldots, n = 30$. Individual growth curves are modelled as

$$(y_i \mid \phi_i, \sigma) \sim N(X_i \mu_i, \sigma^2 I),$$

where X_i is 5×2 design matrix with jth row $(1, x_{i,j})$, and $x_{i,j}$ is the age of rat i at the jth time point, and I is the 5×5 identity matrix. Note that the generality of section 22.3 is not needed here since the individual variance matrices Σ_i are degenerate at $\Sigma_i = I$ and so the corresponding posterior simulation steps are vacuous. Also, $\phi_i \equiv \mu_i$ for each i, and we note that the rats have a common within-group variance σ^2.

Gelfand *et al.* (1990) use the standard conjugate assumptions to model the μ_i, adopting the prior $(\mu_i \mid m, B) \sim N(m, B)$ independently. Their prior for the parameters $\{\sigma, m, B\}$ assumes mutual independence and conditional conjugacy, with

$$\sigma^2 \sim IG\left(\frac{u}{2}, \frac{uU}{2}\right)$$

$$m \sim N(a, A)$$

$$B^{-1} \sim W(c, (cC)^{-1}),$$

where IG denotes the inverse gamma distribution, and the parameters u, U, a, A, c and C determining these priors will be specified in any analysis.

To sample the posterior distribution, Gelfand et al. performed a Markov chain Monte Carlo defined by the following conditional distributions:

$$(\sigma^2 \mid Y, \mu) \sim IG\left(\frac{u+5n}{2}, \frac{1}{2}\left[uU + \sum_{i=1}^{n}(y_i - X_i\mu_i)'(y_i - X_i\mu_i)\right]\right),$$

$$(\mu_i \mid Y, m, B, \sigma) \sim N(D_i(B^{-1}m + \sigma^{-2}X_i'Y_i), D_i),$$

$$(m \mid \mu, B) \sim N(V(A^{-1}a + B^{-1}\sum_{i=1}^{n}\mu_i), V),$$

$$(B^{-1} \mid m, \mu) \sim W\left\{c + n, \left[cC + \sum_{i=1}^{n}(\mu_i - m)(\mu_i - m)'\right]^{-1}\right\},$$

where $D_i = (B^{-1} + \sigma^{-2}X_i'X_i)^{-1}$ and $V = (A^{-1} + nB^{-1})^{-1}$. Note that the conditioning of all distributions here explictly recognizes only those quantities that are relevant; for example, the full, required distribution for $(\sigma^2 \mid Y, \mu, m, B)$ does not depend on m and B so they are not written in the conditioning. This implicit simplication is also used throughout the following discussion.

Consider uncertainty in the distributional form of the prior on μ_i modelled by assuming the above baseline prior within a Dirichlet process framework. Then $\phi_i = \mu_i$, $\gamma = \{m, B\}$, and $\{\theta_1, \ldots, \theta_k\}$ represents the k distinct values among the full set of μ_i's. The parameter σ^2 is common across all observations and we will continue to use an inverse gamma prior for this parameter. From section 22.3 and using equations (5)–(7), we deduce the specific forms of relevant conditional posterior distributions detailed below. On a point of notation, we are replacing ϕ by μ throughout (e.g. in equations (5) and following) so that

$$\mu^{(i)} = \{\mu_1, \ldots, \mu_{i-1}, \mu_{i+1}, \ldots, \mu_n\}.$$

We then have configurations generated according to chances $q_{i,j}$ in equation (7). In this specific context, these are as follows:

- $q_{i,0}$ is proportional to α times the probability density function of a multivariate normal distribution, evaluated at y_i with mean X_im and covariance matrix $X_iBX_i' + \sigma^2 I$;
- for $j > 0$, $q_{i,j}$ is proportional to the probability density function of a multivariate normal, evaluated at y_i, with mean $X_i\mu_j$ and covariance matrix $\sigma^2 I$;
- the chances are normalized so that $q_{i,0} + \sum_{j \neq i} q_{i,j} = 1$.

Given a configuration and the corresponding k distinct means θ, the remaining conditional posteriors are as follows.

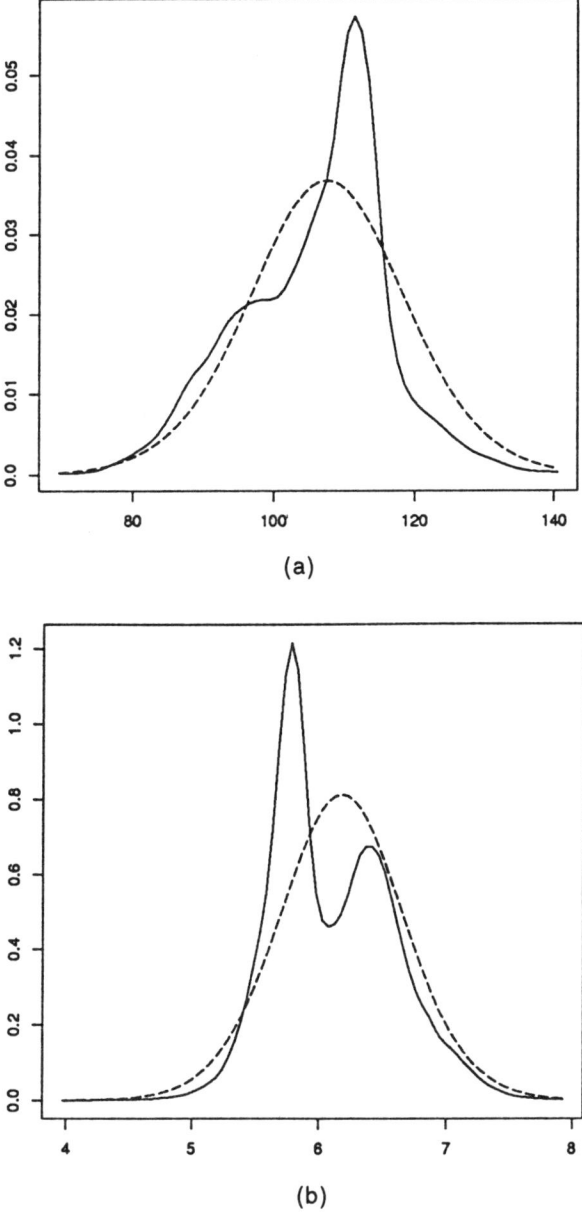

Figure 22.1 Marginal predictive densities for the two elements of the regression vector $\mu_{n+1} = (\mu_{n+1,1}, \mu_{n+1,2})'$ of the growth curve for a future rat. The full lines represent densities from the mixture analysis, to be compared with the dashed curves from the traditional or baseline analysis of Gelfand *et al* (1990). (a) $p(\mu_{n+1,1} | Y)$; (b) $p(\mu_{n+1,2} | Y)$

- σ^2 is sampled from just the inverse gamma detailed above—though the μ_i cluster into k distinct values, the sum of squares term in $p(\sigma^2 \mid Y, \mu)$ does not explicitly recognize the fact;
- $(m \mid \mu, B) \sim N(V^*(A^{-1}a + B^{-1}\sum_{j=1}^{k}\theta_j), V^*)$, with $V^* = (A^{-1} + kB^{-1})^{-1}$; and
- $(B^{-1} \mid \mu, m) \sim W\{c + k, [cC + \sum_{j=1}^{k}(\theta_j - m)(\theta_j - m)']^{-1}\}$.

In extending from the traditional baseline analysis of Gelfand et al. to the Dirichlet process analysis, note that there is no change to the conditional distribution of σ^2, and that the conditional distributions of m and B just have the n assumedly distinct μ_i values replaced by k actually distinct θ_j values.

In connection with the data analysed in Gelfand et al. we assume the same, rather diffuse priors for σ, m, and B as in that paper. Specifically, these are given by

$$u = 0, \quad A^{-1} = 0, \quad c = 2 \quad \text{and} \quad C = \begin{pmatrix} 100 & 0 \\ 0 & 0.1 \end{pmatrix}.$$

The prior for α is taken as gamma with shape and scale both unity, very diffuse. In comparing the Dirichlet process prior method with the more traditional baseline analysis, one of the most surprising findings with analysis of this data set is the strong indication of multimodality in the population of μ_i, evidenced in Figure 22.1. Here we display Monte Carlo[†] estimates of the univariate margins of the posterior expectation $E(G(\mu) \mid Y)$; this is computed as the average of sampled densities $E(G \mid \phi, \mathscr{S}, k)$ from equation (3), each a mixture of a small number of bivariate normals. The displayed densities therefore represent the marginal predictive distributions for the elements of a further μ vector, and so indicate predictions for the growth curve parameters for a further rat from the same population. These two densities appear in the figure as full lines; the dashed lines are corresponding densities from a repeat of the baseline analysis of Gelfand et al. In the latter, these margins are constrained to be smooth and unimodal, but embedding within the 'semi-parametric' Dirichlet family 'around' the baseline prior allows for and apparently supports rather strong heterogeneity in the rat population.

22.5 A MULTIVARIATE DENSITY ESTIMATION EXAMPLE

Dirichlet mixture models have been considered as frameworks for non- (or semi-) parametric density estimation for over 20 years, though only recently

[†] Gibbs sampling here is based on 20 000 consecutive iterations performed, with the conditional densities averaged to give Figure 22.1 computed on every 20th iteration, beginning at the 20th from the start. The simulation was initialized by setting the μ_i to least-squares values for each subject, and the hyperparameters m and B to their least-squares values assuming the initialized μ_i.

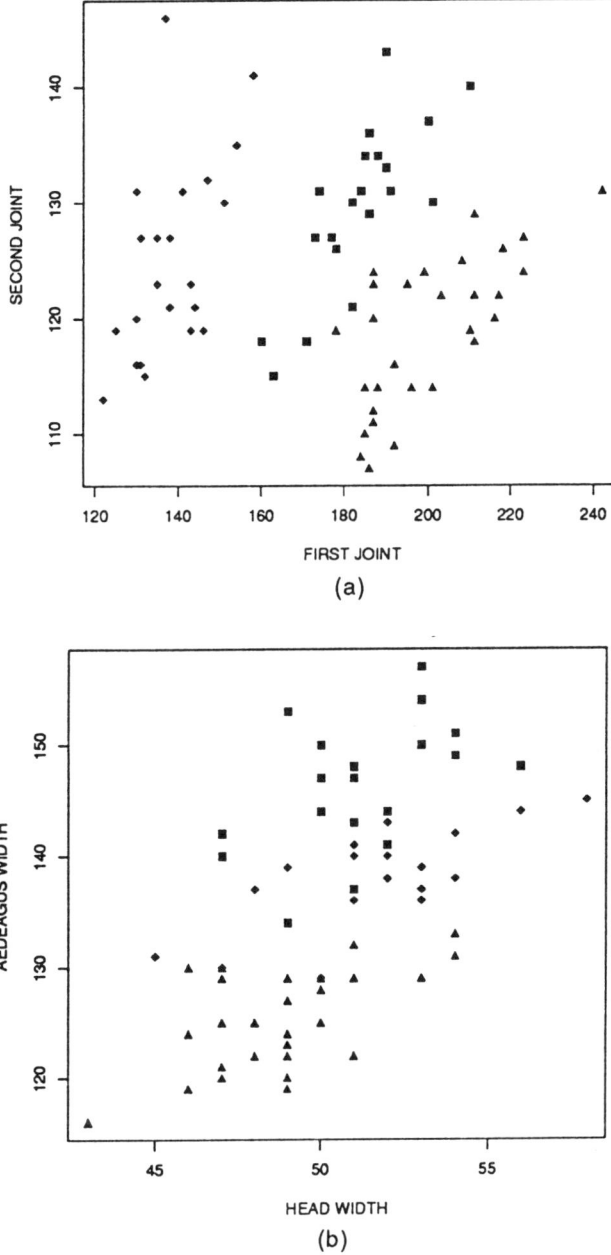

Figure 22.2 Scatterplots of x_1, x_2 and x_3, x_4 from the beetle data. For information, the three different species are marked by diamonds, squares and triangles (though the classification information is not used in analysis).

has routine application become computationally feasible. Restricting the modelling framework of sections 22.2 and 22.3 to the multivariate random sampling context leads directly to density estimation. We illustrate this here with a five-dimensional data set from Lubischew (1962), providing five measurements of physical characteristics of male insects of the species *Chactocnema concina, C. heikertinger* and *C. heptapotamica*. For any beetle, we denote the vector y of five measurements by $y = (x_1, \ldots x_5)$, so that the ith beetle has measurements

$$y_i = (x_{i,1}, \ldots x_{i,5})' \quad (i = 1, \ldots n);$$

the data set has $n = 74$ cases. The five variables are width of the first joint (x_1), width of the second joint (x_2), head width (x_3), aedeagus width (x_4), and aedeagus side width (x_5). Figure 22.2 shows scatterplots of the first two pairs of variables.

One simulation analysis is partly summarized in the remaining figures. The random sampling context has $y_i \sim N(\mu_i, \Sigma_i)$ with $\phi_i = \{\mu_i, \Sigma_i\}$; there is no common parameter σ in this model. We assume a baseline prior G_0 of the non-conjugate form in section 22.3.4, equation (11). The analysis illustrated assumed the mutually independent hyperprior on the parameters $\pi = \{\alpha, m, B, S\}$ in which $p(\alpha)$ is gamma with shape parameter a_0 and scale parameter b_0, $m \sim N(a, A)$, $B^{-1} \sim W(c, (cC)^{-1})$ and $S \sim W(q, q^{-1})$. The specific priors chosen for this illustration have $c = q = 9$, $s = 15$, and the matrix parameters A, C and are diagonal matrices with variances simply reflecting the scale of the problem. The gamma prior for α has shape $a_0 = 5$ and scale $b_0 = 0.5$, therefore supporting a diffuse range of reasonably large values consistent with possibly large values of k. In this case of $n = 74$ observations, the implied prior

$$p(k \mid n) = \int p(k \mid \alpha, n) p(\alpha) d\alpha$$

can be (numerically) evaluated, and supports a wide range of k values between about $k = 8$ and $k = 35$ ($p(k \mid n) > 0$ for all k, of course, though is very small outside this range of k values) with mean/mode near $k = 20$. Such a prior is consistent with the idea that we are concerned with local smoothing, not overduly constraining the model to favour a small number of normal components. Notice that we do not suggest that the model, and this over-smoothing prior, are in any way designed to address issues of clustering or discrimination; the focus is simply one of density estimation. Examining prior predictive densities and other features of predictive distributions allows for assessment of implications of chosen priors. For example, taking m, B and Σ_i at their prior means implies that the conditional prior predictive distribution for any single y_i is just $N(m, B + \Sigma_i)$; contours of one bivariate margin of this density are plotted, at fractions of 0.8, 0.6, 0.4 and 0.2 of the modal value, in Figure 22.5(a). This is quite

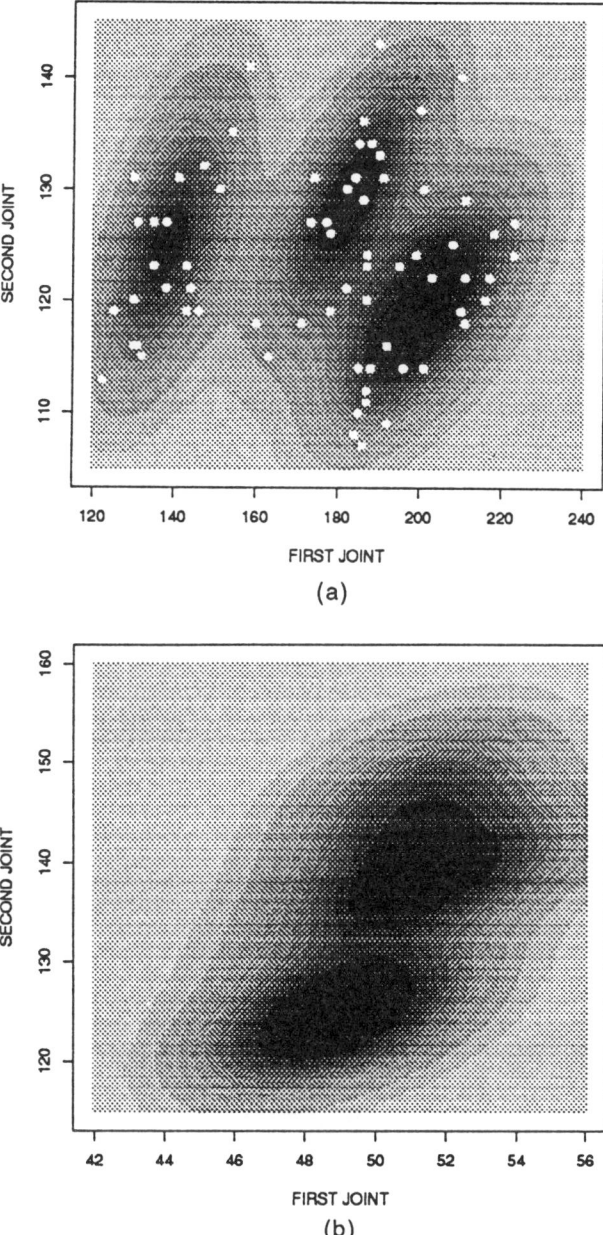

Figure 22.3 Predictive densities $p(x_1, x_2 \mid Y)$ and $p(x_3, x_4 \mid Y)$. Both densities are two-dimensional marginals of the estimated five-dimensional distribution $p(y_{n+1} \mid Y)$. Two of the three clusters are overlapping in the (x_3, x_4)-projection. The estimated clustering corresponds exactly to the three species. (a) $p(x_1, x_2 \mid Y)$; (b) $p(x_3, x_4 \mid Y)$

diffuse over the data range for these two variables, as it is for the other variables, and clearly does not anticipate specific departures from normality.

Analysis reported here is based on 5000 iterations of the Gibbs sampling scheme with summarized predictive densities computed by averaging the conditional mixtures of normals (4) across these all draws. Figure 22.3 shows two of the bivariate projections of the full five-dimensional predictive density estimate. A concentration into three clearly distinguished components is apparent, and, in fact, these correspond precisely with the three species of beetle, though, of course, no information about species of individuals is used in the analysis. Indeed, the analysis is remarkably accurate in reproducing the species discrimination; the Monte Carlo estimate of the posterior for k gives approximate probabilities 0.01, 0.48, 0.30, 0.15, 0.02 and 0.01 for values of $k = 2, \ldots, 8$, respectively, with the mode at $k = 3$, 'truth' in this case. The data substantially overrides the prior on k which favoured much larger values of k. It may be thought that the appreciable posterior mass on $k = 4$ and 5 is a result of this over-smoothing prior, though the usual 'over-fitting' aspects of mixture modelling are likely also to contribute. That is, whatever the prior for k, posteriors in these (and other) mixture models will typically give residual mass to larger values

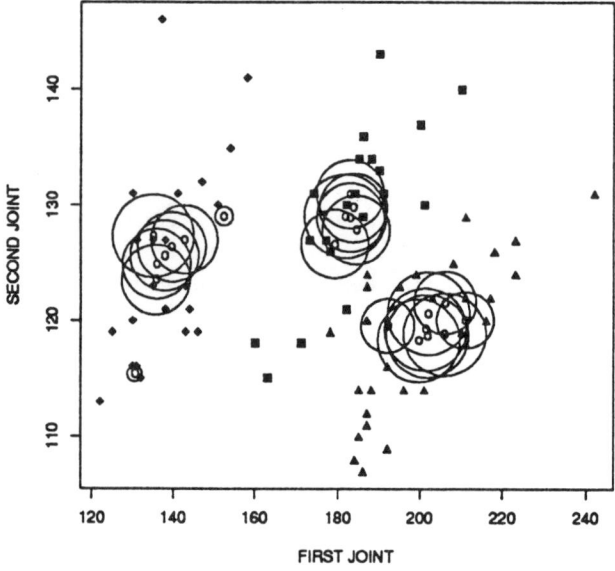

Figure 22.4 Some elements of a few simulated draws $\mu_i \sim p(\mu_i | Y)$. The figure shows the first two elements of the sampled values of μ_i, $i = 1, \ldots, k$, after $t = 100$, 1000, 2000, 3000, 4000 and 5000 iterations of the Gibbs sampler. The centre of each circle corresponds to the value of some μ_i. The areas of the circles show the corresponding sizes n_i of the clusters.

of k corresponding to possible configurations in which some groups contain just a very few observations. To illustrate in this example, simply ignoring simulated groups with fewer that seven observations (i.e. conditioning the posterior on appreciable sized groups) leads to approximate posterior chances of 0.01, 0.92 and 0.06 on $k = 2, 3$ and 4 respectively, perfectly highlighting the three beetle populations. With a much less severe, and perhaps more appropriate and realistic, restriction to groups sizes of at least three observations, the conditional posterior chances are approximately 0.01, 0.71 and 0.25 on $k = 2, 3$ and 4, respectively.

Posteriors for the parameters ϕ_i could be estimated and summarized in similar fashions. We simply note here that the posterior reflects very little uncertainty about the locations μ_i of the suggested three clusters; Figure 22.4 provides some insight by displaying the first two components of just a few of the sampled μ_i.

Some applications of these models are more concerned with smooth regression estimation (Erkanli *et al.* 1992) rather than density estimation *per se*. Regression estimation simply involves summarizing inferences about conditional predictive distributions, and so is trivially obtained within the existing mixture analysis. For example, the conditional predictive expectations $E(x_2 | Y, x_1)$, deduced directly from the Monte Carlo estimate of the bivariate distribution, appears in Figure 22.6(a) with the corresponding data points superimposed. This frame also presents approximate 66% credible bands computed pointwise. Conditional modes, rather than means, appear in Figure 22.6(b), superimposed on the predictive density contours; note the bifurcation induced across the range where the estimated conditional density $p(x_2 | Y, x_1)$ is bimodal.

Some technical comments about implementation of the Gibbs sampling scheme are in order. It is well known that Markov chain Monte Carlo methods generally may become 'trapped' in local modes of posterior distributions. An incidence of this can be observed in the current example. While the posterior gives most weight to $k = 3$ or 4 clusters, there is some remaining posterior probability (estimated at 0.01) for $k = 2$, but a transition from $k = 3$ to $k = 2$ is most unlikely in the Gibbs sampling scheme. Since only one observation at a time is being changed in the sampling of configurations in our simulation routine, such a transition would require the Gibbs sampler to pass through many rather unlikely configurations (when k is still 3, but some observations are 'misclassified') before settling on $k = 2$. However, both states, $k = 2$ as well as $k = 3$, are easily reached from the configuration $k = n$, which was used to initialize the Gibbs scheme. For this reason we implemented the Gibbs sampling iterations by 'reinitializing' the configurations at $k = n$ (though not the hyperparameter values) after each 1000, rather than simply running one longer chain.

All of the above discussion is based on the full five-dimensional density estimate and its bivariate projections. The picture changes slightly when

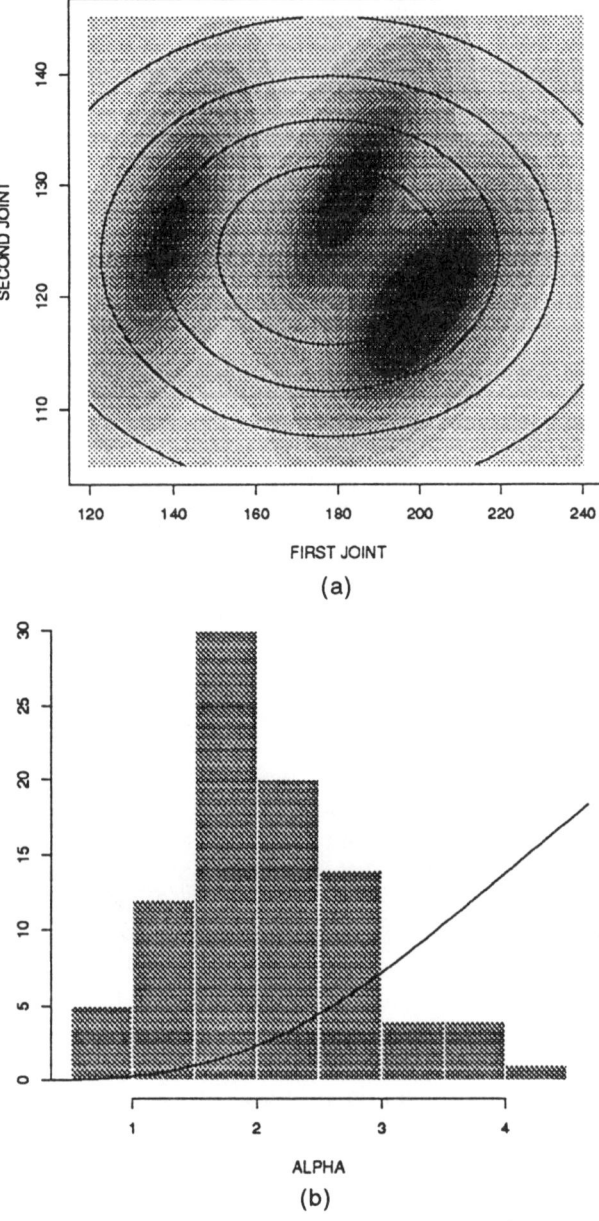

Figure 22.5 Panel (a) compares posterior predictive $p(y_{n+1} | Y)$ (grey shades) and prior predictive $p(y_{n+1})$ (contours). Panel (b) shows the gamma prior $p(\alpha)$ (line) and the estimated posterior $p(\alpha | Y)$ (histogram). The implication is that inference in this example is mostly driven by the data, which partially justifies the use of some default choices for hyperprior parameters. (a) Prior/posterior predictive distributions; (b) prior/posterior for α.

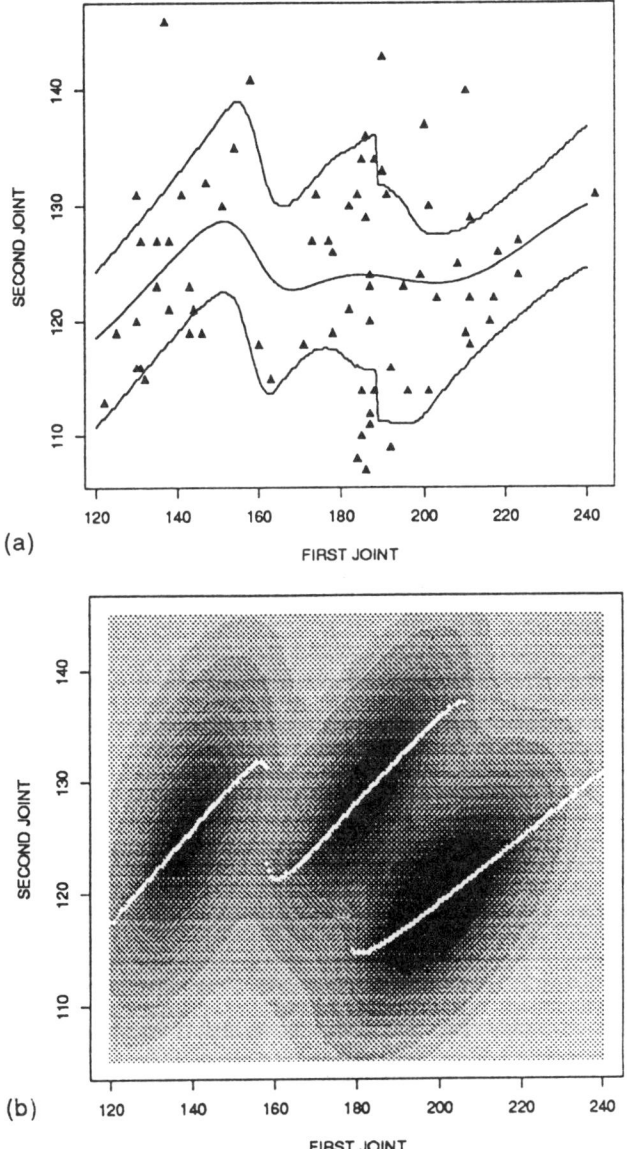

Figure 22.6 Taking predictive conditional means $E(x_2 | x_1, Y)$ leads to the non-linear regression line shown in (a). Alternatively, modes of the conditional predictive distribution $p(x_2 | x_1, Y)$ can be used to obtain modal regression traces shown in (b). The non-linear regression line is shown together with 66% credible bands computed pointwise—for each x_1-value, the margins give a prediction interval for x_2, and the intervals are found by taking the shortest interval of conditional posterior probability 0.66 which contains the conditional mode (note that highest density regions become disconnected when the conditional density becomes biomodal). (a) Conditional means; (b) conditional modes.

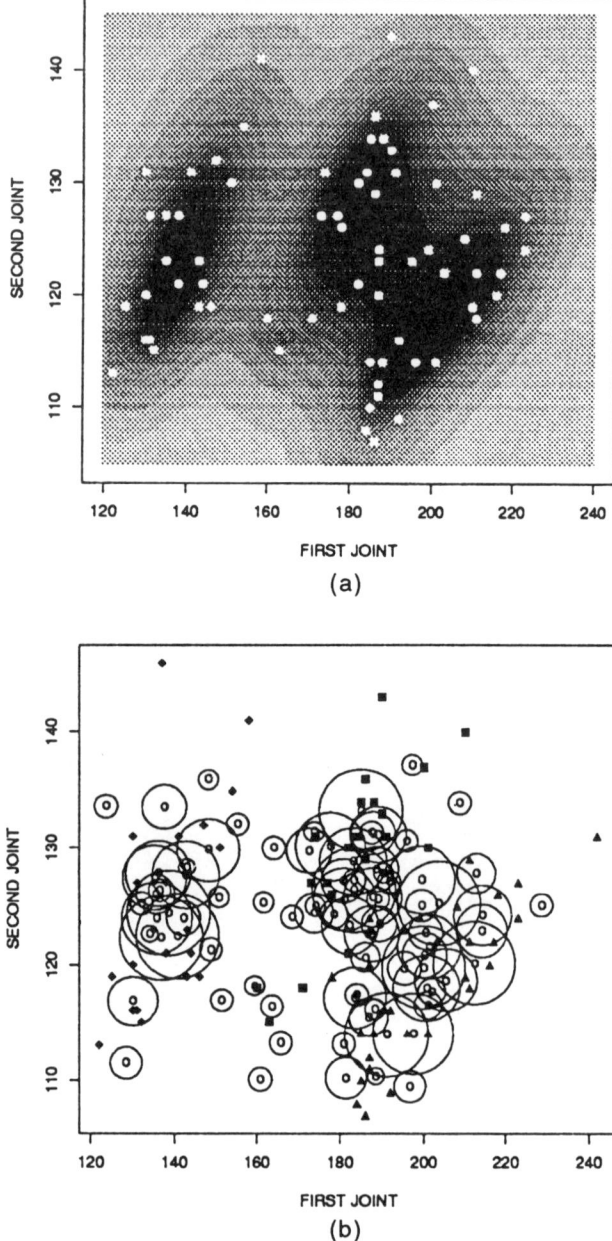

Figure 22.7 Predictive density $p(x_1, x_2 \mid Y)$ and a flavour of the posterior distribution for elements of the μ_i parameters in the bivariate model. While the density estimate shows only little change from Figure 22.5(a), the posterior on μ_i displays much more uncertainty about the clusters. (a) $p(x_{n+1,1}, x_{n+1,2} \mid Y)$; (b) some sampled μ_i.

deriving the bivariate density estimate from only the marginal data set of x_1 and x_2. Reanalysis in just the two-dimensional space is briefly summarized in Figure 22.7, this provides the bivariate density estimate and draws μ_i from the posterior distribution as in the original five-dimensional analysis. While the density estimate has changed only a little, the posterior on the model parameters exhibits greater uncertainty. Indeed, the estimated posterior probabilities for the number of clusters k in this case are considerably different with mass diffusely spread over the range $k = 9, \ldots, 27$, consistent with the prior; the data is rather uninformative about k and configurations. This greater diffuseness of posterior inference for k (which is also reflected in increased uncertainty about the cluster means μ_i shown in Figure 22.7(b)) indicates that crucial information on the discrimination of the three species is lost when considering only this two-dimensional slice of the full five-dimensional data set.

ACKNOWLEDGEMENTS

Discussions with Cliff Litton and Steve MacEachern contributed materially to the final revision of this chapter. Mike West was partially supported by the National Science Foundation under grant DMS 90–24793. Michael D Escobar was partially financed by National Cancer Institute No. RO1-CA54852-01 and a National Research Service Award from NIMH Grant No. MH15758.

REFERENCES

Antoniak, C. E. (1974) Mixtures of Dirichlet processes with applications to nonparametric problems, *Ann. Statist.*, **2**, 1152–74.
DuMouchel, W. M. and Harris, J. E. (1983) Bayes methods for combining the results of cancer studies in humans and other species (with discussion), *J. Amer. Statist. Assoc.*, 78, 293–315.
Erkanli, A., Müller, P., and West, M. (1992) Curve fitting using Dirichlet process mixtures. *Discussion Paper #92–A09*, Institute of Statistics and Decision Sciences, Duke University.
Escobar, M. D. (1988) Estimating the means of several normal populations by nonparametric estimation of the distribution of the means, unpublished PhD dissertation, Yale University.
Escobar, M. D., and West, M. (1991) Bayesian density estimation and inference using mixtures, Invited revision for *J. Amer. Statist. Assoc.*
Escobar, M. D. and West, M. (1992) Computing Bayesian nonparametric hierarchical models. *ISDS Discussion Paper 92–A20*, Duke University.
Gelfand, A. E. and Smith, A. F. M. (1990) Sampling based approaches to calculating marginal densities. *J. Amer. Statist. Assoc.*, **85**, 398–409.

Gelfand, A. E., Hills, S. E., Racine-Poon, A., and Smith, A. F. M. (1990) Illustration of Bayesian inference in normal data models using Gibbs sampling, *J. Amer. Statist. Assoc.*, 85, 972–85.

Kuo, L., and Smith, A. F. M. (1992) Bayesian computations in survival models via the Gibbs sampler, in *Survival Analysis: State of the Art*, J. P. Klein and P. K. Goel (eds.) Kluwer.

Lindley, D. V. (1972) *Bayesian Statistics, a Review*, Philadelphia, SIAM.

Lindley, D. V. and Smith, A. F. M. (1972) Bayes' estimates for the linear model (with discussion), *J. Roy. Statist. Soc.*, **B34**, 1–41.

Lubischew, A. (1962), On the use of discriminant functions in taxonomy, *Biometrics*, 18, 455–77.

MacEachern, S. M. (1992) Estimating normal means with a conjugate style Dirichlet process prior. Technical report No. 487, Department of Statistics, The Ohio State University.

MacEachern, S. M. and Müller, P. (1994) Estimating mixture of Dirichlet process models, *ISDS Discussion Paper*, Duke University.

O'Hagan, A. (1988) Modelling with heavy tails, in *Bayesian Statistics 3*, J. M. Bernardo, M. H. De Groot, D. V. Lindley and A. F. M. Smith (eds), Oxford.

Smith, A. F. M. (1983) Comment on article by DuMouchel and Harris, *J. Amer. Statist. Assoc.* 78, 293–315.

Turner, D. A. T. and West, M. (1993) Statistical analysis of mixtures applied to postsynpotential fluctuations. *Journal of Neuroscience Methods* (to appear).

West, M. (1984) Outlier models and prior distributions in Bayesian linear regression, *J. Roy. Statist. Soc.*, **B46**, 431–9.

West, M. (1985) Generalised linear models: scale parameters, outlier accommodation and prior distributions, in *Bayesian Statistics 2*, J. M. Bernardo, M. H. De Groot, D. V. Lindley and A. F. M. Smith (eds), North-Holland, Amsterdam.

West, M. (1990) Bayesian kernel density estimation, *ISDS Discussion Paper 90–A02*, Duke University.

West, M. (1992) Hyperparameter estimation in Dirichlet process mixture models, *ISDS Discussion Paper 92–A03*, Duke University.

West, M. and Cao, G. (1993) Assessing mechanisms of neural synaptic activity, in *Bayesian Statistics in Science and Technology: Case Studies*, C. Gatsonis, J. Hodges, R. Kass and N. Singpurwalla (eds), (to appear).

Institute of Statistics and Sciences
Duke University
Durham
NC 27708–0251
USA

Department of Statistics
Baker Hall 232-212
Carnegie-Mellon University
Pittsburgh
PA 15213
USA

Index

admissibility and boundedness 225–31, 242–4
adversarial approach in law 265
annealing and Bayesian clustering 70–1
Armitage, P., Dennis Lindley: the first 70 years 1–12
auditors, Bayesian sampling schemes for 51–65
axiomatizations of probability, utility and SEU 172–4

Barlow, R. E., M. B. Mendel, Operational Bayesian approach 19–28
Barnard, G. A., Pivotal inference illustrated on the Darwin maize data 29–39
Bartholomew, D. J., Bayes' theorem in latent variable modelling 41–50
baseball batting averages 193–4
Bather, J. A., and P. J. Browne, Bayesian sampling schemes for auditors 51–65
Bayes–Stein estimators 289–93
Bayesian analysis
 Bayes' theorem in latent variable modelling 41–50
 Bayesian decision theory and the legal structure 261–6
 Bayesian sampling schemes for auditors 51–65
 categorical data — a selective review 283–310
 Darwin maize data 30, 38
 design of clinical experiments 331–48
 estimation of the spectrum of a time series 125–33
 fully Bayesian hierarchical analysis for exponential families via Monte Carlo computation 181–99
 heavy-tailed Bayesian modelling 311–15
 inference for a covariance matrix 78–9
 Lindley's first interest 5–6, 9–11
 older controversies 225, 254–7
 operational Bayesian approach 19–28
 optimizing prediction with hierarchical models: Bayesian clustering 67–76
 plural decision analysis 96–8
 probability of guilt 160–4
 residual analysis and outliers in Bayesian hierarchical models 149–57
 utility: probability's younger twin? 173–9
beetle data 377–80
Bernardo, J. M., Optimizing prediction with hierarchical models: Bayesian clustering 67–76
Berry, D. A., and G. Parmigiani, Applications of Lindley information measure to the design of clinical experiments 329–48
beta-binomial set-up 195–6
binary data and latent variable modelling 46–8
biometry, the signature as a covariate in reliability and 119–47
bivariate distributions, heavy-tailed 311–27
Brown, P. J., N. D. Le and J. V. Zidek, Inference for a covariance matrix 77–92

Brown, R. V., The role of statistical theory in decision aiding: measuring decision effectiveness in the light of outcomes 93–117
Browne, P. J. and J. A. Bather, Bayesian sampling schemes for auditors 51–65

Cambridge University
  Lindley at 2, 4–6
  Lindley's lectures 349
Campodónico, S. and N. D. Singpurwalla, The signature as a covariate in reliability and biometry 119–47
categorical data, Bayesian analysis of 283–310
Chaloner, K., Residual analysis and outliers in Bayesian hierarchical models 149–57
classic theorems involving the characteristic function 349–62
clinical experiments 329–48
clustering, Bayesian 70–1
coin tossing 203–6
conflicting information and a class of bivariate heavy-tailed distributions 311–27
contingency tables 283–310
cost function in dynamic programming 54–7
covariance matrix, inference for a 77–92
covariate in signature analysis, spectrum as a 133–7
Cramer's theorem 349, 359–61
crime
  and evidence 159–70
  judgements of guilt 262–4

Darwin maize data, pivotal inference illustrated on 29–39
datelessness of knowledge (DoK) 32, 37
Dawid, A. P., The island problem: coherent use of identification evidence 159–70
decision theory
  Bayesian, and the legal structure 261–6

experimental design and 278–82
Lindley information measure 329–48
optimizing prediction with hierarchical models 69–70
statistical theory in decision aiding: measuring decision effectiveness in the light of outcomes 93–117
subjective expected utility 171–80
Deely, J., and F. Lad, Experimental design from a subjective utilitarian viewpoint 267–82
design of experiments 267–82
diabetes, measurements in 268–82
Dirichlet process priors 372–6
duration and follow-up in clinical experiments 336–41
dynamic programming and sampling schemes for auditors 54–64

educational testing 46–8
Eggleston, R. and the island problem 160–2
Einstein, Albert 224
election forecasting and prediction with hierarchical models 71–5
electrocardiogram, signatures of 124–5
engineering apprentices data 288
Escobar, M. D., M. West and P. Müller, Hierarchical priors and mixture models with application in regression and density estimation 363–87
evidence
  admission of all cost-free 264–5
  the island problem: coherent use of identification 159–70
exchangeable beliefs: subjectivist foundations for the inductive argument 201–22
experimental design from a subjective utilitarian viewpoint 267–82

factor analysis model 45–6
Feller, W. 227
finite/infinite problem and formalism 223–60
Fisher, R. A. 4–5, 29–30, 267

French, S., Utility: probability's younger twin? 171–80
frequency domain methods 120
fully Bayesian hierarchical analysis for exponential families via Monte Carlo computation 181–99

Galton, Francis 34
gamma-Poisson set-up 185, 188, 194–5
gene frequency data 287–9
generalized inverted Wishart distribution (GIW) 82–9
George, E. I., U. E. Makov and A. F. M. Smith, Fully Bayesian hierarchical analysis for exponential families via Monte Carlo computation 181–99
Gibbs sampler 187–90, 194, 295, 367–8, 381
GIW (generalized inverted Wishart distribution) 82–6
Goldstein, M., Revising exchangeable beliefs: subjectivist foundations for the inductive argument 201–22
guilt
  abolition of judgements of 262–4
  probability of 160–4

heavy-tailed Bayesian modelling 311–15
hierarchical modelling
  for exponential families via Monte Carlo computation 181–99
  multicentre clinical trials 341–5
  optimizing prediction with 67–76
  priors and mixture models applied to regression and density estimation 363–87
  residual analysis and outliers in Bayesian 149–57
Hill, B. M., On Steinian shrinkage estimators 223–60
Hsu, J. S. J., and T. Leonard, The Bayesian analysis of categorical data—a selective review 283–310

identification evidence 159–70
indifference principle in operational Bayesian approach 20–3
inductive argument, exchangeable beliefs: subjectivist foundations for the 201–22
inference for a covariance matrix 77–92
information measure, Lindley 329–48
innocence, presumption of 168–70
inverted Wishart distribution (IW) 77–81
island problem: coherent use of identification evidence 159–70
IW (inverted Wishart distribution) 77–81

James–Stein estimator 230–1, 233–42

Kadane, J. B., Bayesian decision theory and the legal structure 261–6
kinetic energy and operational Bayesian approach 21, 24–5

Lad, F. and J. Deely, Experimental design from a subjective utilitarian viewpoint 267–82
latent variable modelling, Bayes' theorem in 41–50
Le, Huiling and A. O'Hagan, Conflicting information and a class of bivariate heavy-tailed distributions 311–27
Le, Nhu D., P. J. Brown and J. V. Zidek, Inference for a covariance matrix 77–92
least squares estimation
  Bayes estimator and 250
  of the spectrum 120–3
legal structure and Bayesian decision theory 261–6
Leonard, T. and J. S. J. Hsu, The Bayesian analysis of categorical data—a selective review 283–310
life distributions and operational Bayesian approach 25–6
life model, accelerated 134–6

Lindeberg central limit theorem 349, 350–9
Lindley, Dennis V
  abolition of the adversarial approach in law 265–6
  Bayesian decision theory 261, 267
  bibliography 12–18
  biography 1–12
  evidence of guilt 164–5, 167–8
  hierarchical modelling 363
  information measure 329–48
  lectures in Cambridge 349
  measure theory 6
  prescriptive decision research 95–7
linear structural relations models 48–9
LISREL 48–9, 78
locomotive traction motors, service life 137–41
logit-multivariate normal distribution 305–6

maize data 29–39, 34
Makov, U. E., E. I. George and A. F. M. Smith, Fully Bayesian hierarchical analysis for exponential families via Monte Carlo computation 181–99
Mendel, M. B, R. E. Barlow, Operational Bayesian approach 19–28
Mexico state elections 73–5
Monte Carlo methods 181–99
Müller, P., M. West and M. D. Escobar, Hierarchical priors and mixture models with application in regression and density estimation 363–87
multicentre clinical trials 341–5
multinomial-Dirichlet distribution 304–5
multivariate distributions
  heavy-tailed 313–15
  multivariate density estimation 376–85

normal linear model and latent variable modelling 45–6
number of observations, optimal 267–8

O'Hagan, A. and Huiling Le, Conflicting information and a class of bivariate heavy-tailed distributions 311–27
oil discovery data 194–5
operational Bayesian approach 19–28
optimizing prediction with hierarchical models: Bayesian clustering 67–76
outliers 149–57

paradoxes, Steinian 249–54
Parmigiani, G. and D. A. Berry, Applications of Lindley information measure to the design of clinical experiments 329–48
path analysis and latent variable modelling 48–9
pivotal inference 29–39
plural evaluation methodology 95–103
power spectrum (the signature) 119–23
prediction with hierarchical models 67–76
prescriptive decision research 94–9
prosecutor's fallacy 159–60, 161

random effects model and Steinian estimators 244–9
Rasch model 47–8
realized errors in non-hierarchical linear models 149–50
reference experiment and decision-making 175–8
references
  applications of Lindley information measure to the design of clinical experiments 346–8
  Bayes' theorem in latent variable modelling 49–50
  the Bayesian analysis of categorical data—a selective review 306–10
  Bayesian decision theory and the legal structure 266
  Bayesian sampling schemes for auditors 65
  conflicting information and a class of bivariate heavy-tailed distributions 326–7

INDEX

Dennis Lindley: the first 70 years 12
exchangeable beliefs: subjectivist foundations for the inductive argument 222
experimental design from a subjective utilitarian viewpoint 281–2
fully Bayesian hierarchical analysis for exponential families via Monte Carlo computation 196–8
hierarchical priors and mixture models with application in regression and density estimation 385–7
inference for a covariance matrix 90–2
island problem: coherent use of identification evidence 170
the operational Bayesian approach 27–8
optimizing prediction with hierarchical models: Bayesian clustering 75–6
pivotal inference illustrated on the Darwin maize data 39
residual analysis and outliers in Bayesian hierarchical models 156–7
signature as a covariate in reliability and biometry 146–7
statistical theory in decision aiding: measuring decision effectiveness in the light of outcomes 115–17
on Steinian shrinkage estimators: the finite/infinite problem and formalism in probability and statistics 258–60
on two classic theorems involving the characteristic function 361–2
utility: probability's younger twin? 179–80
regression
 estimation of the spectrum 120–3, 126–9
 and hierarchical modelling 363–87

reliability and biometry, the signature as a covariate in 119–47
residual analysis and outliers in Bayesian hierarchical models 149–57
risk funtions and Steinian estimators 238–42
Russell, Bertrand 224

sampling schemes for auditors, Bayesian 51–65
Savage, L. J. Bayesian approach to decision analysis 172–4
school examination results 293–5
SEU (subjective expected utility) 171–80
signature as a covariate in reliability and biometry 119–47
Simpson's paradox in three-way tables 296–304
Singpurwalla, N. D. and S. Campodónico, The signature as a covariate in reliability and biometry 119–47
Smith, A. F. M., E. I. George and U. E. Makov, Fully Bayesian hierarchical analysis for exponential families via Monte Carlo computation 181–99
Smith, W. L., On two classic theorems involving the characteristic function 349–62
sphere and Steinian estimators 231–42
statistical theory in decision aiding: measuring decision effectiveness in the light of outcomes 93–117
Steinian shrinkage estimators 223–60
subjective expected utility (SEU) 171–80
subjectivist foundations for the inductive argument, revising exchangeable beliefs 201–22
submarine combat aid 96–7
survival analysis 226–8
 exponential data 334–6

toxoplasmosis data 195–6

University College, London, Lindley at 7–9

utility
    conceptions of, as information 277–8
    experimental design from a subjective utilitarian viewpoint 267–82
    plural decision analysis 102–13
    probability's younger twin? 171–80

Valencia elections 72
vibrations of rotating machinery
    defects in 137–41

signature of 123–4
von Winterfeldt, Detlof, and plural estimates 98

Wales, University College of, Lindley at 6
Weibull model, generalized 24–5
West, M., P. Müller and M. D. Escobar, Hierarchical priors and mixture models with application in regression and density estimation 363–87
Wishart distribution, inverted 77–89

Zidek, J. V., P. J. Brown and N. D. Le, Inference for a covariance matrix 77–92

Index compiled by Dr. M. P. M. Merrington

Applied Probability and Statistics (Continued)
  BROWN and HOLLANDER · Statistics: A Biomedical Introduction
  BUCKLEW · Large Deviation Techniques in Decision, Simulation and Estimation
  BUNKE and BUNKE · Nonlinear Regression, Functional Relations and Robust
    Methods: Statistical Methods of Model Building
  CHATTERJEE and HADI ·Sensitivity Analysis in Linear Regression
  CHATTERJEE and PRICE · Regression Analysis by Example, *Second Edition*
  CLARKE and DISNEY · Probability and Random Processes: A First Course with
    Applications, *Second Edition*
  COCHRAN · Sampling Techniques, *Third Edition*
 *COCHRAN and COX · Experimental Designs, *Second Edition*
  CONOVER · Practical Nonparametric Statistics, *Second Edition*
  CONOVER and IMAN · Introduction to Modern Business Statistics
  CORNELL · Experiments with Mixtures, Designs, Models and the Analysis of
    Mixture Data, *Second Edition*
  COX · A Handbook of Introductory Statistical Methods
 *COX · Planning of Experiments
  CRESSIE · Statistics for Spatial Data
  DANIEL · Applications of Statistics to Industrial Experimentation
  DANIEL · Biostatistics: A Foundation for Analysis in the Health Sciences, *Fifth
    Edition*
  DAVID · Order Statistics, *Second Edition*
  DEGROOT, FIENBERG, and KADANE · Statistics and the Law
 *DEMING · Sample Design in Business Research
  DILLON and GOLDSTEIN · Multivariate Analysis: Methods and Applications
  DODGE and ROMIG · Sampling Inspection Tables, *Second Edition*
  DOWDY and WEARDEN · Statistics for Research, *Second Edition*
  DRAPER and SMITH · Applied Regression Analysis, *Second Edition*
  DUNN · Basic Statistics: A Primer for the Biomedical Sciences, *Second Edition*
  DUNN and CLARK · Applied Statistics: Analysis of Variance and Regression,
    Second Edition
  ELANDT-JOHNSON and JOHNSON · Survival Models and Data Analysis
  EVANS, PEACOCK, and HASTINGS · Statistical Distributions, *Second Edition*
  FISHER and VAN BELLE · Biostatistics: A Methodology for the Health Sciences
  FLEISS · The Design and Analysis of Clinical Experiments
  FLEISS · Statistical Methods for Rates and Proportions, *Second Edition*
  FLEMING and HARRINGTON · Counting Processes and Survival Analysis
  FLURY · Common Principal Components and Related Multivariate Models
  GALLANT · Nonlinear Statistical Models
  GROSS and HARRIS · Fundamentals of Queueing Theory, *Second Edition*
  GROVES · Survey Errors and Survey Costs
  GROVES, BIEMER, LYBERG, MASSEY, NICHOLLS, and WAKSBERG ·
    Telephone Survey Methodology
  HAHN and MEEKER · Statistical Intervals: A Guide for Practitioners
  HAND · Discrimination and Classification
 *HANSEN, HURWITZ, and MADOW · Sample Survey Methods and Theory:
    Volume I: Methods and Applications
 *HANSEN, HURWITZ, and MADOW · Sample Survey Methods and Theory:
    Volume II: Theory
  HEIBERGER · Computation for the Analysis of Designed Experiments
  HELLER · MACSYMA for Statisticians
  HOAGLIN, MOSTELLER, and TUKEY · Exploratory Approach to Analysis of
    Variance
  HOAGLIN, MOSTELLER, and TUKEY · Exploring Data Tables, Trends and Shapes
  HOAGLIN, MOSTELLER, and TUKEY · Understanding Robust and Exploratory
    Data Analysis
  HOCHBERG and TAMHANE · Multiple Comparison Procedures
  HOEL · Elementary Statistics, *Fifth Edition*
  HOGG and KLUGMAN · Loss Distributions
  HOLLANDER and WOLFE · Nonparametric Statistical Methods
  HOSMER and LEMESHOW · Applied Logistic Regression

* Now available in a lower priced paperback edition in the Wiley Classics Library.

Applied Probability and Statistics (Continued)
IMAN and CONOVER · Modern Business Statistics
JACKSON · A User's Guide to Principal Components
JOHN · Statistical Methods in Engineering and Quality Assurance
JOHNSON · Multivariate Statistical Simulation
JOHNSON and KOTZ · Distributions in Statistics
                          Continuous Univariate Distributions—1
                          Continuous Univariate Distributions—2
                          Continuous Multivariate Distributions
JOHNSON, KOTZ, and KEMP · Univariate Discrete Distributions, *Second Edition*
JUDGE, GRIFFITHS, HILL, LÜTKEPOHL, and LEE · The Theory and Practice of Econometrics, *Second Edition*
JUDGE, HILL, GRIFFITHS, LÜTKEPOHL, and LEE · Introduction to the Theory and Practice of Econometrics, *Second Edition*
KALBFLEISCH and PRENTICE · The Statistical Analysis of Failure Time Data
KASPRZYK, DUNCAN, KALTON, and SINGH · Panel Surveys
KISH · Statistical Design for Research
KISH · Survey Sampling
LAWLESS · Statistical Models and Methods for Lifetime Data
LEBART, MORINEAU, and WARWICK · Multivariate Descriptive Statistical Analysis: Correspondence Analysis and Related Techniques for Large Matrices
LEE · Statistical Methods for Survival Data Analysis, *Second Edition*
LePAGE and BILLARD · Exploring the Limits of Bootstrap
LEVY and LEMESHOW · Sampling of Populations: Methods and Applications
LINDVALL · Lectures on the Coupling Method
LINHART and ZUCCHINI · Model Selection
LITTLE and RUBIN · Statistical Analysis with Missing Data
MAGNUS and NEUDECKER · Matrix Differential Calculus with Applications in Statistics and Econometrics
MAINDONALD · Statistical Computation
MALLOWS · Design, Data, and Analysis by Some Friends of Cuthbert Daniel
MANN, SCHAFER, and SINGPURWALLA · Methods for Statistical Analysis of Reliability and Life Data
MASON, GUNST, and HESS · Statistical Design and Analysis of Experiments with Applications to Engineering and Science
McLACHLAN · Discriminant Analysis and Statistical Pattern Recognition
MILLER · Survival Analysis
MONTGOMERY and PECK · Introduction to Linear Regression Analysis, *Second Edition*
NELSON · Accelerated Testing, Statistical Models, Test Plans, and Data Analyses
NELSON · Applied Life Data Analysis
OCHI · Applied Probability and Stochastic Processes in Engineering and Physical Sciences
OKABE, BOOTS, and SUGIHARA · Spatial Tessellations: Concepts and Applications of Voronoi Diagrams
OSBORNE · Finite Algorithms in Optimization and Data Analysis
PANKRATZ · Forecasting with Dynamic Regression Models
PANKRATZ · Forecasting with Univariate Box-Jenkins Models: Concepts and Cases
RACHEV · Probability Metrics and the Stability of Stochastic Models
RÉNYI · A Diary on Information Theory
RIPLEY · Spatial Statistics
RIPLEY · Stochastic Simulation
ROSS · Introduction to Probability and Statistics for Engineers and Scientists
ROUSSEEUW and LEROY · Robust Regression and Outlier Detection
RUBIN · Multiple Imputation for Nonresponse in Surveys
RYAN · Statistical Methods for Quality Improvement
SCHUSS · Theory and Applications of Stochastic Differential Equations
SCOTT · Multivariate Density Estimation: Theory, Practice, and Visualization
SEARLE · Linear Models
SEARLE · Linear Models for Unbalanced Data
SEARLE · Matrix Algebra Useful for Statistics